Principles and Applications of Stereochemistry

MICHAEL NORTH
Department of Chemistry
University of Wales, Bangor
Bangor
UK

Stanley Thornes (Publishers) Ltd

First published in 1998 by:
Stanley Thornes (Publishers) Ltd
Ellenborough House
Wellington Street
CHELTENHAM
GL50 1YW
United Kingdom

98 99 00 01 02 / 10 9 8 7 6 5 4 3 2 1

A catalogue record for this book is available from the British Library

ISBN 0-7487-3994-7

Typeset by Blackpool Typesetting Services Ltd, Blackpool, UK
Printed and bound in Great Britain by T.J. International Ltd, Padstow, Cornwall

Contents

Foreword

The shapes of molecules control almost every aspect of our lives. They dictate the differences in taste between spearmint chewing gum and caraway seed cake; the difference in elasticity between a rubber band and the gutta percha coating of a golfball and between the toxicity and therapeutic benefit of a drug. Ultimately, they control our heredity.

Since Pasteur's experiments in the 1840s laid the foundations of modern stereochemistry, a huge amount of sophisticated research has been carried out and with it has grown the necessary and complicated jargon required to express the symmetry of molecules and the spatial relationships between the groups that they possess.

Dr North's book has been designed to introduce this area of chemistry to the stereochemical novice and to lead the student with plenty of diagrams and cross referencing from the simplest molecular form to quite sophisticated understanding of topics such as diastereotopicity and catalytic asymmetric synthesis.

Dr North rightly does not flinch from including discussion of energy differences which determine the shapes and ground state populations of flexible and cyclic molecules and he makes constant use of spectroscopic and other quantitative information which not only illuminates his material but has the invaluable add on of making connections to other aspects of molecular behaviour and molecular construction.

E.L. Eliel's volume *The Stereochemistry of Organic Compounds* is the current reference book in this field but at more then 1200 pages does not claim to be a handy student text. Dr North's book fills a wider gap more succinctly and it will be of great value to students from the start of their university courses as well as to their teachers. The inclusion in each chapter of a set of problems is an essential device for checking the understanding of the text and a good set of models will encourage the familiarity essential for a confident grasp of this vital aspect of chemistry.

C.J.M. Stirling, FRS

Preface

Stereochemistry is the relationship between the three dimensional shape of a molecule and its chemistry. It is a topic that falls between the three classical subdivisions of chemistry into organic, inorganic and physical chemistry. Most of the subject is usually taught as part of an organic chemistry course, with molecular symmetry often being split off into a separate physical chemistry course. The stereochemistry of inorganic compounds has been largely neglected in undergraduate chemistry courses, although this is now changing mainly because of the widespread use of organometallic compounds as catalysts for organic reactions. Thus in this text I have tried to bring together the organic, inorganic and physical chemistry aspects of the subject. The book is still dominated by the stereochemistry of organic compounds but the coverage of inorganic stereochemistry probably exceeds the content of most current undergraduate courses.

This book has its origin in a 10 lecture course given to undergraduate chemistry students at the University of Wales, Bangor. Increasingly, students are coming to university to study chemistry and related subjects without having studied mathematics in the preceding years. It is the need to present the topic in as non-mathematical a way as possible that has largely determined the order in which the material is presented. Thus Chapters 1 to 5 are essentially descriptive and require no mathematical ability beyond multiplication and division. The relationship between stereochemistry, symmetry, and group theory is thus delayed until Chapter 6, by which time the reader should have a thorough understanding of the key stereochemical concepts and of why stereochemistry is important. The first eight chapters deal with the stereochemistry of individual molecules and the final two chapters then expand this to show how a knowledge of stereochemistry can be used to predict the outcome of a variety of chemical reactions.

Stereochemistry is sometimes perceived as being a dry and boring aspect of chemistry, probably because of the number of arbitrary definitions and conventions that are associated with stereochemical nomenclature and representations. Throughout this book I have tried to counter this by including numerous real life examples of the applications of stereochemistry, and by including topics such as the origin of enantiomerically pure compounds in nature that would not normally be covered in an undergraduate course. Numbers are used in many different contexts in stereochemistry, and to avoid confusion I have adopted a system of using numbers in a shadow font (1, 2, 3 etc.) to number atoms according to chemical nomenclature, and numbers in a standard font for stereochemical purposes. Much of stereochemistry is concerned with the three dimensional arrangement of groups. This is often not easily seen in two dimensional representations and the reader would be well advised to purchase a ball and stick type molecular modelling kit such as the inexpensive ORBIT kit. This will allow a three dimensional model of any of the structures discussed in this book to be constructed rapidly, and will greatly aid both the visualization and understanding of the material.

At the end of each chapter, I have included a list of sources of further information which, wherever possible, are books or reviews. A particularly useful, comprehensive and up to date more advanced text is Eliel's *Stereochemistry of Organic Compounds*, and the first reference at the end of each chapter will direct the reader to the appropriate chapter(s) of this book. Also at the end of each chapter are a series of problems based on material covered in the chapter. These problems are of two types: the first being drill type questions usually concerned with arbitrary aspects of nomenclature that can only be memorized by practice. The second type of question are more problem based and designed to test understanding of, and ability to apply, the material covered in the chapter. Answers to these questions have not been included since I feel that this makes it too tempting for students to look at the answers and decide that they could have answered the question without actually doing so. Answers to all of the problems, along with explanations of the answers and two and three dimensional colour versions of many of the diagrams, are, however, available via the internet. Links to the appropriate web pages will be found at: http://www.bangor.ac.uk/ch/mnhome.htm.

This book is not a comprehensive coverage of all of chemistry so it has been assumed that the reader has some previous chemical knowledge. In particular, to get the most out of this book, the reader should have previously studied chemical bonding (valence bond and molecular orbital approaches), basic organic reactions, and interpretative NMR spectroscopy. However, where topics such as these which are not immediately related to stereochemistry are required in the text, then a brief review of the area is given and references to further reading are included at the end of the chapter.

A number of my colleagues in Bangor (Professor M.S. Baird, Dr P.K. Baker and Dr J.N. MacDonald) along with external reviewers have assisted in the preparation of this book by reading all or part of the manuscript at various stages of completion and making helpful suggestions as to its improvement. I thank them for their efforts in removing errors from the text, and accept responsibility for any remaining typographical or scientific errors. It would, however, be greatly appreciated if readers could let me know if they find any errors so that these can be corrected in future versions of the book.

Michael North, CChem FRSC
Bangor 1997

1 Molecular structure and bonding

1.1 Isomerism

Isomers are different compounds that share the same molecular formula. Common examples include ethanol **1.1** and dimethyl ether **1.2**, both of which have the molecular formula C_2H_6O; and 1,2-dimethylbenzene **1.3**, 1,3-dimethylbenzene **1.4**, and 1,4-dimethylbenzene **1.5**, all of which share the molecular formula C_8H_{10}. In both of these cases, the connectivity of the atoms is different, i.e. in compound **1.1** the order in which the carbon and oxygen atoms are bonded is C–C–O, whilst in **1.2** the order is C–O–C. Because they differ in the connectivity of their atoms, compounds **1.1** and **1.2** are said to be constitutional isomers (also sometimes called structural isomers), as are compounds **1.3–1.5**.

$$H_3C-CH_2-OH \qquad\qquad H_3C-O-CH_3$$

$$\textbf{1.1} \qquad\qquad\qquad \textbf{1.2}$$

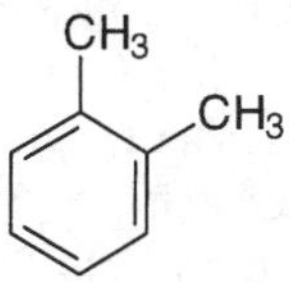

1.3 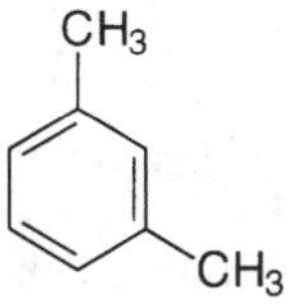   **1.4**    **1.5**

Constitutional isomers have different chemical and physical properties. This is markedly apparent in the case of isomers **1.1** and **1.2**, since ethanol is a liquid at room temperature and undergoes all the chemical reactions that would be expected for a primary alcohol (dehydration to an alkene, oxidation to an aldehyde or acid etc.). Dimethyl ether **1.2** on the other hand is a gas at room temperature and is unaffected by the reaction conditions and reagents used to dehydrate or oxidize ethanol. This difference in reactivity is to be expected since the two compounds contain different functional groups.

Isomers **1.3–1.5** are more closely related since they contain the same functional groups (two methyl groups attached to a benzene ring), and differ only in the relative positions of these functional groups. Such isomers are often called

positional isomers, but note that positional isomerism is a subclass of constitutional isomerism. Once again the three isomers have different physical properties, for example the boiling points are 144°C, 139°C and 138°C for **1.3**, **1.4** and **1.5** respectively. In general, these three isomers will undergo the same types of chemical reaction (benzylic oxidation for example), but will do so at different rates, again illustrating their different chemical properties.

$$\underset{\textbf{1.6}}{\overset{\displaystyle \underset{\overset{1}{H_3C}}{\underset{\underset{H\ 2\quad 3\ H}{}}{\diagdown}}\,\overset{4}{CH_3}}{C=C}}\qquad\qquad \underset{\textbf{1.7}}{\overset{\displaystyle \underset{\overset{1}{H_3C}}{\underset{\underset{H\ 2\quad 3\ CH_3}{\underset{4}{}}}{\diagdown}}\,H}{C=C}}$$

There is, however, another type of isomerism, one in which all of the atoms in the two isomers **do** have the same connectivity. A familiar example is found in 1,2-disubstituted alkenes such as compounds **1.6** and **1.7**. In both of these isomeric compounds, the order in which the carbon atoms are joined together is C1–C2=C3–C4 and the only difference between them is that in isomer **1.6** the two methyl groups are on the same side of the double bond, whilst in isomer **1.7** the two methyl groups are on opposite sides of the double bond. Any pair of isomers which have the same connectivity of their atoms but which differ in the relative orientation of those atoms are called **stereoisomers**.

Stereoisomers are the topic of this book and the following chapters will investigate the different structural features which are responsible for stereo-isomerism, and discuss the chemical, biological and physical consequences of the formation of stereoisomers. Both organic and inorganic compounds can exhibit stereoisomerism, and examples of each will be found throughout this book. Essentially, stereochemistry is concerned with the shapes of molecules, and the consequences of a molecule adopting a particular shape.

Later in this chapter, the way in which the shape of a molecule may be predicted using Valence Shell Electron Pair Repulsion Theory (VSEPR) will be introduced, and the nature of the bonding found in the most common chemical structures will be discussed. At the end of this chapter, the various classifications of stereoisomers will be introduced and these will be discussed in more detail throughout the remainder of this book. However, many of the structures seen later in this chapter are three dimensional, and before they are discussed it is necessary to understand the conventions used when representing three dimensional structures on a two dimensional piece of paper.

1.2 Drawing three dimensional chemical structures

Chemistry is dominated by two structures: the tetrahedron and the octahedron. A tetrahedral structure is found whenever a carbon atom is bound to four groups via four single bonds, whilst many transition metal complexes are octahedral.

Since both of these structures are three dimensional they cannot easily be drawn on a two dimensional piece of paper. To overcome this problem, chemists have developed a number of conventions that will be introduced here.

The usual way of drawing tetrahedral and octahedral structures is as shown in structures **1.8** and **1.9** respectively. In structure **1.8**, two of the bonds are drawn with normal (–) symbols, and represent bonds in the plane of the paper. The other two bonds, however, are drawn with a wedge (➤) and a hash (·····ııı) respectively. The wedge represents a bond in the direction drawn, which is also coming out of the paper towards you, whilst the hash represents a bond in the direction drawn and going into the paper away from you. The same symbols are used in structure **1.9** to illustrate the orientation of the bonds in an octahedral structure.

1.8   **1.9**

When used properly, this system of representing three dimensional structures with wedges and hashes provides a very simple and effective way of illustrating the structures of molecules. However, the correct use of the system requires some care. Structures **1.8** and **1.9** can be rotated to give a number of equivalent representations as shown in **Figure 1.1a** for a tetrahedral structure. It is very

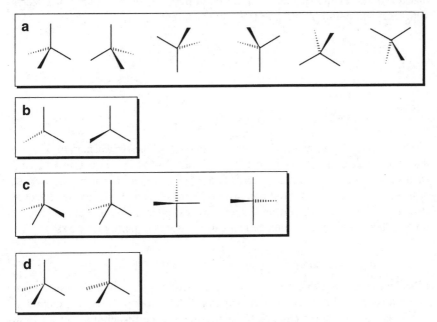

Figure 1.1 Correct (**a** and **b**) and incorrect (**c** and **d**) representations of a tetrahedron.

important, however, that a tetrahedral structure is drawn with the wedge and the hash on the same side of the structure, and that the bond angles are drawn to resemble the 109° 28′ angles found in a perfect tetrahedron. A number of nonsense structures are shown in **Figure 1.1c**.

Another point to note when drawing structures with wedges and hashes is that the wedge and the hash are meant to convey the perspective of the structure; that is the wide end of the wedge/hash should be towards the reader. This is usually not a problem for wedges, but hashes are often drawn the wrong way round or with no change in width as illustrated in **Figure 1.1d**. Such structures will not be used in this text but they are common in the chemical literature.

Another frequently used variation of structures drawn with wedges and hashes is to omit either the wedge or the hash when it represents a bond to a hydrogen atom. Thus only three bonds are drawn as in **Figure 1.1b**, and the reader is left to insert the fourth bond in the appropriate place. This can be a very useful convention, particularly for large, complex structures where the structural diagram can be appreciably simplified and clarified. However, it will not be employed in this book.

One final variation on the use of wedges and hashes, is the use of a curly bond (∿) to represent a bond of unknown direction. Thus in structure **1.10**, the structure represents the situation where either the C–F bond or the C–H bond may be pointing towards the reader. This curly bond is used in two different situations: when the three dimensional structure is unknown or when it is known that a mixture of the species with the bond pointing forwards and the species with the bond pointing backwards are both present.

1.10

1.11

The wedge/hash system can easily be extended to the representation of more than one tetrahedral centre within a molecule, as illustrated in structure **1.11**, and such structures are sometimes referred to as 'flying wedge' formula. The wedge/hash system is the most common way of representing the three dimensional structure of a molecule. However, there are two other systems: Fischer projections and Newman projections, which can be useful in particular cases.

1.2.1   Fischer projections

When drawing complex three dimensional structures, it is sometimes convenient to use a simplified type of structure called a **Fischer projection**, an example of which is shown in structure **1.12**, alongside the corresponding wedge/hash structure **1.13**. When viewing Fischer projections, it is important to remember

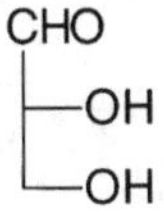

Fischer
Projection

1.12

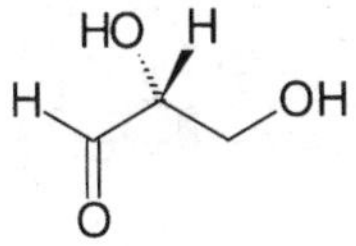

1.13

the conventions used which are that bonds to hydrogen atoms are omitted, and bonds drawn horizontally are coming out of the paper towards you whilst bonds drawn vertically are going into the paper away from you. The latter of these conventions can be remembered via the mnemonic:

HORIZONTAL
OUT
VERTICAL
I N
S YSTEM

To convert a Fischer projection back into a wedge/hash structure, first add hydrogen atoms to complete the valence at each carbon atom, then draw vertical bonds with hashes, and horizontal bonds with wedges. Finally, rotate the structure to minimize the number of wedges and hashes which are needed. This is illustrated in **Figure 1.2** for structure **1.12**. The most difficult stage in this process is the last step, rotating the structure without changing the three dimensional orientation of the substituents, and a molecular model can be of assistance in this respect.

To convert a wedge/hash structure into a Fischer projection, rotate the structure so that the main carbon chain is vertical and the carbon atom which would be numbered 1 in naming the compound is at the top. Next rotate the vertical bonds until all the horizontal bonds are coming out of the page, then redraw all wedges and hashes as normal bonds and delete bonds to hydrogen atoms to obtain the Fischer projection as shown in **Figure 1.3** for compound **1.14**. Again, this can be a difficult process and is much simplified by the construction of a molecular model. Nowadays, Fischer projections are normally only used in carbohydrate chemistry. The Fischer projection of the naturally

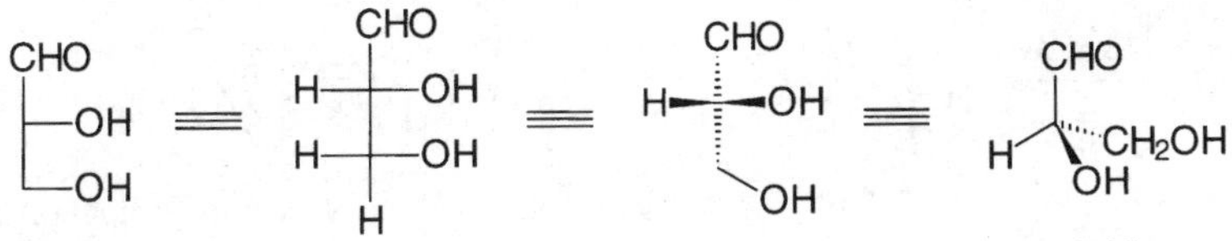

1.12

1.13

Figure 1.2 Conversion of Fischer projections into wedge/hash structures.

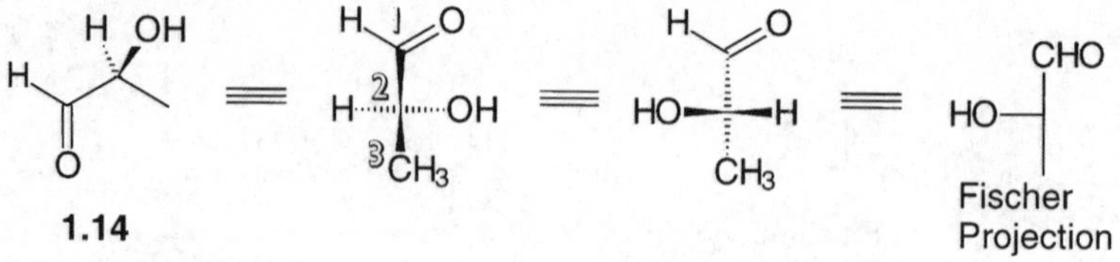

1.14

Figure 1.3 Conversion of a wedge/hash structure into a Fischer projection.

occurring stereoisomer of glucose is shown in structure **1.15**. Most carbohydrates, including glucose, exist predominantly as one or more cyclic hemiacetals (cf. Chapter 8, section 8.10.2), so a Fischer projection, which has to be of the acyclic form, actually represents only a minor component of the carbohydrate structure.

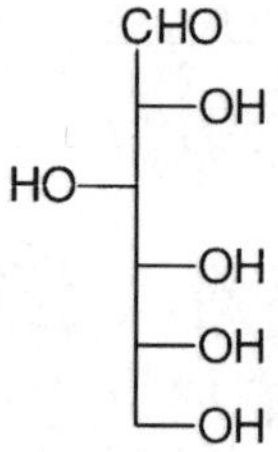

Glucose

1.15

1.2.2 Newman projections

Another useful way of representing the three dimensional arrangement of the atoms in a molecule is the Newman projection. In this representation, the molecule is viewed along the bond about which the arrangement of the atoms is being considered, an example is shown in **Figure 1.4** for two different stereoisomers of 1-bromo-2-chloroethane **1.16a,b**. In this example, the molecule is

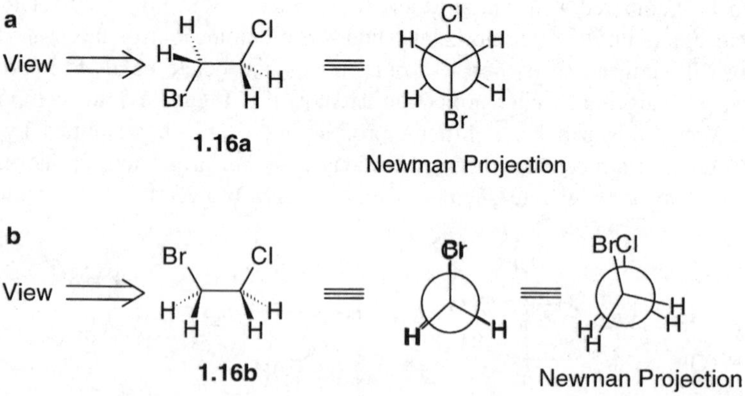

1.16a

1.16b

Figure 1.4 Wedge/hash and Newman projections of two stereoisomers of 1-bromo-2-chloroethane.

viewed along the carbon–carbon bond, so that the rear carbon atom is hidden from view by the front carbon atom which is represented by the circle. The bonds from the front carbon atom are then drawn as lines emanating from the centre of the circle, whilst bonds to the rear carbon atom are drawn so that they stop at the circumference of the circle representing the front carbon atom. When the bonds to the atoms connected to the rear carbon atom would be hidden by the bonds to the front carbon atom, as in **Figure 1.4b**, then the bonds are offset slightly as shown so that they remain visible.

Newman projections are particularly useful for viewing the relative arrangements of atoms around a single bond. It is immediately apparent from the Newman projections that for example in **Figure 1.4a** the two halogens are arranged at 180° to one another whilst in **Figure 1.4b** the torsional angle between the two halogens is 0°.

1.3   VSEPR and the shape of molecules

Atoms are composed of electrons, protons and neutrons, with the protons and neutrons forming the nucleus of the atom and the electrons orbiting around the nucleus. Electrons are negatively charged, whilst protons carry a positive charge and neutrons are uncharged. The electrons can occupy different orbits (or atomic orbitals), the energy of which increases as the mean distance of the orbital from the nucleus increases, and which are occupied by electrons starting with the lowest energy orbitals. The lowest energy orbital is called the 1s orbital and can hold two electrons, with the 2s orbital being next highest in energy, closely followed by the 2p orbitals. There are three 2p orbitals each of which can hold two electrons, so giving a maximum of eight electrons occupying the second level of orbitals (two in the 2s orbital and six in the 2p orbitals).

It is convenient to group orbitals together into shells, based upon their relative energies. The first shell comprises just the 1s orbital, but the 2nd shell comprises both the 2s and 2p orbitals. Similarly, the 3rd shell comprises the 3s and 3p orbitals, whilst the 4th shell includes the 4s, 4p and 3d orbitals as shown in **Table 1.1**. Only the outermost shell of electrons are involved in chemical bonding and in determining the orientation of the bonds around an atom. This

Table 1.1 Shells of electrons in order of increasing energy

Shell number	Orbitals included	Maximum number of electrons held
1	1s	2
2	2s, 2p	8
3	3s, 3p	8
4	4s, 4p, 3d	18
5	5s, 5p, 4d	18
6	6s, 6p, 5d, 4f	32
7	7s, 7p, 6d, 5f	32

shell is called the **valence shell**, all of the other occupied shells being called **core shells**. In going from atoms to molecules, the atomic orbitals are transformed into a corresponding set of molecular orbitals, with the molecular orbitals retaining the same order of energies as the atomic orbitals from which they are constructed.

Consider the bonding in a simple molecule such as H–F **1.17**. The hydrogen atom is surrounded by just two electrons, the two which form the H–F bond and which occupy a molecular orbital derived from the hydrogen 1s atomic orbital. The fluorine atom, however, is surrounded by a total of 10 electrons, the first two of which occupy the 1s orbital, and the remaining eight occupy the molecular orbitals constructed from the 2s and 2p atomic orbitals of the second shell, which is the valence shell. The two electrons in the 1s orbital are not shown in structure **1.17**, and need not be considered further. Only two of the eight electrons in the second shell are involved in the chemical bond to the hydrogen atom, the remaining six electrons being held as lone pairs of electrons around the fluorine atom. Thus the fluorine atom is surrounded by four pairs of electrons, one bonding pair and three lone pairs. It is the number of bonding and lone pairs of electrons in the valence shell of an atom that determines the orientation of the bonds around that atom. A series of rules (the Valence Shell Electron Pair Repulsion or VSEPR rules) have been developed to allow the shapes of molecules to be predicted; these are given below.

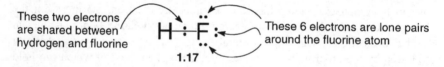

1.17

VSEPR rules

Rule 1   Electron pairs in the valence shell orientate themselves so as to maximize the distance between them and so minimize their mutual repulsion.

Rule 2   Lone pair/lone pair repulsions are stronger than lone pair/bond pair repulsions, which are stronger than bond pair/bond pair repulsions.

Rule 3   The repulsion between electron pairs at 90° or less to one another is significantly more severe than that between electron pairs at angles >90°, so the former is the dominant factor in determining the structure.

Rule 4   Only σ-bonds (σ-bonds are single bonds and one of the bonds in a double or triple bond) are important in determining the basic geometry of a molecule.

Rule 5   The electrons in double and triple bonds occupy more space than those in single bonds and so repel other electrons more strongly than the electrons in single bonds repel each other, although not as strongly as lone pairs of electrons repel other bond pairs.

Table 1.2 Relationship between geometry and number of electron pairs

Electron pairs	Electron pair structure	Possible bond structures
2	Linear	Linear
3	Trigonal planar	Trigonal planar, bent
4	Tetrahedral	Tetrahedral, trigonal pyramidal, bent
5	Trigonal bipyramid	Trigonal bipyramid, seesaw, T-shaped, linear
6	Octahedral	Octahedral, distorted square pyramidal, square planar

As electrons are negatively charged, each pair of electrons will repel all of the other pairs and they will arrange themselves in the geometry that minimizes these electronic repulsions as indicated by rule 1. The preferred geometry depends upon the number of electron pairs as shown in **Table 1.2**. In **Table 1.2**, two different types of structure are given: the electron pair structure, which is most useful for stereochemical purposes, is the geometry adopted by all of the electron pairs around the atom under consideration. The bond structure is the geometry of just the bond pairs of electrons around the central atom, i.e. lone pairs of electrons are not considered in these structures. The bond structure is often used to describe the shape of the molecule.

The geometries given in **Table 1.2** represent the five basic shapes which electron pairs can form around an atom. However, lone pairs of electrons are held nearer to the nucleus of an atom than are bond pairs. This causes electron/electron repulsions involving lone pairs that are larger than the corresponding repulsions between bonding pairs and leads to a distortion of the bond angles in these structures when lone pairs of electrons are involved (rule 2). Therefore, the bond angle between two lone pairs will be larger than that between a lone pair and a bond pair, which again will be larger than that between two bond pairs of electrons.

The repulsion between two electron pairs depends upon the average distance between them, which is related to the angle between them. The repulsion is largest at small angles between the electron pairs and falls off rapidly as the angle between the electron pairs increases. Thus a compound will adopt the structure which minimizes the number of <90° interactions between two electron pairs as stated in rule 3.

When an atom is involved in a double or triple bond, that bond is treated as a single pair of bonding electrons in structure determination, although there are actually two or three pairs of electrons involved in the bond (rule 4). The presence of more than one pair of electrons in the bond is, however, apparent in the geometry of the resulting structure, since the four electrons in a double bond (or six in a triple bond) will repel other bond pairs more strongly than the two electrons in a single bond repel other bond pairs, resulting in a distortion of the bond angles (rule 5).

To see how VSEPR can be used to predict the shapes of molecules, a number of examples will be considered, starting with molecules containing just two bond pairs around an atom, and considering all possibilities up to and including six electron pairs around an atom. Consider first, beryllium dichloride $BeCl_2$ **1.18**: the beryllium atom is surrounded by just two pairs of electrons (in the valence shell), both of which are bond pairs to the chlorine atoms. Hence, the structure will be linear with the two pairs of electrons at 180° to one another, as far apart as it is possible for them to get (rule 1 and **Table 1.2**). Each chlorine atom is, however, surrounded by eight valence electrons, three lone pairs and a bond pair. These four electron pairs will arrange themselves in a tetrahedral arrangement around the chlorine atom (rule 1) but this will be distorted since the lone pairs repel one another more strongly than they repel the bond pair (rule 2), causing the angle between the lone pair of electrons to be >109° and that between the lone pairs and bond pairs to be <109°.

Cl—Be—Cl   H—C≡N(**:**)

1.18   **1.19**

Another example of a linear molecule is hydrogen cyanide HCN **1.19**. This time, the central carbon atom is surrounded by eight valence shell electrons, six in the triple bond to nitrogen and two in the bond to hydrogen. The six electrons shared with the nitrogen atom count as a single bond pair (rule 4), so the carbon is surrounded by two bond pairs, and again the structure is linear. The nitrogen atom is also surrounded by eight valence shell electrons, six in the triple bond to carbon and a lone pair, which again for stereochemical purposes constitute just two electron pairs (rule 4), and so are arranged linearly as shown in structure **1.19**. For clarity, in structure **1.19** only the major lobe of hybridized orbital containing the lone pair of electrons is shown. The same convention will be used throughout this book when molecular orbitals are displayed.

Examples of compounds where one or more atoms are surrounded by three valence shell electron pairs include boron trifluoride BF_3 **1.20**, methanal H_2CO **1.21**, and *N*-methyl methanimine **1.22**. For BF_3 **1.20**, the central boron atom is surrounded by just six valence electrons in three bond pairs. The compound therefore adopts a trigonal planar structure in which all of the bond angles are exactly 120°, this being the furthest that the three electron pairs can get away from one another (rule 1 and **Table 1.2**).

The situation with methanal **1.21** is very similar, the carbon atom is again surrounded by three bond pairs (ignoring the second pair of electrons in the

F
|
F—B—F

1.20

O
‖
H—C—H

1.21

Me—N(**:**)
‖
H—C—H

1.22

double bond; rule 4) and so there will be a trigonal planar arrangement of the atoms around this carbon. In this case, however, the bond angles will not be exactly 120° since there are four electrons in the C=O bond which therefore repel the electrons in the C–H bonds more strongly than the C–H bonds repel one another (rule 5). This leads to the O=C–H angle being slightly larger than 120°, and the H–C–H angle being less than 120°. The oxygen atom in compound **1.21** is also surrounded by three valence shell electron pairs (rule 4), a bond pair and two lone pairs. Thus the distribution of the electrons around the oxygen atom will also be trigonal planar with a slightly larger bond angle between the lone pairs than between the lone pair and bond pair.

The carbon atom in imine **1.22** has a slightly distorted trigonal planar geometry exactly analogous to the situation with the carbon atom of compound **1.21**. The nitrogen atom of **1.22** is also surrounded by three valence shell electron pairs, two bond pairs (ignoring the second pair of electrons in the double bond; rule 4), and a lone pair. Thus the electron pair geometry is again distorted trigonal planar. The bonds around this nitrogen atom are usually described as having a bent geometry, as the lone pair of electrons is not included in bond structures.

1.23   **1.24**   **1.25**

The archetypal example of a molecule in which the central atom is surrounded by four pairs of valence electrons is methane **1.23**. In this case, all four electron pairs are bond pairs and the structure is tetrahedral with all bond angles of 109° 29′ (rule 1 and **Table 1.2**). If one of the bond pairs is replaced by a lone pair of electrons, as is the case in ammonia **1.24**, then the electrons are again distributed tetrahedrally around the nitrogen atom, although this time the tetrahedron will be distorted with a larger bond angle between the lone pair and bond pairs than between the bond pairs (rule 2). The bond structure of ammonia is described as a trigonal pyramid. Replacement of a further bond pair of electrons by a lone pair gives a structure such as found in water **1.25**. Again, the electrons are distributed in a distorted tetrahedron around the oxygen atom, although the bond structure is now planar and bent.

Compounds in which an atom is surrounded by five bond pairs of electrons adopt a trigonal bipyramidal structure as shown for PF_5 **1.26** (rule 1 and **Table 1.2**). Note that in this case the fluorine atoms adopt two different sorts of position around the phosphorus atom, three of the fluorine atoms adopt equatorial positions, whilst the other two fluorines adopt axial positions. If one of the bond pairs is replaced by a lone pair of electrons as in SF_4 **1.27**, then there are now two possible structures since the lone pair can occupy an equatorial **1.27a** or axial **1.27b** position. To decide which of these two possible structures will be formed, the distance (i.e. angle) between the lone pair and the bond pairs is considered (rule 3), since these groups repel one another more strongly than bond pairs repel other bond pairs. In structure **1.27a**, there are two bond pairs at 90° to the lone pair, and two at 120° to the lone pair, whilst in structure **1.27b** there are three bond pairs at 90° to the lone pair and one at 180° to the lone pair. Since structure **1.27a** contains only two bond pairs at 90° to the lone pair, whilst there are three such bond pairs in structure **1.27b**, structure **1.27a** is the lower energy structure and the one that is actually formed. The shape of the bonds in structure **1.27a** is called a seesaw geometry, since if structure **1.27a** is rotated by 90° so that the wedge and hash bonds are towards the bottom of the page, the structure will resemble a seesaw. Note how in structure **1.27a**, the axial fluorines are repelled more by the lone pair than they are by the bond pairs (rule 2), giving a structure where they are slightly off vertical. Similarly, the angle between the two equatorial fluorines is slightly less than 120° (rule 2).

1.26   **1.27a**   **1.27b**

If a molecule contains an atom surrounded by three bond pairs and two lone pairs of electrons, as is the case for $BrCl_3$ **1.28**, then there are three possible structures **1.28a–c** depending on whether zero, one, or two lone pairs adopt axial positions. Structure **1.28b** with one lone pair equatorial and one axial is clearly the least likely to be formed, since the two lone pairs are at only 90° to one another in this structure (rule 3). In both structure **1.28a** and **1.28c**, the lone pairs are >90° from one another (120° and 180° respectively), so the repulsion between the lone pairs is minimal. However, in structure **1.28a** there are only four 90° lone pair/bond pair interactions, whilst there are six such interactions in **1.28c**. Hence, a molecule such as $BrCl_3$ adopts a T-shaped bond structure in which both lone pairs are equatorial as shown in **1.28a**. The structure is however, distorted slightly from an ideal T shape since the lone pairs will repel the axial chlorines more strongly than will the equatorial bond pair.

1.28a   **1.28b**   **1.28c**

With six bond pairs around an atom, an octahedral structure is the geometry which allows the electron pairs to separate as far as possible as shown in structure **1.29** for SF_6 (rule 1 and **Table 1.2**). In an octahedral structure, each bond pair of electrons is orientated at 90° to four other bond pairs and at 180° to the fifth bond pair. Also, in an octahedral structure all six positions are equivalent. If one of the bond pairs is replaced by a lone pair, as in ClF_5, **1.30**, then the electrons still adopt an octahedral structure and the bond structure is now described as a square based pyramid. However, there are now two different positions for the fluorines to occupy, one axial position trans to the lone pair and four equatorial positions at 90° to the lone pair. Also, since lone pair/bond pair repulsions are stronger than bond pair/bond pair repulsions, the four equatorial fluorines will actually be placed such that the angle between them and the lone pair is slightly greater than 90° and, correspondingly, the angle between the axial and equatorial fluorines will be slightly less than 90°. With a structure containing four bond pairs and two lone pairs such as XeF_4 **1.31**, the two lone pairs will adopt positions trans (180°) to one another, leaving the bond pairs to form a square planar structure.

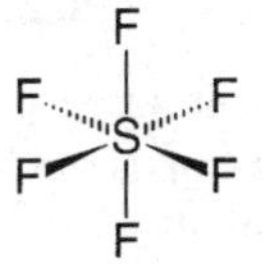

| 1.29 | 1.30 | 1.31 |

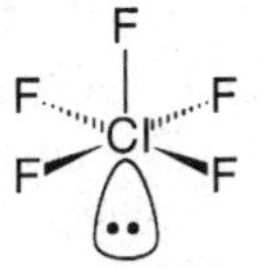

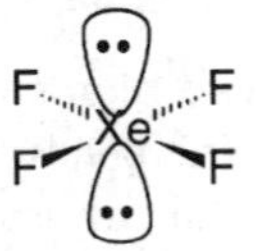

Hence using VSEPR theory, it is possible to predict the geometry of the electrons and/or the bonds around each atom in a molecule and so build up the shape of the molecule as a whole. It should be remembered, however, that VSEPR theory is only a simplified model of the bonding in molecular structures. The majority of compounds contain carbon atoms, and the three possible geometries for the orientation of groups around a carbon atom are shown in **Figure 1.5**. These geometries are very important stereochemically, and will be seen on many occasions throughout this book.

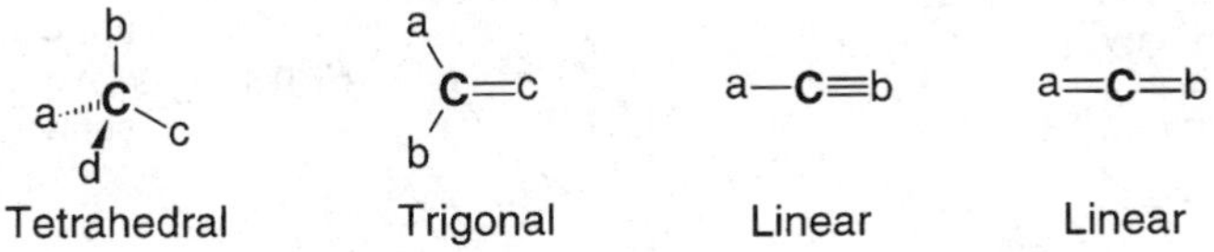

| Tetrahedral | Trigonal | Linear | Linear |

Figure 1.5 Geometries of carbon atoms.

1.4 Structure and bonding

A full description of chemical bonding is beyond the scope of this text, however, some aspects of bonding have important stereochemical consequences so a brief discussion of the salient points is appropriate. Almost always, whenever two atoms are connected by a single bond then this bond will be a σ-bond which is

formed by the overlap of orbitals along the axis of the bond as is shown in
structure **1.32** for the C–C bond of ethane. Since the σ-bond is formed along the
axis of the bond, the extent of orbital overlap is unaffected by rotation around
this bond. Thus in ethane, the two CH_3 groups can rotate around the central C–C
bond, and adopt an infinite number of orientations relative to one another, two
examples of which are shown in the Newman projections **1.32a,b**. One consequence of this rotation around the σ-bond is that, although VSEPR theory can
predict that both carbon atoms will have a tetrahedral geometry, it cannot predict
the relative orientation of these two tetrahedra.

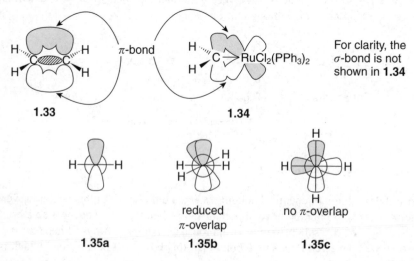

Whenever two atoms are bound together by two or three bonds, then the
second and third bonds are π-bonds rather than σ-bonds. π-Bonds may be
formed by orbital overlap between two p-atomic orbitals as shown for ethene
1.33, by overlap between a p and d-orbital (e.g. ruthenium alkylidene complex
1.34) or occasionally by the overlap of two d-orbitals. Unlike a σ-bond, the
orbital overlap in a π-bond is not equally distributed around the bond axis, but
rather occurs in two well defined lobes. This effectively prevents rotation around
a double or triple bond, since any such rotation would result in a decrease in
orbital overlap, and hence a loss of bonding in the structure as shown in the
Newman projections **1.35a–c** along the C=C of ethene. This lack of rotation is
the reason why alkenes such as **1.6** and **1.7** discussed at the start of this chapter
exist as separate, stable, stereoisomeric species.

For each bonding molecular orbital (σ or π), there will be a corresponding antibonding molecular orbital designated σ^* or π^*. A σ^* orbital is orientated along the bond axis but the lobes of the orbital are orientated away from one another and do not overlap, as shown in structure **1.36**. Similarly, the lobes of a π^* orbital are orientated away from the axis of the bond as shown in structure **1.37**. Antibonding orbitals are higher in energy than bonding orbitals, and for neutral molecules no electrons are placed in antibonding orbitals. Anionic species, however, may contain electrons in antibonding orbitals, and antibonding orbitals are also important in determining the stereochemical consequences of chemical reactions as will be discussed in Chapters 9 and 10.

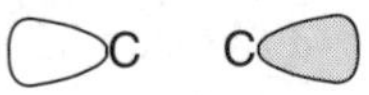

1.36 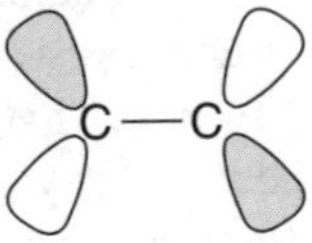　　　　　　　　　　**1.37**

1.5　Classification of stereoisomers

There are two different classification systems for subdividing stereoisomers. These two classification systems (into enantiomers and diastereomers; and into conformational isomers and configurational isomers) are completely independent of one another, both are widely used and, indeed, often used together.

1.5.1　Enantiomers and diastereomers

Consider the structures **1.38** and **1.39** in which the vertical line represents a mirror placed between them. Structure **1.39** is clearly the mirror image of structure **1.38**. Furthermore, these two structures are not superimposable, as diagrams **1.40**–**1.42** all represent the same structure as **1.38** but have each been rotated about the vertical C–Me bond. As can be seen, none of these structures

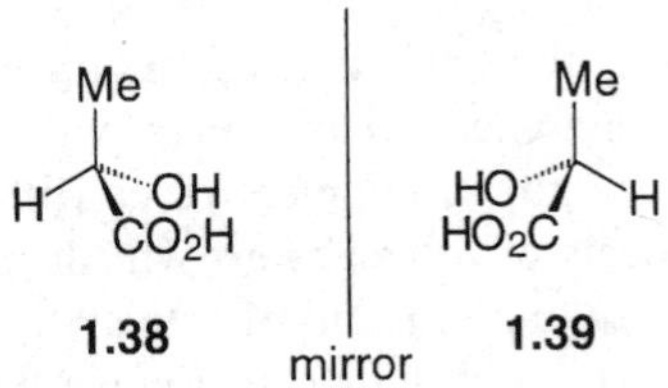

1.38　　mirror　　**1.39**

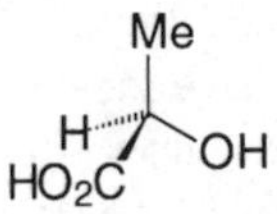

1.40 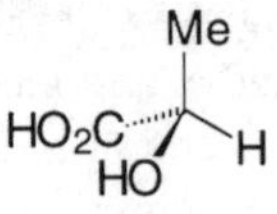　　　　　　**1.41** 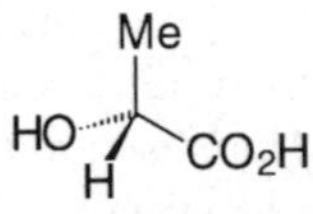　　　　　　**1.42**

is identical to (i.e. superimposable on) structure **1.39**. In each case, two of the four substituents on the central carbon atom are correctly orientated to match structure **1.39**, but the other two substituents are interchanged. No other rotation of structure **1.38** would produce structure **1.39** (a molecular model may be useful in visualizing this), so these two structures represent separate compounds which are non-superimposable mirror images of one another. Structures **1.38** and **1.39** are clearly isomers of one another as they have the same molecular formula, and they are stereoisomers as they have the same connectivity: a carbon atom bonded to a hydrogen, a methyl group, a hydroxyl group and a carboxylic acid group. Stereoisomers that are non-superimposable mirror images of one another are called **enantiomers**.

Enantiomeric structures have identical bond lengths and angles. The torsional angles also have identical magnitudes, the only difference between these two structures (and between any two enantiomers) is in the sign of the torsional angles. In one structure these will be positive, whilst in the other enantiomer they will have the same magnitude but will be negative. This is illustrated for the case of enantiomers **1.43** and **1.44** in **Table 1.3**. The two enantiomers are said to possess opposite **absolute configurations**.

1.43   **1.44**

Table 1.3 Bond lengths and angles for enantiomers **1.43** and **1.44**

Bond length (Å)	1.43	1.44	Bond angle (°)	1.43	1.44	Torsional angle (°)	1.43	1.44
C–H	1.1104	1.1104	H–C–F	110.9	110.9	Br–C–H–Cl	+119.2	−119.2
C–F	1.3974	1.3974	H–C–Cl	107.3	107.3	Cl–C–F–H	−119.1	+119.1
C–Cl	1.7922	1.7922	H–C–Br	107.1	107.1	F–C–H–Cl	−121.0	+121.0
C–Br	1.9538	1.9538	F–C–Cl	110.7	110.7	F–C–Cl–H	+121.3	−121.3
			F–C–Br	109.7	109.7	Br–C–Cl–F	+122.0	−122.0
			Cl–C–Br	111.0	111.0	Br–C–F–H	+118.2	−118.2

Since the magnitudes of the bond lengths, angles and torsional angles are identical for a pair of enantiomers, enantiomers will have the same energies. Therefore, whenever a pair of enantiomers react with another chemical, the transition states and products of the reactions will themselves be enantiomeric and so of equal energy, resulting in identical activation energies (E_a) and Gibbs free energies of reaction (ΔG) as shown in **Figure 1.6**, where the solid and broken lines represent the reaction pathways for the two enantiomers. Hence, the reactions will occur at equal rates, and enantiomers will have identical chemical properties. There is one important exception to this rule, that is when a pair of enantiomers react with a chemical which can itself exist as a pair of enantiomers. This will be discussed in more detail in Chapter 3, section 3.

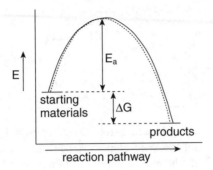

Figure 1.6 Reaction pathway/energy diagram for reaction of enantiomers.

Another consequence of enantiomers having identical bond lengths and bond angles, is that most physical techniques are incapable of distinguishing between them. Thus, for example, the infrared spectrum of a compound is determined by the bond lengths, bond angles and bond strengths as well as the atoms that are present in the molecule. Since these are identical in a pair of enantiomers, then the enantiomers will have identical infrared spectra. Similarly, enantiomers will also have identical NMR and UV spectra, etc.

Furthermore, enantiomers have identical molecular volumes and shapes which are identical except for being mirror images of one another. As a result, the way in which the molecules pack together and the magnitude of their intermolecular interactions will be identical for a pair of enantiomers. Thus any physical property that depends on molecular packing and/or intermolecular interactions will also be unable to distinguish between enantiomers (samples which are not composed of a single enantiomer of a compound may have different physical properties as will be discussed in Chapter 3, section 3.4). Examples of such techniques include melting point, boiling point and density. Thus enantiomers not only have identical chemical properties (unless reacting with other enantiomers as will be discussed in Chapter 3, section 3.3), they also have identical physical properties, with one important exception which will be introduced in Chapter 3, section 3.4.1.

$$H_3C \underset{H}{\overset{}{\diagdown}} C = C \overset{}{\underset{H}{\diagup}} CH_3 \qquad H_3C \underset{H}{\overset{}{\diagdown}} C = C \overset{}{\underset{CH_3}{\diagup}} H \qquad F \underset{H}{\overset{}{\diagdown}} C = C \overset{}{\underset{H}{\diagup}} F \qquad F \underset{H}{\overset{}{\diagdown}} C = C \overset{}{\underset{F}{\diagup}} H$$

1.6   **1.7**   **1.45**   **1.46**

The two stereoisomers **1.6** and **1.7** introduced at the beginning of this chapter, however, are not mirror images of one another and so cannot be enantiomers. By definition, any pair of stereoisomers which are not enantiomers of one another are called **diastereomers**. **Table 1.4**, lists the bond lengths, bond angles and torsional angles for a pair of diastereomers **1.45** and **1.46**. Comparison of the data in **Table 1.4** with that in **Table 1.3** illustrates the important difference

Table 1.4 Bond lengths and angles for diastereomers **1.45** and **1.46**

Bond length (Å)	1.45	1.46	Bond angle (°)	1.45	1.46	Torsional angle (°)	1.45	1.46
C–F	1.3228	1.3224	H–C–F	118.4	118.8	F–C=C–H	180	0
C–H	1.1023	1.1020	H–C=C	119.9	120.2	F–C=C–F	0	180
C=C	1.3427	1.3418	F–C=C	121.8	121.1	H–C=C–H	0	180

between enantiomers and diastereomers. Diastereomers have different bond lengths, bond angles and torsional angles. Hence they have different energies and will have different chemical properties. Often, diastereomers will have similar chemical properties since they contain the same functional groups, but will undergo reactions at a different rate. Occasionally, however, the diastereomers will have completely different chemical properties as shown in **Scheme 1.1**. Thus alkene **1.47** when heated to 180°C, loses water and forms a cyclic anhydride **1.49**. Alkene **1.48** which is a diastereomer of compound **1.47** however, cannot be dehydrated to a cyclic anhydride.

Diastereomers have different molecular shapes and volumes, as is particularly apparent for isomers **1.6/1.7** or **1.45/1.46**. As a result, they will pack together in different ways and will have different physical properties such as melting point, solubility, infrared spectra, etc. This difference in chemical and physical properties of diastereomers is very important and the consequences of it will be met throughout the rest of this book. It should also be noted that alkene isomers such as **1.6/1.7** are only one example of diastereomers, there are many other ways in which a compound can exist as diastereomers and these will be introduced in Chapters 2–4.

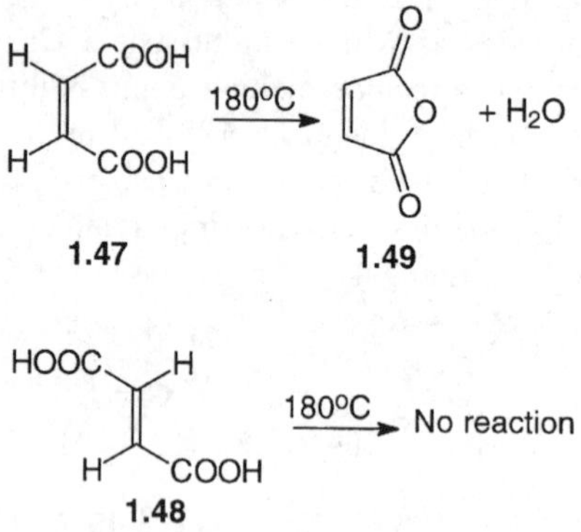

1.47   1.49

1.48

Scheme 1.1

1.5.2 Conformations and configurations

Structures **1.6** and **1.7** represent a pair of stereoisomers as do structures **1.32a** and **1.32b**. There is, however, a fundamental difference between these two pairs of stereoisomers: compounds **1.6** and **1.7** cannot be interconverted unless at least

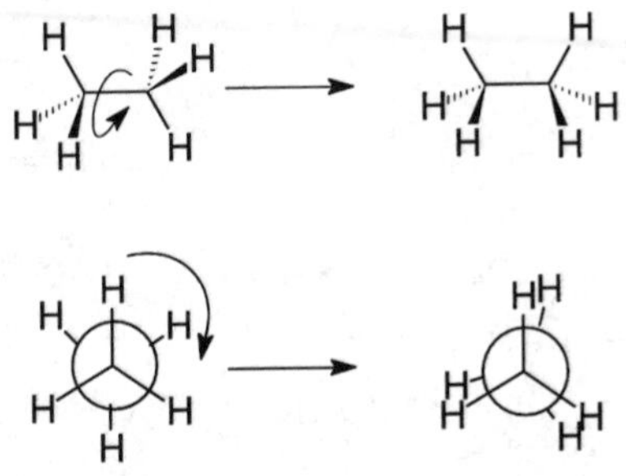

1.32a **1.32b**

Figure 1.7 Interconversion of conformational isomers **1.32a** and **1.32b** by rotation about the C–C single bond.

the π-bond component of the C=C double bond is broken (cf. section 1.4). Stereoisomers **1.32a** and **1.32b**, however, readily interconvert as rotation about the C–C single bond can easily occur as shown in **Figure 1.7**. Stereoisomers such as **1.6** and **1.7** which cannot be interconverted unless at least one chemical bond is broken are called **configurational isomers**, or **configurations**, whilst stereoisomers which can be interconverted without breaking a chemical bond (such as **1.32a** and **1.32b**) are called **conformations**. The subdivision of stereoisomers into conformations and configurations has been widely adopted and will be used throughout this book. However, the method is somewhat arbitrary and not without its limitations. There are two main problems which arise: temperature dependence and the situation with compounds containing delocalized electrons.

Although rotation around a σ-bond generally occurs readily, this does not mean that there is no energy barrier to such a rotation. This topic will be discussed in more detail in Chapter 8 but the energy profile for one of the simplest cases, rotation about the carbon–carbon bond of ethane is shown in **Figure 1.8**. It is apparent from **Figure 1.8**, that the staggered conformations **1.32a** occur at minima on the energy curve, whilst the eclipsed conformations **1.32b** occur at energy maxima. The energy barrier to rotation in this case is 3 kcal/mol (12.6 kJ/mol). As the size of the substituents attached to the σ-bond about which rotation is occurring increases, the energy barrier to be overcome during rotation also increases. The ability to overcome this energy barrier will depend upon the energy that is available to the molecule, and so will be temperature dependent. Thus, two stereoisomers which do not interconvert at a given temperature (and so may be classified as configurations) may interconvert at a higher temperature and so be classified as conformations. In such cases, the classification of the compounds as conformations or configurations is somewhat arbitrary.

Any point along the curve in **Figure 1.8** represents a conformation of ethane, thus there are an infinite number of conformations. There are, however, only three minimum energy conformations which in this case happen to be indistinguishable.

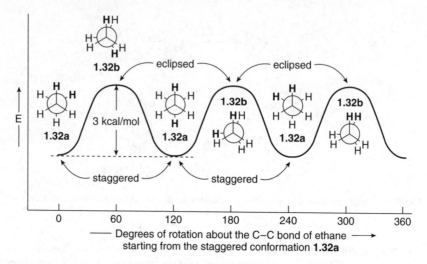

Figure 1.8 Energy profile for rotation about the carbon–carbon bond of ethane.

One hydrogen on each of the two carbon atoms has been emboldened in **Figure 1.8**, to show what happens to their relative orienation during a 360° rotation. Minimum energy conformations are called **conformers** or **conformational isomers**. In general, for any compound with one or more bonds about which rotation can occur, there will be an infinite number of conformations but a finite number of conformers.

The second problem with the classification of stereoisomers as conformations or configurations concerns compounds which possess delocalized electrons. A good example of this is found in amides, since the carbonyl to nitrogen bond is intermediate between a single and double bond due to delocalization of the nitrogen lone pair of electrons into the carbonyl bond as shown in **Figure 1.9**. Hence, stereoisomers **1.50** and **1.51** can be considered as a pair of conformations (structures **1.50a** and **1.51a**) or configurations (structures **1.50b** and **1.51b**). The choice of descriptor is again somewhat arbitrary, although stereoisomers about an amide are

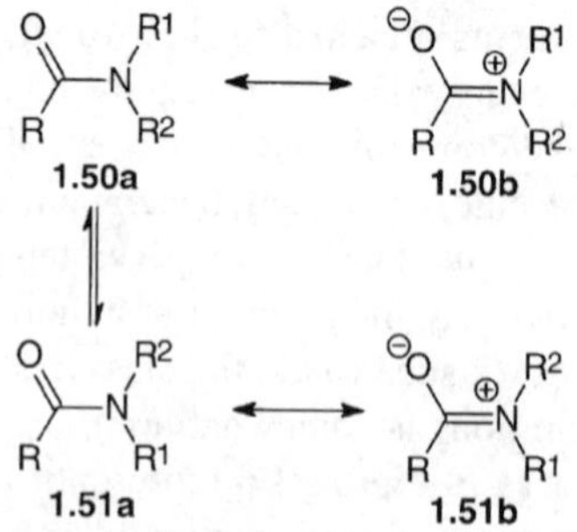

Figure 1.9 Electron delocalization in amides.

usually considered as conformations. A number of other classes of compounds show a similar effect, and these will be met throughout Chapters 2 and 3.

Both conformations and configurations are stereoisomers, and can be distinguished by referring to them as conformational stereoisomers and configurational stereoisomers respectively. It is normal practice, however, to refer to configurational stereoisomers just as stereoisomers. Hence, throughout the remainder of this book, if it is not specified that conformations are being discussed, then the reader should assume that the discussion is restricted to configurations.

1.6 Further reading

General
Stereochemistry of Organic Compounds E.L. Eliel and S.H. Wilen. Wiley: London, 1994, chapters 2 and 3.

VSEPR
Inorganic Chemistry K.F. Purcell and J.C. Kotz. Holt-Saunders: London, 1977, chapters 2 and 4.

Classification of isomers
H. Dodziuk. *Tetrahedron*, 1996, **52**, 12941.

1.7 Problems

1. Are the following pairs of structures identical, constitutional isomers, enantiomers or diastereomers of one another?

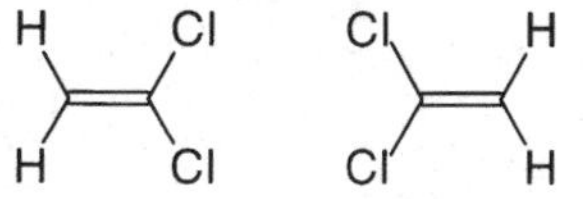

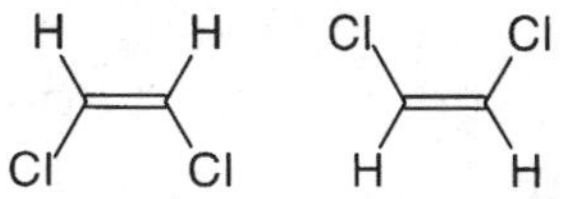

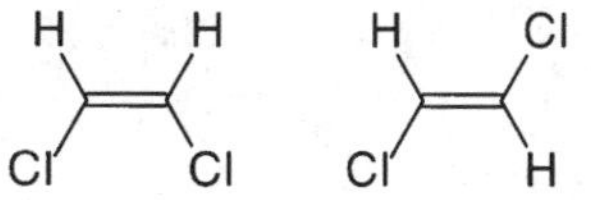

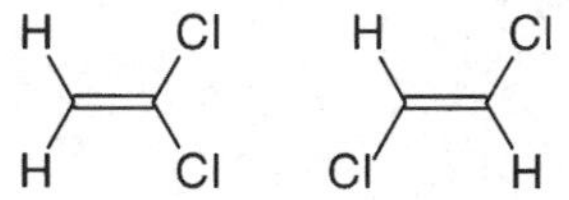

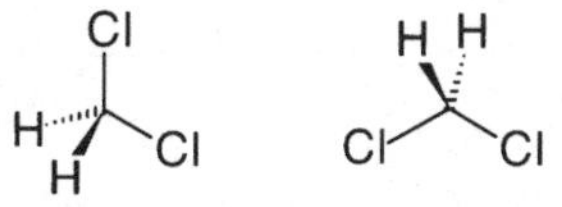

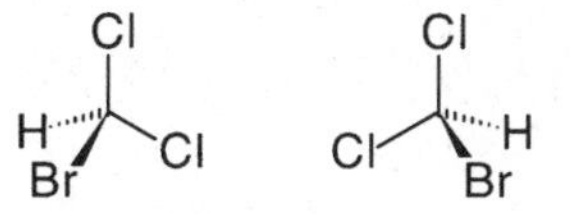

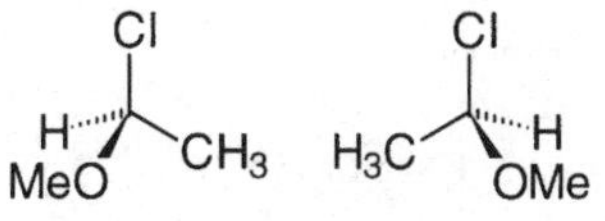

2. Draw all possible isomers of C_4H_6, and for each pair of structures identify whether they are constitutional or stereoisomers. Repeat the exercise for C_4H_8.

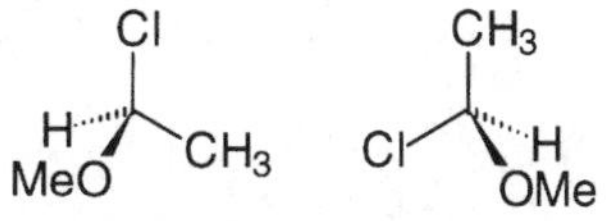

3. Use VSEPR theory to predict the orientation of both the electrons and the bonds around the underlined atom(s) in each of the following compounds. In each case, specify any deviations that would be expected from the 'ideal' structures listed in **Table 1.2**. Draw a representation of each structure, using wedges and hashes where necessary.

a $\underline{C}Cl_4$	b $Me_3\underline{P}$	c $\underline{Be}(OH)_2$	d $H_2\underline{S}$	e $\underline{CH_3OH}$
f $\underline{SOCl_2}$	g $\underline{S}O_2Cl_2$	h $\underline{S}O_2$	i $\underline{C}O_2$	j $\underline{S}O_3$
k $Me_3\underline{O}^+$	l $F-\underline{N}=\underline{N}-F$	m $Me_2\underline{O}$	n $H_2\underline{C}=\underline{C}Me_2$	o $\underline{C}H_3\underline{N}H_2$
p $Me_3\underline{S}^+$	q $[I-\underline{I}-I]^-$	r $[Me_4\underline{P}]^+[PF_6]^-$	s $\underline{I}F_5$	t $[\underline{I}Cl_4]^-$
u $\underline{Xe}F_2$	v $[\underline{S}O_4]^{2-}$	w $\underline{N}O_2$	x $[\underline{N}O_2]^+$	y $[\underline{N}O_2]^-$
z $[\underline{N}O_3]^-$				

4. Borane (BH_3) actually exists as a dimer (B_2H_6) with the structure shown below. Predict the shape of both BH_3 and B_2H_6.

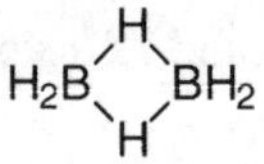

5. Draw a conformation versus energy diagram similar to **Figure 1.8** for rotation around: (a) the C–C bond of C_2Cl_6; (b) the CH_3–CH_2 bond of butane; and (c) the CH_2–CH_2 bond of butane. Specify any differences that you would expect from **Figure 1.8**. Draw a Newman projection corresponding to each maximum and minimum point on the diagram. It may be helpful to build a molecular model.

6. Convert the following wedge/hash structures into Fischer projections.

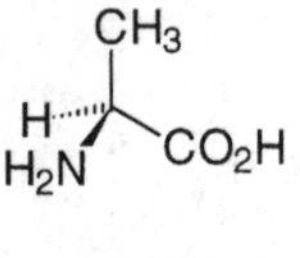

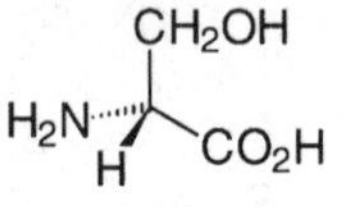

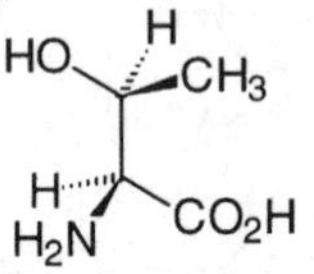

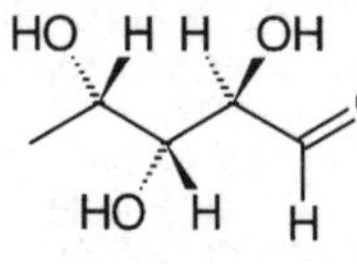

2 Cis–trans isomerism

2.1 Cis-trans isomerism in alkenes

In a discussion of stereochemistry, it is convenient to start with a consideration of cis–trans isomerism as it occurs in alkenes. This provides an area in which some fundamental stereochemical principles can be established using only familiar, easily drawn and easily visualized molecules. Hence, a difficulty in visualizing the molecules in question should not prevent an understanding of the stereochemistry. We will start by discussing the simple alkene, 2-butene **2.1**. It was shown in Chapter 1 (section 1.1) that there are two stereoisomers of compound **2.1** which can be represented by structures **2.2** and **2.3**.

$$\overset{1}{}\overset{2}{}\overset{3}{}\overset{4}{}$$
1 2 3 4
MeCH=CHMe

2.1

2.2

cis-2-butene

2.3

trans-2-butene

The two stereoisomers **2.2** and **2.3** can be distinguished by adding the stereochemical prefixes *cis* or **trans** to their name. The *cis*-isomer of a di-substituted alkene is the isomer in which the two substituents on the double bond (or the two hydrogens attached to the double bond) are adjacent to one another, as in the case of structure **2.2**. The *trans*-isomer is correspondingly the structure in which the two substituents on the double bond (or the two hydro-gens attached to the double bond) are opposite to one another, as in the case of structure **2.3**. The terms *cis* and *trans* are also widely used to describe the relative orientation of two groups. Thus in structure **2.2** the two methyl groups can be said to be *cis* to one another, as can the two hydrogen atoms attached to the alkene. Similarly, in structure **2.3** the two methyl groups (and the two hydrogens) are *trans* to one another.

Isomers such as structures **2.2** and **2.3** which differ only in the arrangement of the atoms around a double bond are referred to as **cis–trans** isomers. Another widely used name for this type of isomerism is **geometrical isomerism**, although the use of this term is now discouraged. Stereoisomers **2.2** and **2.3** are clearly configurational isomers of one another, since they cannot be inter-converted without breaking at least the π-bond component of the alkene double bond (cf. Chapter 1, section 1.4). All stereoisomers are either enantiomers or diastereomers (by definition, cf. Chapter 1, section 1.5.1), and as structures **2.2** and **2.3** are not mirror images of one another they cannot be enantiomers and so must be diastereomers, hence they are configurational diastereomers.

Since the cis–trans isomers of an alkene are diastereomers, they will have different chemical and physical properties just as will any other pair of diastereomers. Thus, for example, the two isomers may have different melting points and boiling points and may give different NMR and infrared spectra. They may also undergo chemical reactions at different rates, or to give different products, as will be discussed in Chapter 9.

Not all alkenes exhibit cis–trans isomerism; for instance there is only one isomer of 1-butene **2.4**. In the general case of an alkene bearing four substituents (a–d) **2.5**, then the molecule will exhibit cis–trans isomerism provided that a ≠ b and c ≠ d. Note that it is perfectly acceptable for 'a' to be identical to 'c' or 'd', as was the case for structures **2.2** and **2.3**, where each end of the double bond was attached to a hydrogen atom and a methyl group. It is only necessary that at **both** ends of the double bond, the two substituents be different to one another.

H   Et   a   c   Ph CO_2H Ph CO_2Me

H   H   b   d   Me CO_2Me Me CO_2H

2.4   **2.5**   **2.6**   **2.7**

Whilst the *cis/trans* nomenclature system is simple to apply, the terms are only appropriate for use with disubstituted alkenes. With more complex systems such as cis–trans isomers **2.6** and **2.7**, the *cis/trans* nomenclature system is not applicable. Hence a new nomenclature system is required, one which can be used to assign unambiguously the stereochemistry of any cis–trans isomer. Just such a system was developed in 1951 by Cahn and Ingold, and later modified and extended by Cahn, Ingold and Prelog using a set of sequence rules that are referred to as the CIP rules after the initials of these three chemists. These rules are extremely versatile and can be used to produce a systematic nomenclature system for many different types of stereochemical systems, not just cis–trans isomers. The rules are given below, and will be utilized extensively throughout this book.

The rules as stated here are a simplification of the current version of the rules, although they are sufficient to cover the vast majority of situations that are likely to be encountered. Indeed, modifications and extensions to these rules are still being proposed to cover rare situations where the six rules given below are inadequate. References both to the original papers on the CIP rules and to recent papers dealing with situations where they are inadequate are included in the further reading.

The CIP rules

Rule 1   Precedence is given to the atom of highest atomic number. Thus if O, H, C, Br and S are being compared then precedence decreases in the order Br > S > O > C > H.

Rule 2   If the first atoms are identical, compare the second atom in the chain and continue until a point of difference is found. Thus $-CH_2CH_3$ takes precedence over CH_3, since in the former case the three second atoms are H, H and C whilst in the latter case all three second atoms are H, and C has a higher priority than H. At each stage where atoms are being compared, the atoms of highest atomic number are compared first.

Rule 3   Double and triple bonds are treated as if they had duplicate or triplicate single bonds, so C=O is treated as (O)–C–O–(C). Aromatic systems are treated as if they contained localized double and single bonds.

Rule 4   A lone pair has the lowest precedence, thus $^+NH_3$ takes precedence over NH_2.

Rule 5   When considering isotopes, the heavier isotope takes precedence, thus D has precedence over H, and ^{13}C takes precedence over ^{12}C.

Rule 6   If two ligands differ only in configuration, then Z takes precedence over E, and R takes precedence over S (this will become clear later).

To show how these rules can be applied to give a nomenclature system for cis–trans isomers, consider the general structure of an alkene **2.5**. Compare first the groups 'a' and 'b' (shown emboldened in structure **2.5** below), i.e. the two groups that are attached to the same carbon atom, and decide which has the higher priority. Then repeat the process for 'c' and 'd' (shown italicized in structure **2.5**). If the groups of higher priority on each of the alkene carbon atoms appear *cis* to one another, then the double bond is given the prefix Z from the German *Zusammen* (together). Conversely, if the two groups of higher priority are *trans* to one another, then the prefix E (*Entgegen*; German for 'across') is used. Thus if in the general case **2.5**, 'a' has a higher priority than 'b', and 'c' has a higher priority than 'd' then the two groups of higher priority are 'a' and 'c', which are *cis* to one another, so this would be the (Z)-isomer of the alkene. If, however, 'a' has a higher priority than 'b', but 'd' has a higher priority than 'c' then the two groups of higher priority would be 'a' and 'd', which are *trans* to one another, and this would correspond to the (E)-isomer of the alkene.

Compounds **2.8–2.10** provide three examples to illustrate the use of the priority rules in assigning the stereochemistry of alkenes. For compound **2.8**, on

2.5

the left-hand side, the two atoms directly bound to the double bond are the C and H shown emboldened. As carbon has a higher atomic number than hydrogen, then the methyl group is the group of higher priority by the appliance of rule 1. The same applies to the right-hand side (atoms shown in italics), so the two groups of highest priority are on the same face of the double bond (*cis* to one another), hence the prefix is *Z* and this compound would be named (*Z*)-2-butene.

        2.8   **2.9**   **2.10**

For compound **2.9**, on the left-hand side C takes precedence over H (rule 1, atoms shown in outline) as discussed for compound **2.8**, so the methyl group is the group of higher priority. On the right-hand side, however, the first atoms are both carbon (shown in bold), so rule 2 has to be invoked. Hence, the atoms attached to the emboldened carbon atoms are compared, these being highlighted in italics. The atoms are H, H, H (in the methyl group) and C, H, H (in the ethyl group). Comparison of these atoms shows that in two cases there is again no difference between the atoms (H being compared with H), but in the remaining case C is compared with H, and as carbon has a higher atomic number than hydrogen, it takes precedence so the ethyl group has a higher priority than the methyl group. Hence the two groups of higher priority are on opposite faces of the double bond and this compound is 3-methyl-(*E*)-2-pentene.

In the case of compound **2.10**, on the left hand side of the double bond, S has precedence over O (rule 1, atoms shown in outline), so the CH_3S group has the higher priority. On the right-hand side, however, the first atoms directly attached to the double bond are both C (shown in bold), so it is necessary to look at the next atoms along the chains (rule 2). In the case of the CH_2OH group, these are the H, H and O shown in italics. The phenyl ring, however, has only two carbon atoms (shown in italics) attached to the carbon joined to the alkene. Therefore rule 3 is invoked and the aromatic ring is treated as if it had alternating single and double bonds, i.e. as $-C=C-C=$, where the central of these three carbons corresponds to the atom shown emboldened in structure **2.10**, and the other two carbons are shown in italics in **2.10**. A third, dummy, carbon atom is then added to the central carbon giving $-C-C(-C)-C$, and the three second atoms become

C, C and C. Comparison of these three atoms with the H, H and O of the CH_2OH group is then carried out starting with the atoms of highest atomic number. Thus C is compared with O, and the oxygen takes precedence (rule 1), so that the CH_2OH has priority over the phenyl ring. Hence the double bond has the Z-geometry since the two groups of higher priority CH_3S and CH_2OH are *cis* to one another. In this case it is important that the comparisons be carried out starting with the atoms of highest atomic number, as the opposite and incorrect result would have been obtained if C of the phenyl ring had been compared with H of CH_2OH.

2.11	**2.12**	**2.13**
trans:trans (*EE*)	*trans:cis* (*EZ*)	*cis:cis* (*ZZ*)
(2*E*,4*E*)-2,4-hexadiene	(2*E*,4*Z*)-2,4-hexadiene	(2*Z*,4*Z*)-2,4-hexadiene

It is apparent from the above examples that the *E/Z* nomenclature system can be applied to any alkene that is capable of exhibiting cis–trans isomerism, including disubstituted alkenes for which the *cis/trans* nomenclature system can also be used. Thus the *cis/trans* nomenclature is not needed. However, *cis* and *trans* are still in very common usage as stereochemical descriptors, and will be used alongside *E* and *Z* throughout this book. The *E/Z* or *cis/trans* nomenclatures can easily be extended to systems with more than one double bond simply by including the atom numbers in front of each descriptor to specify which double bond the descriptor refers to. Structures **2.11**−**2.13** show how this can be applied to the three possible cis–trans isomers of 2,4-hexadiene.

Probably the hardest thing to remember about the *E/Z* nomenclature is which stereochemistry each prefix represents. This is compounded by the fact that a capital E resembles a *cis*-alkene (omit the central horizontal line from E), whilst a capital Z resembles a *trans*-alkene if the diagonal of Z is taken as the double bond. This is unfortunate, since the *E*-prefix actually represents the situation where the two groups of highest priority are *trans* to one another, whilst the *Z*-prefix represents the case where the groups of highest priority are *cis* to one another. Hence, it may be helpful to remember that the *E/Z* prefixes represent the situation that they **do not** resemble!

2.2 Cis–trans isomerism in other systems which contain a double bond

Cis–trans isomerism can be found in systems other than alkenes, and the descriptors *cis/trans* or *E/Z* can be used as appropriate. In general, any system that has a planar (or nearly planar) arrangement of atoms about a bond around which there is restricted rotation will show cis–trans isomerism if suitable substituents are present. Thus imines may exhibit cis–trans isomerism (cf. structures **2.14** and **2.15**) provided R and R^1 are different, and in assigning the stereochemistry the lone pair of electrons on the nitrogen atom is included as a substituent. Thus R is compared with R^1, and R^2 is compared with the lone pair of electrons to determine precedence. Note that R^2 will always have a higher priority than the lone pair since rule 4 states that a lone pair always has the lowest priority. For imines in which R or R^1 is hydrogen, then the *cis/trans* nomenclature can be used, since there will be only one carbon based substituent on each end of the double bond. For other imines, however, the *E/Z* nomenclature is more appropriate. Other compounds which contain a carbon–nitrogen double bond (such as oximes and hydrazones) may similarly exhibit cis–trans isomerism. The prefixes *syn* and *anti* are sometimes used instead of *cis* and *trans* when referring to the cis–trans isomers of imines, oximes and hydrazones, although this is now discouraged.

2.14 **2.15**

Azo compounds contain a nitrogen nitrogen double bond and are always capable of exhibiting cis–trans isomerism (cf. structures **2.16** and **2.17**) since the two substituents on each nitrogen atom are always different: a lone pair of electrons and an R group. Hence, in this case either nomenclature system (*cis/trans*, or *E/Z*) can be used.

2.16 **2.17**

2.3 Cis–trans isomerism in cyclic systems

It is a general rule that a ring and a double bond are stereochemically equivalent, since in both cases substituents are located above and below the plane containing the π-orbital or the ring atoms: thus cis–trans isomerism can occur in cyclic systems. The *E/Z* nomenclature system should not be used to assign the stereochemistry to cyclic compounds, as the ring carbon atoms will generally be sp^3 rather than sp^2 hybridized and so carry three rather than just two substituents. Hence the *cis/trans* nomenclature is still preferred in this case, provided that each carbon atom bears a hydrogen atom as in isomers **2.18** and **2.19**. When considering cyclic compounds, assume that the ring is planar (this is not generally true, see Chapter 8), then two substituents can be on the same face of the ring (*cis*) or opposite faces (*trans*) as shown in structures **2.18** and **2.19**. This applies to any size of ring, not just to the cyclobutanes shown in **2.18** and **2.19**. For more highly substituted cyclic systems (such as cyclopropane **2.20**), the descriptors *cis* and *trans* are no longer unambiguous, however the *R/S* nomenclature system which will be introduced in Chapter 3 or the *l/u* system which will be discussed in Chapter 4 can usually be used to define the stereochemistry of these systems. This will be covered in more detail in Chapter 8.

cis	trans	
2.18	**2.19**	**2.20**

2.4   Cis–trans isomerism in square planar and octahedral metal complexes

Square planar metal complexes which contain two or more different ligands and no more than two of each ligand can exist in two diastereomeric forms which can be described as cis–trans isomers of one another. An example of this is the diaminedichloroplatinum(II) complexes **2.21** and **2.22**. These two complexes have both been prepared and have different chemical, physical and biological properties, as would be expected for any pair of diastereomers. Indeed, the *cis*-isomer **2.21** which has the trivial name of cisplatin is one of the most effective treatments for testicular and ovarian cancer, whilst the *trans*-isomer **2.22** is completely devoid of anticancer activity. Square planar complexes of the form Ma_2b_2, $Mabc_2$, and $Mabcd$ will exhibit cis–trans isomerism provided a–d are all distinct, although in the case of $Mabcd$ complexes there are three rather than just two stereoisomers and the *cis/trans* nomenclature is not appropriate in this case.

cis	trans
2.21	**2.22**

A similar situation exists with octahedral complexes, although there are many more possibilities, since the metal can be attached to up to six different ligands. Complexes of the general structure Ma_2b_4 can exist as a pair of cis–trans isomers (*cis*-**2.23**) and (*trans*-**2.24**) exactly analogous to the situation with square planar complexes. The use of *cis* and *trans* descriptors can also be extended to complexes of general structure Ma_2bcde, although in this case there is more than one *cis* and more than one *trans*-isomer since the ligands b, c, d, and e can also be *cis* or *trans* to one another.

cis	trans	fac	mer
2.23	**2.24**	**2.25**	**2.26**

Complexes of general structure Ma_3b_3 and Ma_3bcd can also exist as a pair of cis–trans isomers, although in this case the descriptors *cis* and *trans* are not appropriate. Instead, the descriptors *fac* (facial) and *mer* (meridonal) are used. An octahedron has eight triangular faces, hence the name octahedron. The *fac*-isomer **2.25** is the isomer in which the three identical ligands all lie on the same face of the octahedron; this is illustrated by the dotted lines in **2.25**. In the *mer*-isomer **2.26** the three identical ligands lie around the equator of the octahedron. The stereochemistry of octahedral complexes will be discussed in more detail in Chapter 3.

2.5 Cis–trans isomerism about single bonds (conformational diastereomers)

Consider butadiene for which there are two conformers (minimum energy conformations) **2.27** and **2.28**, both of which are planar, but which differ in the orientation of the alkylidene groups about the central carbon–carbon bond shown emboldened. Since these two structures can be interconverted without breaking a bond, they are classified as conformers rather than configurations, although in this case since the π-electrons of butadiene are delocalized, leading to the central carbon–carbon bond having a bond order intermediate between a single and double bond, this classification is somewhat arbitrary (cf. Chapter 1, section 1.5.2). However, since the carbon atoms forming the central carbon–carbon single bond are sp^2 hybridized, it is possible to use the *cis/trans* nomenclature system to distinguish between them. In this case, the stereochemical descriptors are modified slightly to *s-cis* and *s-trans* to indicate that the stereochemistry is about a single rather than a double bond. The descriptors can then be applied as shown in structures **2.27** and **2.28**. The nomenclature can be applied equally well to other conjugated systems such as α,β-unsaturated ketones. Whilst conformational diastereomers such as structures **2.27** and **2.28** cannot usually be separated and isolated, the ability or lack of ability of a conjugated system to adopt the *s-cis* or *s-trans* conformation can have important consequences for the reactivity of the system as will be discussed in Chapter 9.

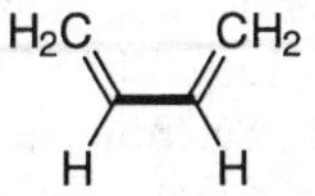

s-cis

2.27

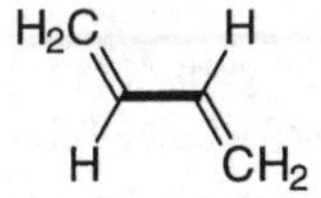

s-trans

2.28

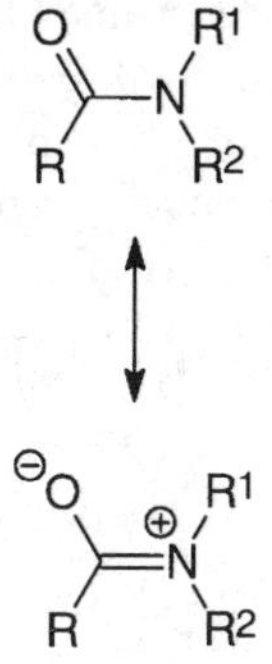

2.29

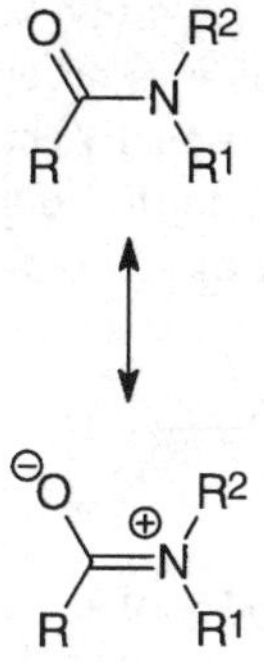

2.30

In amides, the carbonyl to nitrogen bond is also intermediate between a single and double bond due to delocalization of the nitrogen lone pair of electrons into the carbonyl bond (cf. Chapter 1, section 1.5.2). This delocalization results in the atoms around an amide bond being planar, so suitably substituted amides will show cis–trans isomerism as illustrated in structures **2.29** and **2.30**. Either the *cis/trans* (if either R^1 or R^2 is a hydrogen atom) or *E/Z* nomenclatures can then be used to classify these isomers. By convention, stereoisomers of amides are again considered to be conformers rather than configurations.

2.6 Methods for distinguishing cis–trans isomers

A number of methods are available for determining the stereochemistry of compounds which exhibit cis–trans isomerism. Only those techniques which allow the stereochemistry of an unknown compound to be determined are discussed here. For known compounds, since cis–trans isomers are diastereomers, the two isomers will have different physical properties such as melting or boiling point, ^{1}H NMR chemical shifts and infrared absorption maxima, which can be compared with literature data to determine which stereochemistry a particular sample possesses.

2.6.1 NMR spectroscopy

The vicinal coupling constant (^3J) between two vinyl protons depends upon the stereochemistry of the double bond as shown in structures **2.31** and **2.32**. For hydrogens which are *cis* to one another, ^3J is usually in the range 6–12 Hz **2.31**, whilst for *trans* hydrogens ^3J is typically between 14 and 18 Hz **2.32**. This is very characteristic and is a reliable method for determining the stereochemistry of disubstituted alkenes. However, the two vinyl hydrogens must be in different environments (i.e. R and R^1 are different), as coupling is observed only between non-equivalent hydrogens. Furthermore, the difference in environment between H_a and H_b must be sufficient for their ^1H NMR resonances not to overlap. Finally, this method can only be used to determine the stereochemistry of disubstituted alkenes as it is necessary to retain two hydrogens (H_a and H_b) on the double bond. Occasionally, however, another suitable nucleus (e.g. ^{19}F) is attached to the alkene and the coupling to this nucleus can be used to determine the stereochemistry of tri- or tetra-substituted alkenes.

$^3J_{ab}$= 6-12Hz $^3J_{ab}$= 14-18Hz

2.31   **2.32**

Another way in which NMR spectroscopy can be used to determine the stereochemistry of an alkene (or of any other species which exhibits cis–trans isomerism) is by use of the nuclear Overhauser effect (nOe). A full description of the origin of this effect is beyond the scope of this book but essentially whenever an NMR sample is irradiated (i.e. subjected to electromagnetic radiation with a frequency corresponding to the resonance frequency of one of protons in the sample), then the intensity of the signals seen for nearby protons will be perturbed. The nOe falls off rapidly with distance, being proportional to r^{-6} where r is the distance between the two protons. Hence, 'nearby' in this sense refers to protons that are within the same molecule and within about 5 Å of the proton whose resonance frequency is being irradiated. The closer the two protons are together, the larger the nOe that will be observed, and beyond about 5 Å the effect will be too small to detect.

irradiate here

2.33   **2.34**

For a trisubstituted alkene (as shown in structures **2.33** and **2.34**), irradiation at the frequency corresponding to the vinyl hydrogen will result in nOe's being observed at the resonances corresponding to each of the other hydrogens shown in structures **2.33**–**2.34**. The nOe to the hydrogens shown in normal typeface gives no stereochemical information, but the relative intensities of the nOe's to the emboldened and italicized hydrogens allow the stereochemistry of the double bond to be determined. In the case of isomer **2.33**, the nOe to the emboldened hydrogens will be larger than that to the italicized hydrogens since the former are nearer to the vinyl hydrogen (approximately 2.8 Å) than the latter (approximately 3.8 Å). In the case of isomer **2.34**, however, the reverse is true, as the italicized hydrogens are now nearer to the vinyl hydrogen and will show a larger nOe than the emboldened hydrogens.

Similar experiments can be used to determine the stereochemistry of tetra-substituted alkenes (e.g. **2.35**), where irradiation of the resonance corresponding to either set of emboldened hydrogens will show a larger nOe to the vicinal (*cis*) emboldened hydrogens than to the vicinal (*trans*) italicized hydrogens, and *vice versa* if the resonance of one of the sets of italicized hydrogens is irradiated. In this case, the actual distances between the vicinal hydrogens are about 2.4 Å for hydrogens of the same emphasis in structure **2.35**, and approximately 4.0 Å between emboldened and italicized hydrogens. The nOe method can also be applied to other species which exhibit cis–trans isomerism such as imines. The main limitation on any nOe experiment is that the relevant resonances in the ^{1}H NMR spectrum must not overlap. Thus, in the case of isomers **2.33** and **2.34**, all of the emboldened and italicized hydrogens must have different chemical shifts and should not be coincident with any of the other signals in the NMR spectrum.

$$RH_2C \qquad CH_2R2$$
$$R^1H_2C \qquad CH_2R3$$

2.35

2.6.2 X-ray crystallography

Again, a full description of the theory of X-ray crystallography is beyond the scope of this text. The essence of the technique, however, is that a mono-chromatic beam of X-rays is directed at a single crystal of the chemical under investigation. The X-rays interact with, and are diffracted by, the electrons within the sample to produce an X-ray diffractogram. This diffractogram contains sufficient information to reconstruct an electron density map of the mole-cule(s) present in the crystal. From the electron density map, the relative positions of each of the constituent atoms can be determined. The whole analysis from diffractogram to atomic positions is carried out by computer. The results are usually output as a three dimensional picture of the molecule(s) comprising the crystal along with tables of bond lengths, bond angles and torsional angles.

Clearly, X-ray analysis gives a great deal of information about a molecule, and from the bond angles and torsional angles or much more simply just by inspection of the three dimensional molecular diagram, the configuration of any alkene, ring or other structural feature can be determined. The main limitation of X-ray crystallography is the requirement for a single, good quality crystal of the substance to be investigated. Many organic compounds are oils or amorphous solids, and even if the substance is crystalline, determining a suitable solvent system for use in recrystallization to give good quality crystals can be very time consuming. It is possible, however, to obtain X-ray structures at low temperatures so that compounds which only crystallize below room temperature can be studied.

2.6.3   Dipole moments

The presence of electron withdrawing or electron donating substituents on an alkene induces a dipole moment (μ) into the molecule, the magnitude of which can be measured. Symmetrical *trans*-alkenes (e.g. **2.36**) have a dipole moment of zero, whilst the corresponding *cis*-isomers (e.g. **2.37**) have large dipole moments. For unsymmetrically disubstituted alkenes, if the two substituents are both electron withdrawing or both electron donating then the *trans*-isomer will have the smaller dipole moment. If, however, one substituent is electron donating whilst the other is electron withdrawing, as in isomers **2.38** and **2.39**, then the *cis*-isomer will have the smaller dipole moment. In the latter cases both isomers should be available in order to allow their respective dipole moments to be compared.

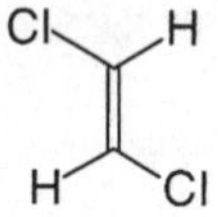

μ= 0.0D	μ= 1.85D	μ= 1.97D	μ= 1.64D
2.36	**2.37**	**2.38**	**2.39**

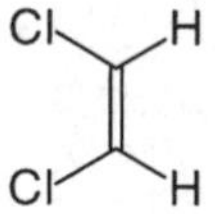

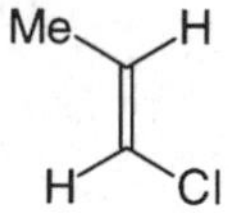

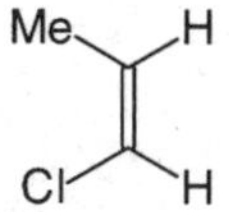

2.7   Further reading

General
Stereochemistry of Organic Compounds E.L. Eliel and S.H. Wilen. Wiley: London, 1994, chapters 9 and 11.

CIP rules and their application
R.S. Cahn and C.K. Ingold. *J. Chem. Soc.*, 1951, 612.
R.S. Cahn, C.K. Ingold and V. Prelog. *Experientia*, 1956, **12**, 81.

R.S. Cahn, C.K. Ingold and V. Prelog. *Angew. Chem., Int. Ed. Engl.*, 1966, **5**, 385.
V. Prelog and G. Helmchem. *Angew. Chem., Int. Ed. Engl.*, 1982, **21**, 567.

Recent papers recommending modifications to the CIP rules

P. Mata, A.M. Lobo, C. Marshall and A.P. Johnson. *Tetrahedron Asymmetry*, 1993, **4**, 657.
M. Perdih, and M. Razinger. *Tetrahedron Asymmetry*, 1994, **5**, 835.
P. Mata and R.B. Nachbar. *Tetrahedron Asymmetry*, 1995, **6**, 693.

nOe and other aspects of NMR

Spectroscopic Methods in Organic Chemistry 5th edn, D.H. Williams and I. Fleming. McGraw-Hill: London, 1996, chapter 3.
Topics in Stereochemistry Vol. 7, R.A. Bell and J.K. Saunders (N.L. Allinger and E.L. Eliel eds). Wiley: London, 1972, chapter 1.
Introduction to Organic Spectroscopy: Oxford Chemistry Primer Number 43, L.M. Harwood and T.D.W. Claridge. Oxford University Press: Oxford, 1997.

2.8 Problems

1. Which of the following compounds would exhibit cis–trans isomerism. For those compounds that do exist as cis–trans isomers, draw each isomer and use any suitable nomenclature system to classify the isomers.

a

b

c

d

e

f

g

h Cl-N=N-Cl

i

j

k

l

m

n $CrCl_3(OH)_3$

o $CoCl_2(H_2O)_4$

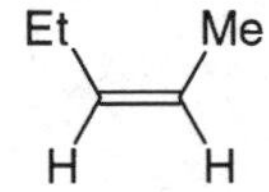

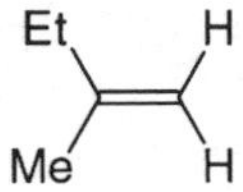

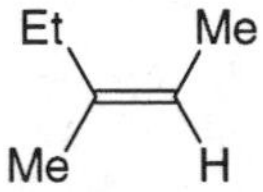

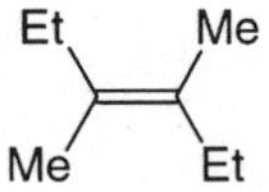

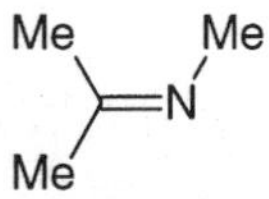

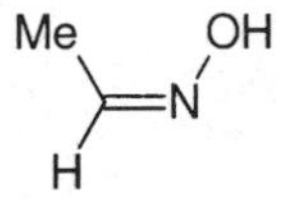

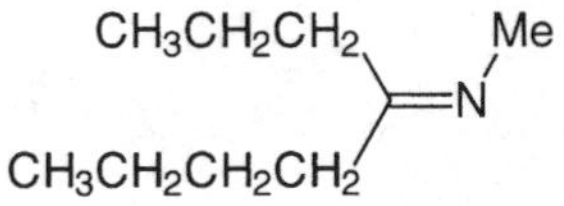

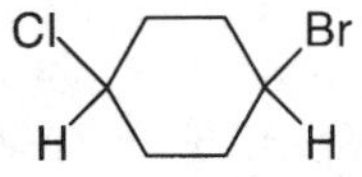

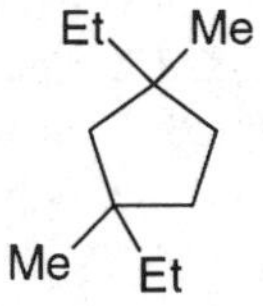

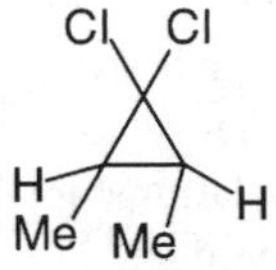

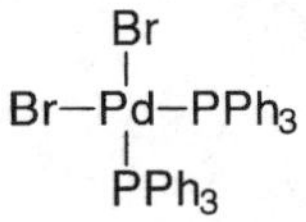

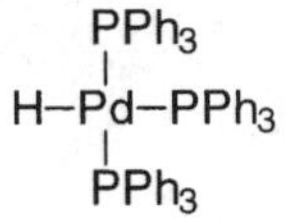

2. Use the E/Z nomenclature system to assign the stereochemistry of the double bond(s) in the following compounds.

a

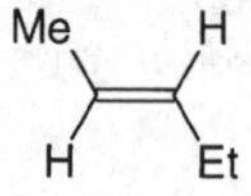

b

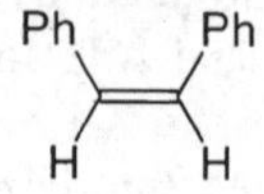

c

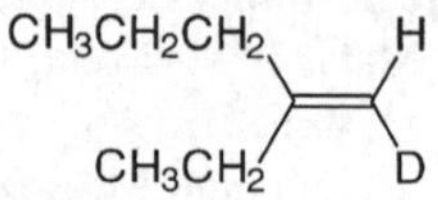

d

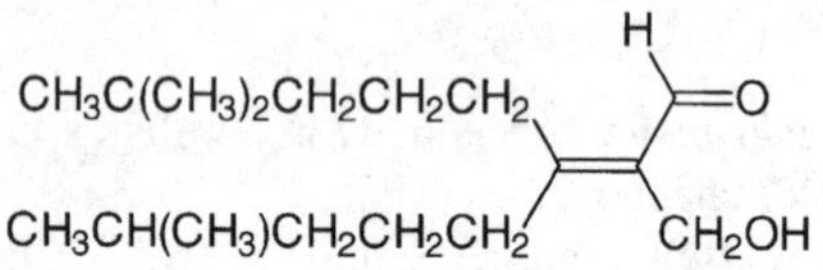

e

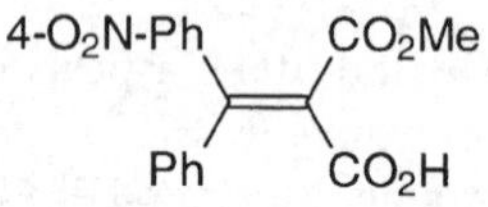

f

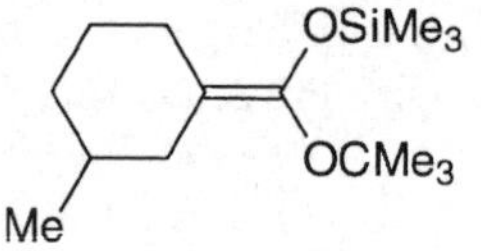

g

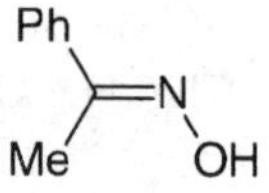

h

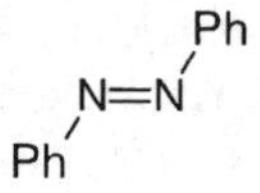

i

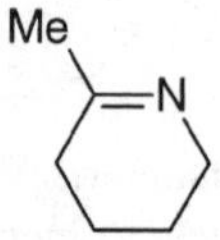

j

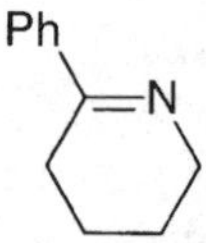

k

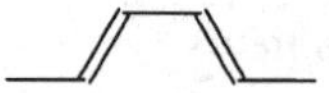

l

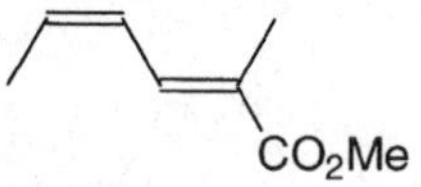

m

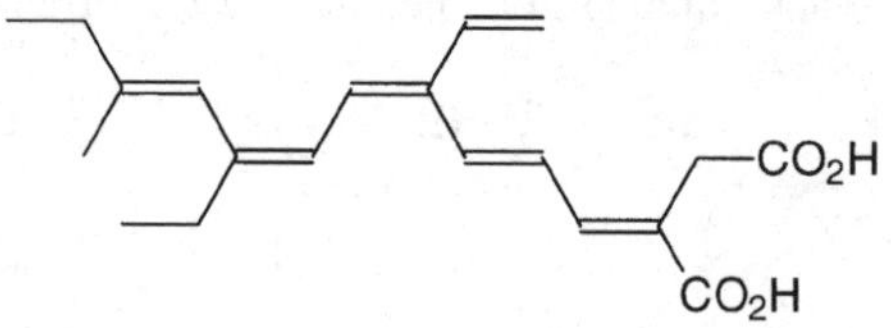

n

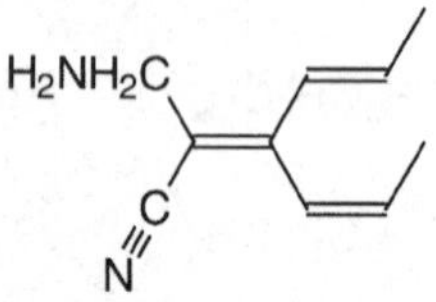

3. The reaction shown below is a Diels–Alder reaction, the stereochemistry of which will be discussed in detail in Chapter 9. Is the alkene being used the (E)- or the (Z)-isomer? What is the conformation of the diene? What is the relationship between the substituents on the cyclohexane ring of the product?

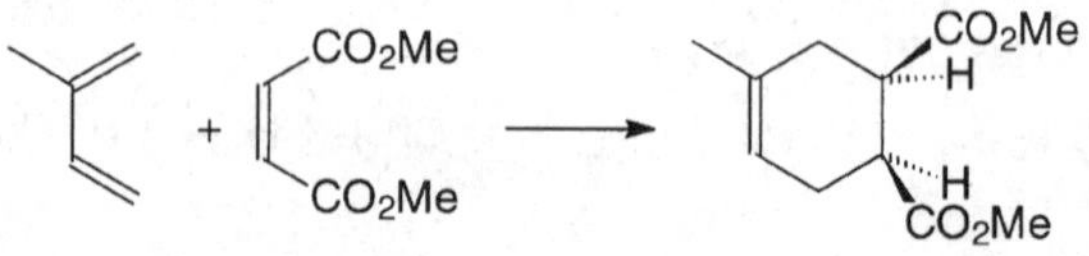

4. Deduce the structure including any stereochemistry of the compound (C_9H_{10}) whose ^1H NMR spectrum shows the following signals: 2.2 ppm (3H, d J 7.0 Hz); 6.4 ppm (1H, dq J 15.5 and 7.0 Hz); 6.8 ppm (1H, d J 15.5 Hz); 7.2–7.3 ppm (5H, m).

5. A chemical synthesis of the vinyl ether shown below produced a mixture of the two possible stereoisomers (A and B) of this compound. Draw the structure of both stereoisomers and name them using the E/Z nomenclature. Irradiation of one of the methyl groups attached to the alkene in isomer A produced a 10% nOe enhancement in the signal corresponding to the other methyl group attached to the alkene. The same experiment on isomer B, however, resulted in an enhancement of just 3%. Determine which of A and B is the (E)-isomer and which is the (Z)-isomer.

$$Me(H)C=C(Me)OEt$$

3 Enantiomers

3.1 Structure of enantiomers

Enantiomers were introduced in Chapter 1 (section 1.5.1) as stereoisomers that are non-superimposable mirror images of one another. This definition of enantiomers based upon the non-superimposability of mirror images was originally proposed by Lord Kelvin and will serve as our definition in this and the next two chapters. An alternative but equivalent definition of enantiomers based upon the symmetry or lack of symmetry in a substance will be introduced in Chapter 6. Any molecule (or other substance) that is not superimposable on its mirror image and so exists as a pair of enantiomers is said to be **chiral** and to exhibit **chirality**. Conversely, any molecule (or other substance) that is superimposable on its mirror image is **achiral**.

3.1

By far the most common origin of chirality is when a tetrahedrally coordinated atom (usually carbon) is bound to four different substituents as shown in structure **3.1**. Indeed, whenever a molecule contains a single atom which is tetrahedrally bound to four different substituents, then two enantiomers are possible. It is, however, important that the four substituents a–d are different to one another, as if any two of them are identical then structure **3.1** would become superimposable on its mirror image and so achiral. The atom X in structure **3.1** is best referred to as a **stereogenic centre** or simply a **stereocentre**. A widely used, although somewhat misleading alternative name for a stereocentre is a **chiral centre**. This name is misleading since it implies that the chirality is localized around the atom X, whereas, in fact, chirality is a property of the molecule as whole that cannot be localized around one atom or a group of atoms. Hence the term 'chiral centre' will not be used in this book.

The presence of a stereocentre is not a requirement for a molecule to exhibit chirality, it is simply the most common cause of chirality. Later in this chapter (section 3.8) other structural features (**stereogenic elements**) which can result in a molecule exhibiting chirality will be introduced. Whilst the presence of a single stereogenic element in a molecule will always cause the molecule to be

chiral, if more than one stereogenic element is present, then the molecule may or
may not be chiral as will be discussed in Chapter 4. Non-superimposability upon
its mirror image rather than the presence of any structural feature is the most
reliable method to determine whether or not a species is chiral, although an
alternative method based on symmetry considerations will be introduced in
Chapter 6.

3.2 Nomenclature for enantiomers (specification of absolute configuration)

There are three common systems in use for defining the absolute configuration
of an enantiomer, i.e. for specifying which enantiomer is being referred to. Two
of these nomenclature systems will be discussed here, and the third will be
introduced in section 3.4.1.

3.2.1 The D/L-nomenclature system

The D,L-nomenclature system is still widely used for amino acids and carbohydrates, but has been superseded for other classes of compounds. It relies on the
fact that the three dimensional structures of the two enantiomers of glyceraldehyde **3.2** and **3.3** are arbitrarily given the stereochemical descriptors L- and

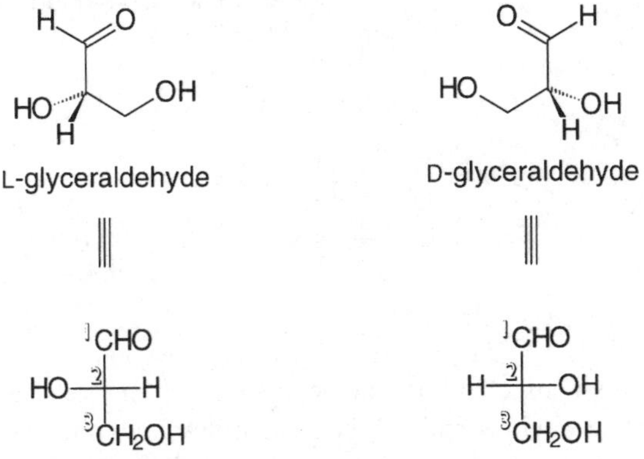

L-glyceraldehyde

D-glyceraldehyde

3.2

3.3

D- respectively. Note that in the Fischer projection of the L-enantiomer, provided
the projection is drawn with Cl at the top then the OH substituent on the
stereocentre is on the left, and conversely for the D-enantiomer the OH substituent will be on the right. This provides a means for extending the nomenclature to other compounds that possess substituents along a chain of carbon
atoms.

To determine whether a chiral species belongs to the D or the L-series, it is
drawn as a Fischer projection (cf. Chapter 1, section 1.2.1) with the first atom of

the longest carbon chain (the atom that would be numbered 1 in standard organic nomenclature) at the top. For carbohydrates other than glyceraldehyde, which usually contain more than one stereocentre, the D,L-nomenclature is then based on the substituent at the stereocentre furthest from C1 as illustrated for D-glucose **3.4**. For amino acids, if the amino group attached to the stereocentre is on the left then the amino acid is in the L-series, and if the amino group is on the right then the structure represents a D-amino acid. This is illustrated for L-α-methyl phenylalanine **3.5**.

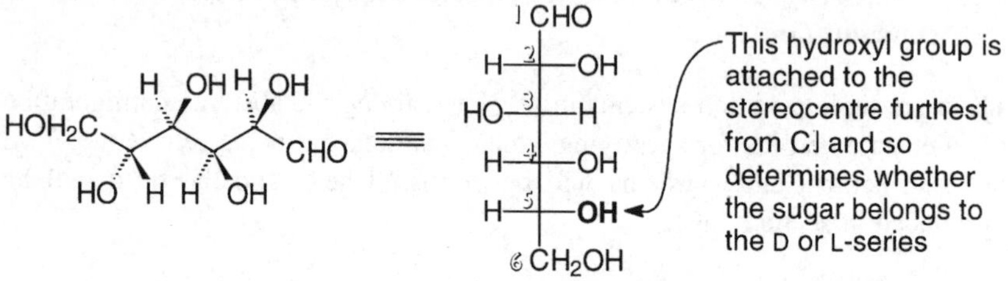

3.4   D-glucose

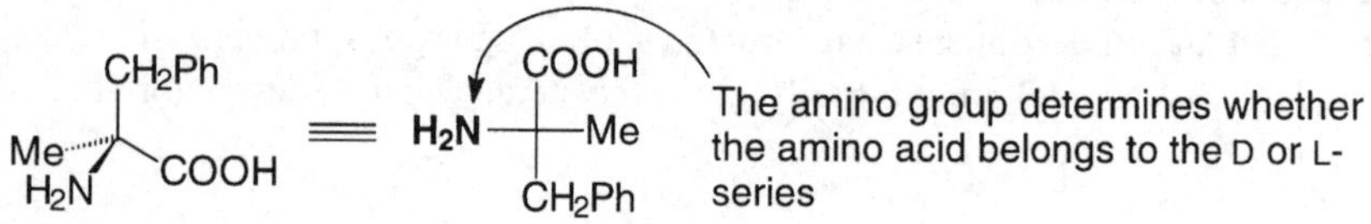

This hydroxyl group is attached to the stereocentre furthest from C1 and so determines whether the sugar belongs to the D or L-series

3.5   L-α-methyl-phenylalanine

The amino group determines whether the amino acid belongs to the D or L- series

L- If this oxygen group is used to assign the stereochemistry

D- If this oxygen group is used to assign the stereochemistry

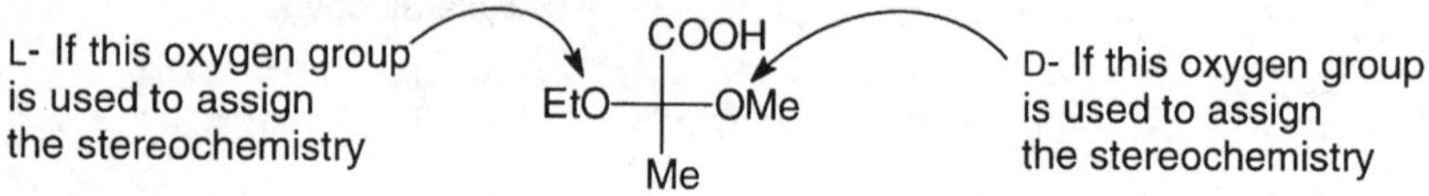

3.6

The main limitations of the D,L-nomenclature system are that not all chiral molecules have an obvious main carbon chain. Furthermore, even if there is a main carbon chain, it may not be obvious which of the substituents attached to the stereocentre should be used in the determination of the D,L-nomenclature as in the case of compound **3.6**. Finally, as will be discussed in section 3.8, many chiral molecules do not even contain a stereocentre. Because of these limitations, a new and more versatile nomenclature system was developed which can be used to name any chiral molecule.

3.2.2   The R/S nomenclature system

The currently preferred nomenclature system for enantiomers whose chirality is caused by the presence of a tetrahedral stereocentre utilizes the same sequence rules described in Chapter 2 for the E/Z nomenclature system. The CIP rules (cf. Chapter 2 section 2.1) are used to define a stereocentre as R or S as described below.

1. Assign the four ligands attached to the stereocentre into an order of priority. Label them as 1 (highest priority) to 4 (lowest priority).
2. View the molecule with the ligand of lowest priority (ligand 4) at the back. If the remaining three ligands decrease in order of priority in a clockwise direction, then the stereocentre is said to possess the (R)-configuration. If, however, the three ligands decrease in priority in an anti-clockwise direction then the stereocentre is said to be of the (S)-configuration.

The correct use of the R and S labels can be easily remembered by the following mnemonic:

CLOCKWISE

R

ANTICLOCKWISE

S

SYSTEM

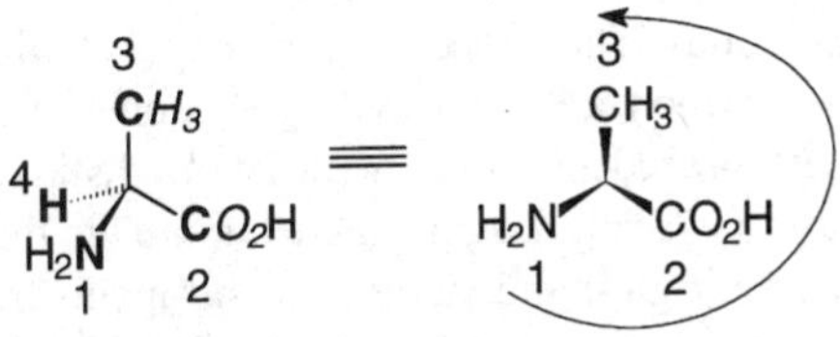

3.7

Compounds **3.7** and **3.8** provide two examples of the way in which the R/S nomenclature system is applied. In structure **3.7**, the four substituents attached to the stereocentre are first assigned a relative order of precedence. CIP rule 1 is used to determine that the hydrogen atom substituent has the lowest priority and that the nitrogen atom of the NH_2 group has the highest priority. The two remaining substituents both start with a carbon atom (shown emboldened), but are distinguished by rule 2 as the second atoms in the acid group are three oxygens (three since the C=O is treated as two C–O bonds according to rule 3; these oxygens are shown in italics) which take priority over the second atoms in the methyl group, which are three hydrogens (also shown in italics). Hence the acid group has precedence over the methyl group. The molecule is then redrawn with the group of lowest precedence at the rear and so hidden from view. The three groups which remain visible are found to decrease in order of precedence in an anti-clockwise direction as shown. Thus the stereochemistry at the stereocentre can be assigned as S, and the molecule named as (S)-alanine, alanine being the trivial name for this amino acid. Alternatively, the compound could be named as (S)-2-aminopropanoic acid using systematic organic nomenclature.

3.8

In the case of compound **3.8**, which is one of the enantiomers of carvone, rule 1 can only be used to assign the precedence of one of the four groups attached to the stereocentre, the hydrogen atom which has the lowest precedence (4), since the other three substituents all start with a carbon atom, shown emboldened. Rules 2 and 3 determine the precedence of these three groups, starting with the use of rule 2 to examine the second atoms (shown italicized) in each substituent. In the case of the isopropenyl substituent, the second atoms are three carbon atoms since, by rule 3, a carbon–carbon double bond is considered as two carbon–carbon single bonds. The other two substituents both start with a carbon attached to one other carbon and two hydrogen atoms. Thus the isopropenyl group has the highest precedence (1). In order to distinguish the other two groups, it is necessary to use rule 2 again and examine the third atoms which are shown in outline. On the right-hand side of the six-membered ring of structure **3.8**, these are two oxygens (apply rule 3 to the C=O) and a carbon, whilst on the left-hand side the three atoms are two carbons (apply rule 3 to the C=C) and a hydrogen. Thus the right-hand side substituent has a higher precedence (2) than the left-hand side (3). The molecule is then redrawn with the hydrogen atom at the back; in this case this entails rotating the molecule through 180°, and the three remaining substituents are found to decrease in precedence in a clockwise direction. Thus structure **3.8** represents the (R)-isomer of carvone.

In the two examples discussed above, the importance of viewing the molecule with the group of lowest precedence at the back was emphasized. If this convention is not followed, then an incorrect assignment of the stereochemistry may result. However, there are two short cuts which can be used once you have some experience of the R/S nomenclature system. The first of these is that the group of lowest precedence need not be directly behind the stereocentre; it is sufficient that this group be drawn with a hashed bond on a correctly drawn tetrahedral stereocentre. (The correct and incorrect ways of representing the three dimensional arrangement of atoms around a tetrahedral centre were discussed in Chapter 1, section 1.2.) Thus in the case of structure **3.7**, the substituents of precedence 1–3 decrease in order of precedence in an anti-clockwise direction in both the left and right-hand representations, even though in the former the hydrogen is not directly behind the stereocentre.

The second short cut is both very useful and potentially very confusing to the inexperienced user. It is that, when the group of lowest priority is pointing forward (i.e. drawn with a wedge) rather than backwards, then the relationship between stereochemical descriptor (R/S) and direction of decrease of precedence can be reversed. That is a clockwise decrease corresponds to the (S)-enantiomer and an anti-clockwise decrease to the (R)-enantiomer. This short cut can be very useful as it means that structures never need to be rotated to place the group of lowest precedence at the rear, a task which can be difficult with large molecules. It is recommended, however, that this short cut *not* be used until the reader is completely familiar with the clockwise = R, anti-clockwise = S system. The use of this short cut can be illustrated with reference to structure **3.8**, since the stereochemistry can now be defined directly from the left-hand structure. Thus the three groups of highest precedence decrease in an anti-clockwise direction, but since the group of lowest priority is at the front rather than at the back, this corresponds to the (R)-enantiomer.

3.3 Chemical properties of enantiomers

Enantiomers have identical chemical properties except when reacting with other chiral species or with achiral species in the presence of chiral additives (e.g. a chiral catalyst or solvent). To see why this is so, consider two reactions of both enantiomers of the chiral alcohol **3.9**; reaction with acetyl chloride **3.10** to give esters **3.11**; and reaction with the (R)-enantiomer of acid chloride **3.12** to give esters **3.13** as shown in **Scheme 3.1**. The reaction of enantiomers with achiral species was briefly introduced in Chapter 1, section 1.5.1.

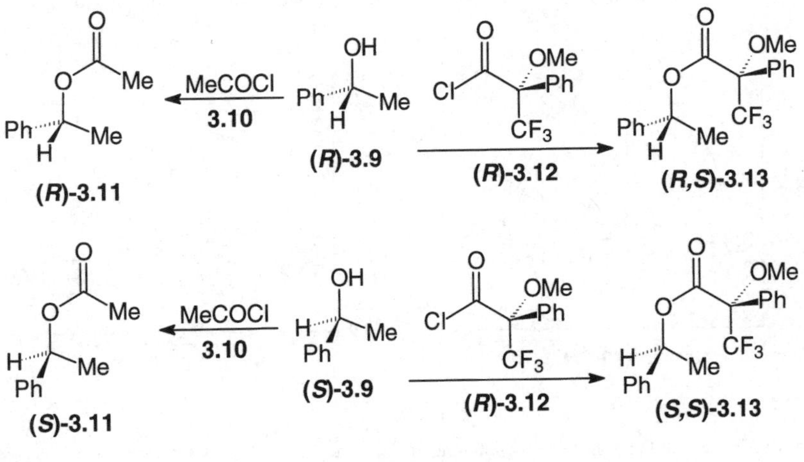

Scheme 3.1

Before the chemistry shown in **Scheme 3.1** is discussed, one nomenclature point needs to be clarified. Acid chloride **3.12** has the (R)-configuration, since the order of priority of the four substituents attached to the stereocentre is OMe > COCl > CF_3 > Ph. In particular, when comparing the CF_3 and COCl substituents, the COCl group has the higher priority since the first atoms (C) are the same in both cases, but the Cl of the acid chloride takes precedence over the F of the trifluoromethyl group. On forming esters **3.13**, however, the order of priority changes to OMe > CF_3 > $CO_2CH(Me)Ph$ > Ph, since there is no longer a chlorine attached to the carbonyl group. This change in order of priority causes the absolute configuration of the stereocentre to be redefined as S. Note, however, that there is no change in the actual stereochemistry of the stereocentre; the change in label from R to S is purely a consequence of the way in which the CIP rules are applied.

Since enantiomers have identical bond lengths and bond angles, and torsional angles of equal magnitude, they have identical energies (cf. Chapter 1, section 1.5.1). Hence in the reaction of the two enantiomers of alcohol **3.9** with acetyl chloride, the energy of the starting materials ((R)- or (S)-**3.9** and **3.10**) will be the same whichever enantiomer is used. The products of this reaction are esters **3.11**, which are also enantiomeric and so of equal energy, and HCl which is achiral. Thus the energy of the products of the reaction will also be equal whichever enantiomer of **3.9** is used. In fact, as the reaction between alcohol **3.9** and acetyl chloride progresses, at any point along the reaction pathway, the species formed from (R)-**3.9** and from (S)-**3.9** will be enantiomeric and hence of equal energy. This is illustrated in **Figure 3.1a**, although this reaction pathway has been simplified from the actual pathway for this reaction as no intermediates are shown. The important point is that the energy of the starting materials, transition state and products are all equal for the two enantiomers. Thus the activation energy (E_a) and Gibbs free energy of reaction (ΔG) will be equal whichever enantiomer is used. This means that the rate of reaction of the two

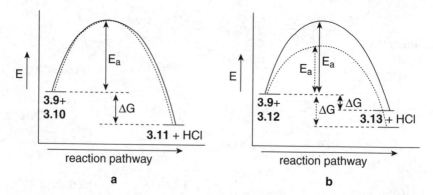

Figure 3.1 Reaction pathway/energy diagrams for the reaction of alcohol **3.9** with: **a**, acid chloride **3.10**; and **b**, acid chloride **3.12**.

enantiomers of **3.9** with acetyl chloride will be identical, and that the equilibrium constant for the two reactions will be identical.

When the two enantiomers of alcohol **3.9** react with the enantiomerically pure acid chloride **3.12**, however, the situation is rather different. The starting materials ((**R**)- or (**S**)-**3.9** and **3.12**) will still be of equal energy, but the products ((**R**,**S**)- or (**S**,**S**)-**3.13**) in this case are not a pair of enantiomers but a pair of diastereomers. The products are not enantiomers since they are not mirror images of one another but they are stereoisomers as they have the same connectivity, so they must be diastereomers of one another (cf. Chapter 1, section 1.5.1). Diastereomers have different bond lengths, bond angles and torsional angles, and hence different energies.

Similarly, at any point along the reaction pathway, the species formed from (**R**)-**3.9** and from (**S**)-**3.9** with acid chloride **3.12** will be diastereomeric rather than enantiomeric and hence have different energy (**Figure 3.1b**). Thus, the energy of the products and of the transition states leading to the products will be different, giving different activation energies (E_a) and Gibbs free energies of reaction (ΔG), and the reactions will occur at different rates and have different equilibrium constants. Depending upon the magnitude of the energy differences between the diastereomeric transition states (or products), this difference in the rate of reaction and/or equilibrium constant may be very small or very large. In the latter case, it is possible that under a given set of reaction conditions, one enantiomer will have completely reacted whilst the other remains unreacted.

Hence the two enantiomers of alcohol **3.9** (or the two enantiomers of any other chiral species) will react identically with an achiral species but differently with a chiral species. The cause of this difference is whether the structures formed along the reaction pathway are enantiomeric or diastereomeric. This is an extremely important concept which will be used to explain much of the remaining material in this and subsequent chapters.

3.4 Physical properties of enantiomers

As was discussed in Chapter 1, section 1.5.1, enantiomers generally have identical physical properties such as melting point, boiling point, infrared absorptions and NMR spectra. It is important to realize, however, that whilst the melting point, etc. of one enantiomer will be identical to that of the other enantiomer, the melting point (or any other bulk property) of a mixture of the two enantiomers may be different. This is because the intermolecular interactions between opposite enantiomers (i.e. between the R and S enantiomers) may be different to those between like enantiomers (i.e. between two molecules both of R or both of S stereochemistry) as discussed in section 3.3.

The one class of physical techniques that can distinguish between the two enantiomers of a compound are chiroptical techniques, the most common of which is optical rotation. The chiroptical properties of a molecule are determined not just by the bond lengths and angles but also by the sign and magnitude of the torsional angles, the sign of the torsional angles being the one difference between enantiomers (cf. Chapter 1, section 1.5.1).

3.4.1 Optical rotation

A light beam can be described by two orthogonal and sinusoidally varying vectors representing the electrical and magnetic components of the wave respectively. When a beam of light passes through a solution of a compound, the electrical and magnetic fields within the light beam will be refracted by the electrons of the solute (and solvent) molecules. If the beam of light is initially plane polarized, i.e. all of the electrical and magnetic vectors are aligned, then the net result of this refraction will be to rotate the plane of the polarization. The direction and magnitude of this rotation will depend upon the orientation of the molecule, and if in a particular orientation a molecule rotates the plane by $\theta°$, then in the mirror image orientation the same molecule will induce a rotation of $-\theta°$.

For an achiral solute, as the light beam passes through the solution it will interact with millions of different molecules. For every molecular orientation encountered, the mirror image orientation will also be encountered, so the rotations will cancel out and no net rotation of the plane of polarization will occur. However, in the case of a single enantiomer of a chiral solute, the mirror image orientation cannot be present (if it were, the molecule would be superimposable on its mirror image and so achiral), so the optical rotations do not cancel out and a measurable rotation builds up. This is shown in **Figure 3.2**, although for clarity only one of the two vectors that constitute the light wave is shown.

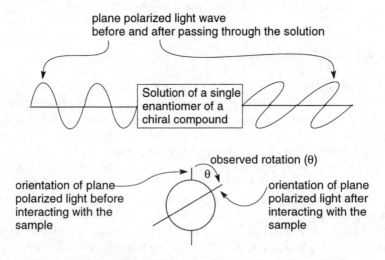

Figure 3.2 The interaction between plane polarized light and an enantiomerically pure solution.

If one enantiomer of a compound rotates the plane of polarized light through an angle $\alpha°$, then its enantiomer measured under identical conditions will rotate the plane through $-\alpha°$; i.e. through the same angle but in the opposite direction. This phenomenon was discovered in 1848 by Louis Pasteur (of pasteurization fame) whilst he was studying salts of tartaric acid. It was already known that many solids such as quartz could rotate the plane of polarized light, but Pasteur discovered that certain solutions also caused this phenomenon. Pasteur concluded that this must be due to dissymmetry (lack of symmetry, cf. Chapter 6) within the molecule. This was an outstanding deduction since at the time the structure of tartaric acid was unknown, and it was not until 1874 that Van't Hoff suggested that tetracoordinate carbon atoms were tetrahedral. An apparatus for measuring optical rotations is called a **polarimeter** and consists of three parts as shown in **Figure 3.3**:

1. A monochromatic source of plane polarized light. This can be obtained from a monochromatic light source (usually a sodium lamp) by passing the light through a special filter called a polarizing filter which allows the light to pass through the filter only when it is orientated in a particular direction.
2. A sample compartment where the monochromatic light can pass through a solution of the sample.
3. A way of detecting the plane of the light after it has passed through the filter. This can be achieved by a second polarizing filter which can be rotated. A beam of light will only be detected through the second filter when it is orientated to match the plane of the polarized light coming from the sample.

The optical rotation depends upon the following factors:

- the wavelength of the light;
- the temperature;
- the concentration of the solution;
- the path length (the length of the sample tube);
- the solvent.

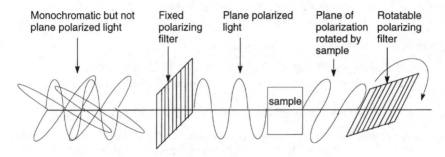

Monochromatic but not plane polarized light

Fixed polarizing filter

Plane polarized light

Plane of polarization rotated by sample

Rotatable polarizing filter

sample

Figure 3.3 A schematic representation of a polarimeter.

The dependence of the optical rotation upon the path length is linear and in many cases the dependence upon concentration is also linear, so the **specific rotation [α]** can be defined as in **Equation 3.1** to remove these variables. The specific rotation of a compound is the value usually quoted in the chemical literature, as this can easily be reproduced provided the wavelength and temperature are kept constant. The factor of 100 which is introduced into the specific rotation is so that most compounds will have specific rotations between +100 and −100. The units of specific rotation are 10^{-1} deg cm^2 g^{-1}, though by convention these are usually omitted when values for specific rotations are quoted.

$$[\alpha]_\lambda^t = \frac{100\alpha}{c \times l} \tag{3.1}$$

α = measured optical rotation;
c = concentration in g per 100 ml;
l = path length in dm (1 dm = 10 cm);
t = temperature;
λ = wavelength – usually 589.3 nm (the Na D line), just reported as D.

A sample composed of equal amounts of the two enantiomers of a chiral compound is referred to as a **racemic mixture** (or a **racemate**) and will not rotate the plane of polarized light. This is because, as the light beam passes through the sample, it will interact with a large number of molecules, half of which will rotate the plane of polarization in a clockwise direction, whilst the other half will rotate the plane of polarization by an equal amount but in an anti-clockwise direction. Thus overall there is no net rotation of the plane of polarization of the light. The specific rotation of a sample is a bulk property, that is a property of the sample as a whole and not of the individual molecules, whilst chirality is a molecular property. A racemate is composed entirely of chiral molecules, yet it does not rotate the plane of polarization of plane polarized light.

The direction in which a sample rotates the plane of polarized light, provides the basis for the third nomenclature system for enantiomers (cf. section 3.2). Thus the sample can be referred to as the (+) or (−)-enantiomer as defined below.

(+) = Sample rotates the plane of polarized light in a clockwise direction
(−) = Sample rotates the plane of polarized light in an anti-clockwise direction

Often the d and l descriptors are used instead of +/−

d = dextrorotatory = rotates polarized light in a clockwise direction
l = laevorotatory = rotates polarized light in an anti-clockwise direction.

This is very simple nomenclature system but it has three major disadvantages:

1. The sign of the specific rotation is not related to the absolute configuration of the molecule. Hence it is not possible to draw the three-dimensional structure of an enantiomer given just the sign of its optical rotation.

2. Sometimes the optical rotation changes sign when the solvent, temperature or wavelength of light is changed, so (+) and (−) become ambiguous.
3. Some compounds have specific rotations which are so small that they cannot be measured by most polarimeters.

A racemic mixture is described with the prefixes DL, *RS*, or ±, depending upon which of the three nomenclature systems is being used.

3.4.2 Other chiroptical methods (optical rotary dispersion and circular dichroism)

In a polarimeter, optical rotations are measured at a single wavelength, usually 589.3 nm (the sodium D-line), for convenience. However, both the sign and magnitude of the optical rotation of a sample are wavelength dependent and this phenomenon is referred to as **optical rotary dispersion (ORD)**. The specific rotation generally increases in magnitude as the wavelength of the light decreases. Typical rotary dispersion curves for a chiral species and its enantiomer are shown in **Figure 3.4**. It should be noted that the two enantiomers have specific rotations of equal magnitude but opposite sign at every wavelength of light.

The specific rotation of a compound is essentially a measure of the ability of the sample to refract light; however, light of an appropriate wavelength may also be absorbed by the sample. Plane polarized light can be considered to be composed of two oppositely rotating but equally intense beams of circularly polarized light (**Figure 3.5**). A chiral compound may show different absorptions for these two circularly polarized light beams, an effect which is known as **circular dichroism (CD)**.

One effect of circular dichroism is to distort the plane curves that would otherwise be produced by ORD into curves with maxima and minima, an effect known as the **Cotton effect**. The Cotton effect can be either positive or negative. A positive Cotton effect (**Figure 3.6a**) occurs when the maximum occurs at a

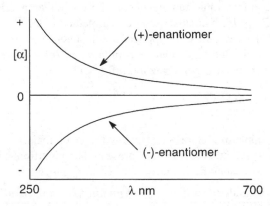

Figure 3.4 Typical ORD curves for a chiral compound and its enantiomer.

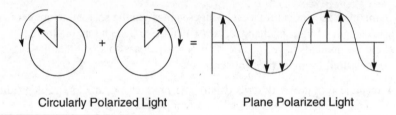

Circularly Polarized Light   Plane Polarized Light

Figure 3.5 The relationship between circularly and plane polarized light.

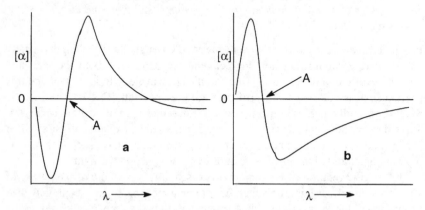

Figure 3.6 a, A positive Cotton effect; **b**, a negative Cotton effect.

higher wavelength than the minimum, and a negative Cotton effect (**Figure 3.6b**) occurs when the maximum occurs at a lower wavelength than the minimum.

The wavelength of the point where a Cotton effect curve crosses the $[\alpha] = 0$ axis (labelled A in **Figure 3.6**) corresponds to the wavelength of the absorption that is causing the Cotton effect. Circular dichroism can also be measured directly in a suitably modified spectrophotometer designed to measure and record the difference in absorption of left and right circularly polarized light. Thus the same information can be obtained from either a refraction based technique (ORD) or from an absorption based method (CD).

ORD and CD have a number of applications in stereochemistry, including the determination of absolute and relative configuration (cf. Chapter 5), and the determination of molecular conformation (cf. Chapter 8).

3.5   Enantiomeric excess

Sometimes, a sample of a chiral substance will contain mainly one enantiomer, but with some of the other enantiomer present. The **enantiomeric excess (ee)** (sometimes called enantiomeric purity) is used to quantify this:

$$ee = \frac{(\text{Amount of major enantiomer} - \text{Amount of minor enantiomer}) \times 100}{\text{Amount of major enantiomer} + \text{Amount of minor enantiomer}}$$

An alternative way of writing this definition is:

ee = %Major enantiomer − %Minor enantiomer

For example, a sample which contains five molecules of one enantiomer for every 95 of the other would have an ee of:

$$ee = \frac{(95 - 5) \times 100}{95 + 5} = \frac{90 \times 100}{100} = 90\%$$

Note that a racemic mixture contains equal amounts of the two enantiomers and so will have an enantiomeric excess of zero. Another term that is often used to define the composition of a mixture of enantiomers is **optical purity**. The optical purity is defined in terms of specific rotations (section 3.4.1) measured by polarimetry, thus:

$$\textbf{Optical purity} = \frac{\textbf{Observed specific rotation} \times \textbf{100}}{\textbf{Specific rotation of enantiomerically pure sample}}$$

For example, a sample which showed an $[\alpha]_D^{20}$ of +20, whilst the literature value for an optically pure sample was +60, would have an optical purity of:

$$\text{Optical purity} = \frac{20 \times 100}{60} = 33.3\%$$

In many cases, the optical purity and enantiomeric excess of a sample are numerically equal. This will be the case whenever the specific rotation of a sample varies linearly with the enantiomeric excess as indicated by the hashed line in **Figure 3.7**. However, cases are known in which the specific rotation does not vary linearly with enantiomeric excess, with both positive and negative deviations being possible as shown by the curves in **Figure 3.7**. In these cases, the enantiomeric excess and optical purity will be different.

Polar compounds (alcohols, acids etc.) are particularly prone to exhibiting a non-linear dependence of specific rotation on enantiomeric excess, since these compounds can form hydrogen bonded dimers or oligomers which, being different species to the monomers, may have different specific rotations. During the formation of such multimolecular species, two (or more) chiral molecules

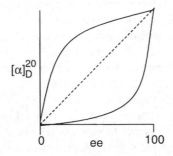

Figure 3.7 Three possible variations of specific rotation with enantiomeric excess.

react together, and the propensity for reaction of a given molecule with the two enantiomers present may be different (section 3.3). Consider, for example, the case of a chiral acid where formation of a hydrogen bonded dimer between two molecules of a particular enantiomer is disfavoured for steric reasons, but formation of such a dimer between two different enantiomers is favourable. In an enantiomerically pure sample, only the monomeric species will be present. However, as the enantiomeric excess is lowered and the other enantiomer is introduced into the solution, formation of a dimer can occur. The concentration of this dimer will increase as the enantiomeric excess of the solution decreases and may cause a non-linear dependence of specific rotation on enantiomeric excess.

3.6   Biological properties of enantiomers

Many natural products which contain a stereocentre exist in nature as a single enantiomer. This is particularly important for amino acids and carbohydrates, since all common, naturally occurring, chiral amino acids have the L-configuration, and all common, naturally occurring carbohydrates have the D-configuration (cf. structures **3.4** and **3.5** in section 3.2.1). Amino acids and carbohydrates are the monomer units from which proteins and polysaccharides are constructed, and nucleic acids also contain carbohydrates as part of their structure (cf. Chapter 8, section 8.10). Hence, all living things (including humans) are composed of chemicals which are present as a single enantiomer.

Since enantiomers react differently with other chiral compounds (section 3.3), whenever a chiral compound is eaten, drunk, smelt, ingested, etc. then the two enantiomers may have different biological effects as they will interact differently with the chiral proteins, polysaccharides and nucleic acids they encounter. One example of this is the naturally occurring terpene called carvone **3.8**. The (*R*)-enantiomer of this compound smells of spearmint, whilst the (*S*)-enantiomer smells of caraway. However, both enantiomers have identical boiling points (98–100°C), NMR spectra, IR spectra, etc. and react identically with achiral chemicals. They smell different because the receptors in the nose are composed of proteins which are built from single enantiomers of naturally occurring amino acids, so the two enantiomers of carvone interact differently with them.

A more serious example of the differing biological properties of enantiomers is illustrated by thalidomide **3.14**. In the 1960s, this pharmaceutical was manufactured as a racemic mixture of the two enantiomers and was sold in this form as a sedative and antinausea agent which was especially effective in treating morning sickness during pregnancy. It is now known that one enantiomer of thalidomide possesses the sedative properties whilst the other enantiomer causes birth defects. Thus it is important to produce any chiral chemical which will be ingested by an animal or plant as a single enantiomer, or at least to test the biological properties of both enantiomers. Processes to achieve this are a major

3.8

(*R*)-carvone
(spearmint)

(*S*)-carvone
(caraway)

3.14

(*S*)-Thalidomide causes
birth defects

(*R*)-Thalidomide
sedative

area of current chemical research and will be discussed in Chapters 5 and 10. The situation with thalidomide is slightly more complex, however, as it has been shown that the two enantiomers are interconverted *in vivo*, so even if the pure (*R*)-enantiomer was administered as a drug, the unwanted (*S*)-enantiomer would also be present.

The differing biological properties of enantiomers are commonly explained by a three point binding model. In this model, the biological receptor for the chiral compound is assumed to have a fixed geometry and to contain groups capable of interacting with the chiral compound. These groups or binding sites

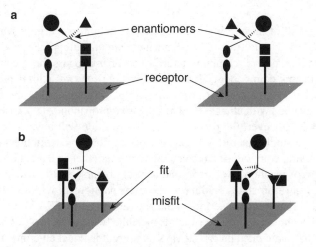

Figure 3.8 Interaction of enantiomers with a receptor containing: **a**, two; and **b**, three binding sites.

within the receptor may interact with the chiral compound sterically or electronically, for example, by hydrogen bond formation or by π–π interactions between aromatic rings. If the receptor contains only one or two groups that can interact with the chiral molecule, then both enantiomers of the chiral chemical will be able to bind to the receptor as shown in **Figure 3.8a**. In this case, the two enantiomers will exhibit the same biological properties. If, however, the receptor contains three or more groups that can interact with the chiral chemical, then only one enantiomer of the chemical will be able to bind to all three groups and so 'fit' into the receptor and elicit a biological response (**Figure 3.8b**). In this case, the two enantiomers will have different biological properties.

3.6.1 Origin of enantiomerically pure compounds in nature

The most widely accepted theory for the origin of the Universe is that it came into existence as the result of a 'big bang'. This big bang produced exotic subatomic particles which as the Universe expanded and cooled eventually produced protons, neutrons and electrons. Only later were these able to combine to form atoms of hydrogen and helium. Much later in the history of the Universe, supernova explosions produced atoms of the heavier elements, and these were able to react to form simple molecules such as hydrogen cyanide, water and ammonia. These simple molecules are often referred to as primordial species since they can be used as starting materials for the synthesis of the essential chemicals required by living species.

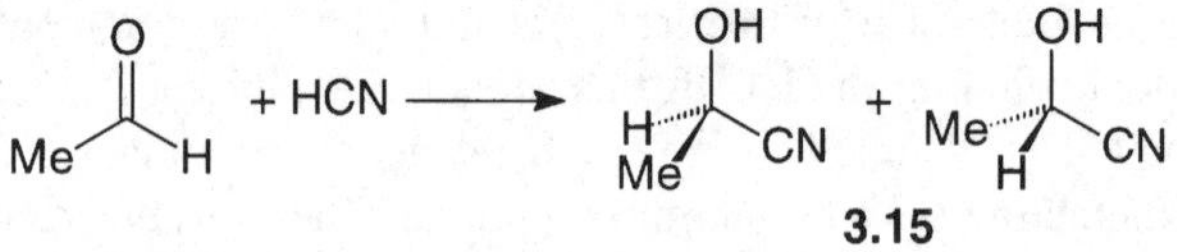

3.15

Scheme 3.2

However, all of these primordial species are achiral (superimposable on their mirror images), yet all naturally occurring proteinogenic amino acids have the L-configuration, and all common naturally occurring sugars the D-configuration, so how did this come about? It is quite possible for two achiral molecules to react together to give a chiral product, a simple example being the reaction of hydrogen cyanide with ethanal to give 2-hydroxypropanonitrile **3.15** as shown in **Scheme 3.2**. However, the product of such a reaction will always be a racemic mixture not a single enantiomer. This is because the two enantiomers of **3.15** (or any other chiral compound) have equal energies and hence have the same Gibbs free energy of reaction (ΔG) from the achiral precursors (**Figure 3.9**). Furthermore, the transition states leading to the two enantiomers will also be enantiomeric and so have equal energy, so the activation energy (E_a) for the formation of the two enantiomers will be equal. Hence, the two enantiomers of **3.15** will be formed at the same rate and will always be present in equal amounts. It is always the case, that if two (or more) achiral molecules react together in an achiral

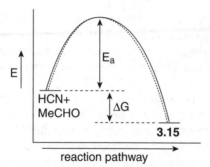

Figure 3.9 Energy diagram for the formation of enantiomeric products from achiral precursors.

environment (i.e. no chiral solvent, catalyst or other chiral influence present) then the product must be either achiral or racemic.

It has been shown (cf. Chapter 10, section 10.3.2) that chemical processes are capable of taking a very small enantiomeric excess of a compound and increasing it to the stage where one enantiomer exists exclusively. There remains, however, the problem of how did even a small enantiomeric excess come about in the first place. A number of possibilities have been proposed, although it is not known which (if any) of these was actually involved. The possibilities include:

1. **Chance**: The primordial sea (i.e. a solution of the primordial chemicals) was completely racemic and it just happened that the first living (i.e. replicating) molecule contained a stereocentre. This stereocentre just happened to have the absolute configuration found in life today (50% chance) and, since the life event only happened once, all life has the same form of chirality.
2. **Non-physical chiral influences**: The Earth just happens to be rotating on its axis and around the Sun in an anti-clockwise direction. Because the Earth has a magnetic field associated with it, this is also rotating and it has been shown that the interaction of a rotating magnetic field with high energy particles which are found in cosmic showers (X-rays, electrons, etc.) results in the formation of circularly polarized light (this is an example of a chiral helix, section 3.8.3). It has also been demonstrated that when chemical reactions occur in the presence of circularly polarized light, asymmetric induction can occur; an example is shown in **Scheme 3.3**. In this example, the racemic

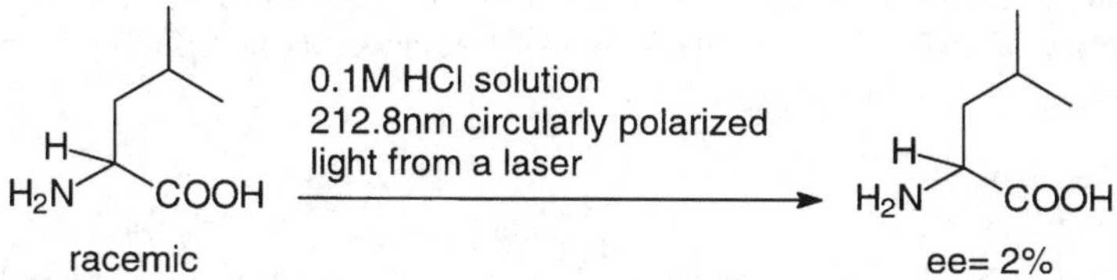

H_2N $COOH$ 0.1M HCl solution
H—   212.8nm circularly polarized
                    light from a laser   H—
                    $\longrightarrow$ H_2N $COOH$

racemic   ee= 2%

Scheme 3.3

starting material undergoes decomposition under the reaction conditions. However, the circularly polarized light decomposes one enantiomer more rapidly than the other, so if the reaction is stopped before all of the starting material is destroyed then the recovered starting material is no longer racemic.

3. **Fundamental fact of nature**: It was shown in the 1950s that certain sub-atomic events were asymmetric; for instance, the β-decay of ^{60}Co produced more electrons with left-handed spin than with right. This has been traced to a fundamental asymmetry in the weak nuclear force and could cause the formation of a small enantiomeric excess in two ways:

   (a) The spinning electrons produce circularly polarized light which causes asymmetric induction as above.

   (b) The asymmetric nature of the weak nuclear force could cause (R)- and (S)-enantiomers of a compound to have slightly different energies; this would affect their stability and reactivity. The effect would be far too small to measure but may have caused the small asymmetry that life processes then amplified.

3.7   Chirality due to stereocentres at atoms other than carbon

So far, all of the chiral molecules in this chapter have been chiral because of the presence of a carbon atom tetrahedrally bound to four different groups. However, whilst this is by far the most common structural feature leading to chirality, it is not the only one. The only requirement for a molecule to be chiral is that it is not superimposable on its mirror image. Indeed, as structure **3.1** suggested, the tetrahedrally bonded carbon atom can be replaced by any other atom capable of being tetrahedrally coordinated to four different groups. This situation can occur for a number of elements in the Periodic Table, and in sections 3.7.1–3.7.4 the more common of these will be surveyed. Also in section 3.7.4 it will be shown that a tetrahedral geometry around the stereocentre is not a prerequisite for chirality. Finally, in section 3.8 a number of ways in which a molecule can be chiral without possessing a stereocentre are introduced.

Sections 3.7 and 3.8 are not intended to be comprehensive but to provide illustrations of the various different structural features which can cause a molecule to be chiral, and to highlight a few unusual situations. For all of the compounds discussed in section 3.7, the absolute configuration of a particular enantiomer can be defined by the R- and S-descriptors as introduced for carbon based stereocentres in section 3.2.2.

3.7.1   Sulphur

A number of sulphur compounds are known in which the sulphur is bonded to three substituents and also retains a lone pair of electrons. These types of compounds can in suitable cases exhibit chirality, since the VSEPR theory predicts a tetrahedral electron structure for these compounds.

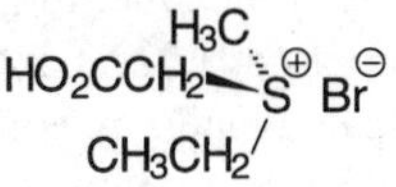

3.16   **3.17**

Sulphonium salts are usually described as having a trigonal pyramidal bond structure. However, if the lone pair of electrons on the sulphur atom is counted as the fourth group, then the structure can be considered as tetrahedral and will be chiral provided R^1, R^2 and R^3 are all different as shown in structure **3.16**. A real example of such a compound is shown in structure **3.17**.

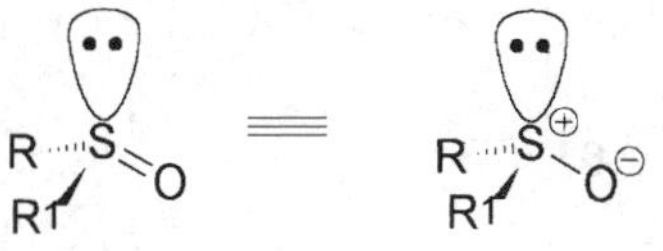

3.18   **3.19**

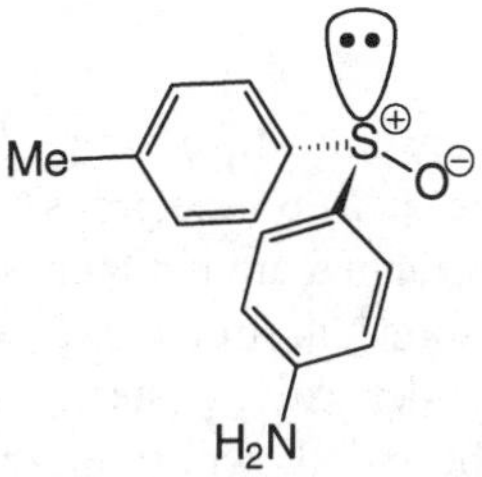

$$[\alpha]_{280}^{25} = +0.71$$

3.20 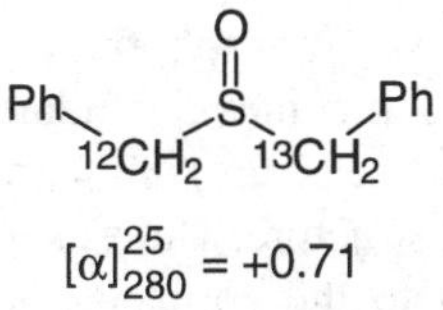   **3.21**

Sulphoxides **3.18** also have a structure with three different substituents (provided R and R^1 are different) and a lone pair of electrons around a sulphur atom, and so can exhibit chirality. If the sulphoxide is drawn in the zwitterionic structure **3.19** (which is a major resonance form of a sulphoxide), then the close relationship between sulphoxides and sulphonium salts **3.16** becomes apparent. Two real examples of chiral sulphoxides are shown in structures **3.20** and **3.21**. The first of these is a straightforward example where the two carbon based substituents are obviously different, whilst the second **3.21** was prepared to demonstrate that even the presence of a single neutron difference (^{12}C and ^{13}C) in the two carbon based substituents is enough to turn the sulphur into a stereocentre and make the molecule chiral. Note that the specific rotation of compound **3.21** had to be determined at 280 nm rather than the more common 589 nm in order for the magnitude of the specific rotation to be large enough to be measured (section 3.4.2).

Sulphur compounds in which the sulphur is attached to four substituents are also tetrahedral and so can exhibit chirality. Examples include the sulphone **3.22** and the sulphoximine **3.23**. Sulphones are normally achiral, since two of the groups attached to the sulphur atom are identical oxygen atoms; however, in compound **3.22**, isotopic labelling was used to make the two oxygen atoms non-equivalent. Compound **3.22** shows the same effect of isotopic substitution in otherwise identical ligands as is seen in compound **3.21**. In this case, the two oxygen atoms differ by two neutrons and the specific rotation of the compound could be measured at 589 nm.

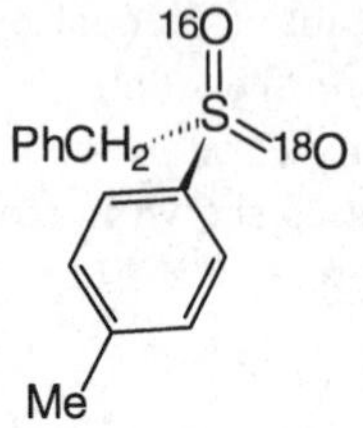

$[\alpha]_D^{20}$ = -0.16°

3.22

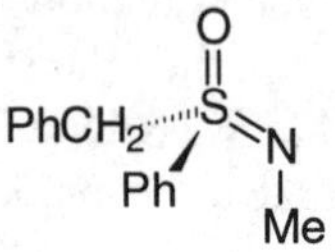

3.23

3.7.2 Phosphorus

Various classes of phosphorus compounds are known in which four different substituents (sometimes including the lone pair of electrons in a trigonal pyramidal bond structure) are tetrahedrally arranged around a phosphorus atom, giving a chiral structure. Three examples are shown in structures **3.24–3.26** which are analogous to the corresponding sulphur derivatives (section 3.7.1). Note, however, that phosphines (e.g. **3.24**) can be chiral, whilst the corresponding sulphides are always achiral since there are two lone pairs attached to the sulphur atom of a sulphide but only one on the phosphorus atom of a phosphine.

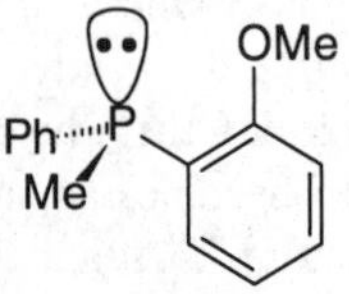

Phosphine   Phosphine Oxide   Phosphonium Salt

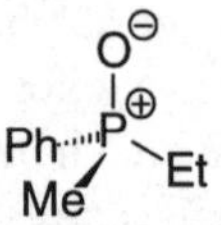

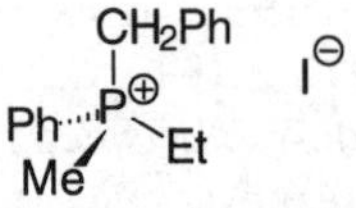

3.24   **3.25**   **3.26**

3.7.3 Nitrogen

Simple amines have a trigonal pyramidal bond structure, as do phosphines (section 3.7.2), and hence are chiral provided the three substituents are different (cf. structures **3.27** and **3.28**). However, unlike phosphines, at room temperature amines undergo a very rapid 'umbrella inversion' (**Scheme 3.4**) which has the effect of interconverting the two enantiomeric structures. Since this inversion occurs without the breaking of a bond, structures **3.27** and **3.28** are best considered as conformational rather than configurational isomers. Hence, simple amines represent a class of compounds in which, although the individual molecules are chiral, a bulk sample can never display optical activity or be separated into its constituent enantiomers.

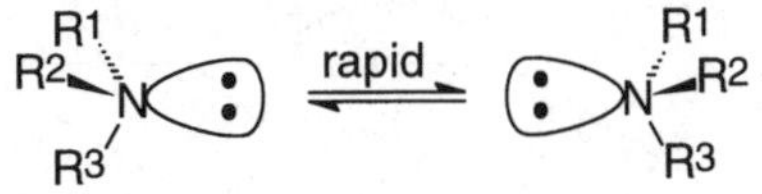

3.27　　　　　　　**3.28**

Scheme 3.4

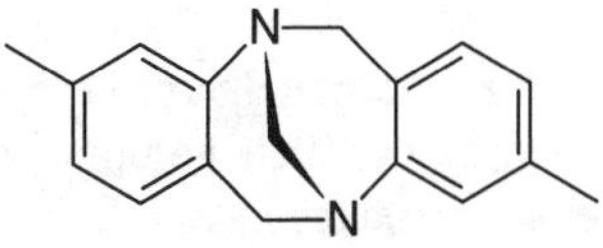

Troger's base **3.29**
Both nitrogen atoms are stereocentres

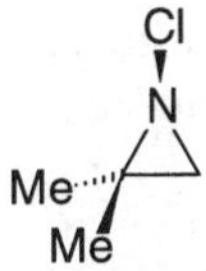

3.30 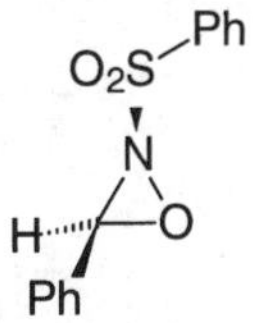　　　　　**3.31** 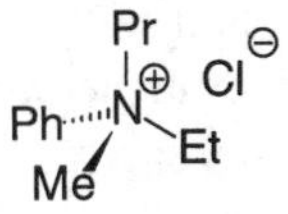　　　　　**3.32**

In bicyclic amines such as Troger's base **3.29**, however, this rapid 'umbrella inversion' cannot occur, since it would result in a highly strained structure in which the lone pairs of the nitrogen atoms were pointing into the centre of the eight-membered ring. Hence, such amines can be obtained as optically active single enantiomers. If the nitrogen atom is part of a three-membered ring, then the barrier to inversion can also be sufficiently high that single enantiomers of these compounds can be obtained. Thus optically active aziridines **3.30** and oxaziridines **3.31** can be obtained. Tetrasubstituted ammonium salts such as **3.32** also do not undergo an 'umbrella inversion', so these can also be obtained in an enantiomerically pure form.

3.7.4 Metal complexes

Metal complexes can adopt a wide variety of different geometries depending upon the coordination number of the metal. However, the three most common are octahedral for six-coordinate species and tetrahedral or square planar for four-coordinate species. Only these three geometries will be considered here.

Square planar complexes being flat are always superimposable on their mirror images and hence are achiral as shown for structure **3.33**. The only way in which a square planar metal complex can be chiral is if there is a stereocentre in one of the four ligands a–d. Square planar complexes can, however, show cis–trans isomerism (cf. Chapter 2, section 2.4).

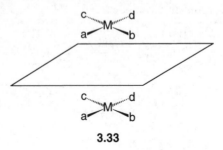

3.33

In a tetrahedral metal complex, the metal will be a stereocentre if the four ligands are all different; however, no such complex has yet been prepared. Tetrahedrally coordinated metal complexes in which the metal is attached to two bidentate ligands can also be chiral, and these have been prepared. These complexes are best considered as possessing a stereogenic axis rather than a stereocentre, however, and are discussed in section 3.8.1

An octahedrally coordinated metal attached to six different monodentate ligands will be chiral and the metal a stereocentre as illustrated in structure **3.34**. This is the first example we have seen of a chiral molecule which does not

3.34

contain a tetrahedrally coordinated atom with four different groups attached to it, and this reinforces the point that a tetrahedral centre of chirality is not necessary for a molecule to be chiral. Very few examples of transition metal complexes octahedrally coordinated to six different monodentate ligands have been prepared, and none has been stereochemically characterized, one example being the platinum(IV) complex **3.35**.

$$Pt(pyridine)(NH_3)(NO_2)(Cl)(Br)(I)$$

3.35

For a tetrahedral stereocentre, there are only two ways in which the four substituents can be attached to the stereocentre, and these are enantiomers of one another. For an octahedral stereocentre, however, there are many more stereo-isomers. For structures of the form Mabcdef, there are a total of 30 stereo-isomers consisting of different arrangements of the ligands around the central metal. In structure **3.34** for example, ligand f is *trans* to ligand a; however, any of the other ligands b–e could occupy this position, giving structures which would be diastereomers of **3.34**. In total, there are 15 such diastereomers, each of which exists as a pair of enantiomers, giving 30 stereoisomers in total.

For hexacoordinated metals with monodentate ligands, it is not necessary for all six ligands to be different for the complex to be chiral. It is sufficient that there are at least three different ligands and no more than two of each of these. For complexes of the form $Ma_2b_2c_2$ there are a total of six stereoisomers; two of these stereoisomers are enantiomers **3.36** and **3.37**; the other four stereoisomers are diastereomeric, cis–trans isomers of structures **3.36** and **3.37**, and of one another. An example of such a complex is the diaminedichlorodinitroplatinum complex **3.38**. If more than three different ligands are present, then there are

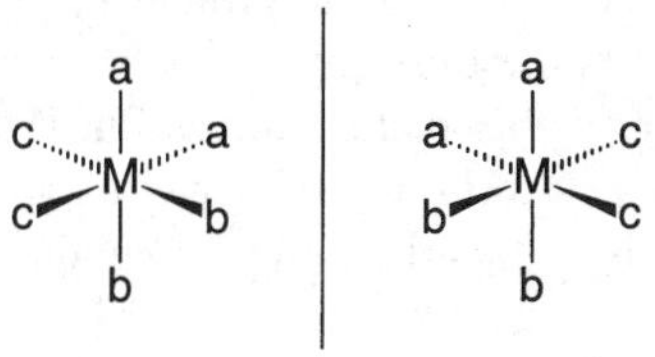

3.36   **3.37** 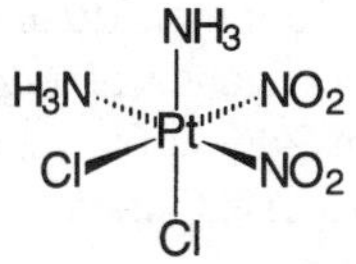   **3.38**

more possible chiral structures; the total number of stereoisomers and pairs of enantiomers is given in **Table 3.1**.

The metal can also be octahedrally coordinated to one or more polydentate ligands, and again many of these complexes are chiral. There is, however, a large number of possible situations depending upon how many different poly-dentate and monodentate ligands are present. The important test of chirality remains, however, is the molecule superimposable on its mirror image? If not then the molecule will be chiral.

The stereochemistry of octahedral systems can be described using the $+/-$ (or d/l) prefixes based upon the direction of rotation of plane polarized light as

Table 3.1 Stereoisomers in octahedral metal complexes with monodentate ligands

Formula	Stereoisomers	Pairs of enantiomers
Ma_6	1	0
Ma_5b	1	0
Ma_4b_2	2	0
Ma_3b_3	2	0
Ma_4bc	2	0
Ma_3b_2c	3	0
Ma_3bcd	5	1
$Ma_2b_2c_2$	6	1
Ma_2b_2cd	8	2
Ma_2bcde	15	6
$Mabcdef$	30	15

described in section 3.4.1. An alternative nomenclature system for chiral octahedral complexes has also been developed utilizing the Cahn, Ingold, Prelog
priority rules (cf. Chapter 2 section 2.1). The nomenclature uses the prefixes *C*
(clockwise) and *A* (anti-clockwise), which are applied using the following
rules:

1. Each of the ligands attached to the octahedral centre is assigned to an order
   of priority (1 to 6).
2. A principal axis is then defined as being the axis containing the ligand of
   highest priority, the octahedral centre, and the ligand *trans* to the ligand of
   highest priority. If two ligands of equal highest priority are present, then the
   principal axis is chosen so that the ligand *trans* to the ligand of highest
   priority has the highest possible priority.
3. The ligands lying in the plane perpendicular to the principal axis are then
   viewed from the side of the ligand of highest priority on the principal axis. If
   the three ligands of highest priority in this plane decrease in priority in a
   clockwise direction, the complex is said to be the (*C*)-enantiomer, whilst if
   they decrease in an anti-clockwise direction, the compound is the
   (*A*)-enantiomer.

To see how these rules are applied, consider the general case of metal
complex **3.34**, in which the order of priority of the ligands is assumed to be
a > b > c > d > e > f. The principal axis is that containing the ligand of
highest priority a. The ligands b, c, d and e form the plane perpendicular to the
principal axis, and the three ligands of highest priority within this plane b, c and
d decrease in order of priority in a clockwise direction when viewed from the
face of the molecule bearing atom a. Hence, this is the (*C*)-enantiomer of the
compound. The *C*/*A* nomenclature can be applied to any octahedral, chiral
species, but if the system contains two or more bidentate ligands, it is more
common to treat the compound as a chiral helix and to use a nomenclature
system designed for helical molecules (section 3.8.3).

3.34

3.8   Other stereogenic elements which can produce molecular chirality

Up to now, all the chiral molecules discussed in this chapter have contained a
stereocentre. There are, however, other structural features that can cause a
molecule to exhibit chirality. Many of the structural features discussed in the

following section are only clearly seen in three dimensions, and the reader will find a molecular modelling kit very useful in visualizing these structures.

3.8.1 Stereogenic axis

Allenes have the general structure **3.39** in which the two π-systems are orthogonal giving the three dimensional structure **3.40**. If viewed end on (along the C=C=C bond) this becomes structure **3.41**, and the molecule resembles a tetrahedron. It is clear therefore that the molecule will be chiral provided R, R^1, R^2 and R^3 are different; an example is compound **3.42**. In fact, as an allene is actually a stretched tetrahedron it has a lower symmetry (cf. Chapter 6) than a true tetrahedral structure, and an allene will be chiral provided R and R^1 are different, and R^2 and R^3 are different (i.e. R and R^2 etc. can be identical). Hence allene **3.43** is also chiral but allene **3.44** is achiral.

3.39   **3.40**   **3.41**

3.42   **3.43**   **3.44**

In general, compounds with cumulative double bonds will show cis–trans isomerism (cf. Chapter 2) if there are an odd number of double bonds involved, and enantiomerism if there are an even number of double bonds in the cumulated π-system. Because a ring and a double bond are stereochemically equivalent (cf. Chapter 2, section 2.3), one or both of the double bonds in an allene can be replaced by a ring (of any size), and the resulting *exo*-alkylidene monocyclic or spiro-bicyclic compounds will exhibit enantiomerism, examples being structures **3.45** and **3.46**. In these compounds, the ring is usually not flat as drawn in structures **3.45** and **3.46** (cf. Chapter 8, sections 8.4–8.7), but this is not important when deciding whether the structures are chiral or not. Another class of compounds that contain a stereogenic axis is tetrahedral metal complexes in which the metal is attached to two bidentate ligands, an example of which is shown in structures **3.47** and **3.48**.

Biphenyls (or more generally biaryls) are compounds which contain two aromatic rings joined together by a carbon–carbon single bond as shown in structure **3.49**. There is usually free rotation about the bond joining the two rings, with the conformation in which the two aromatic rings are coplanar, **3.50a**, being electronically preferred as this maximizes electron delocalization

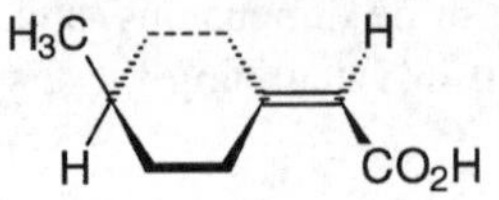

3.45

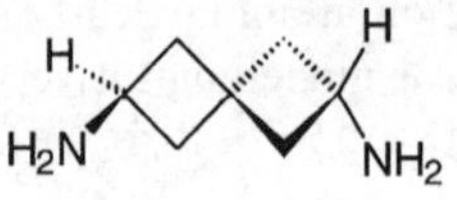

3.46

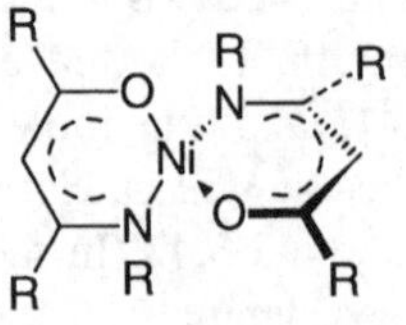

3.47

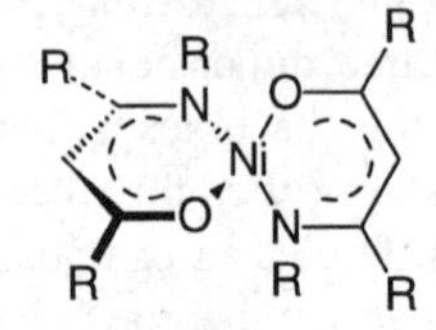

3.48

between the two aromatic rings. However, in structure **3.50a** the *ortho*-substituents on the two rings are brought into close proximity, and to avoid unfavourable steric interactions (shown by the double headed arrows in structure **3.50a**), the minimum energy conformation for these compounds has the two aromatic rings oriented at about 45° to one another as shown in structure **3.50b**. The value of 45° for the angle between the two aromatic rings is a compromise between steric repulsions (maximum at 0°, minimum at 90°) and favourable electron delocalization which is also maximized at 0° and minimized at 90°.

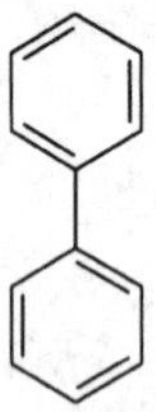

Biphenyl
3.49   **3.50a** 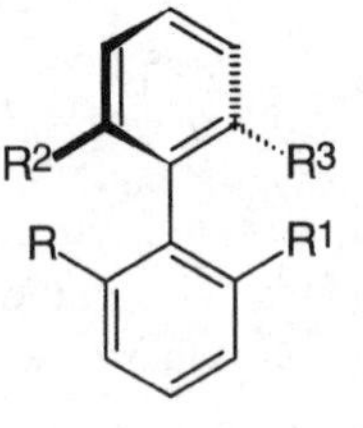   **3.50b** 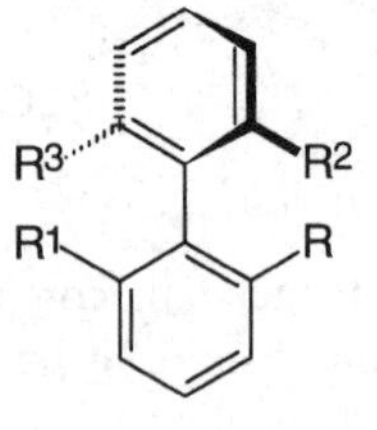   **3.50c**

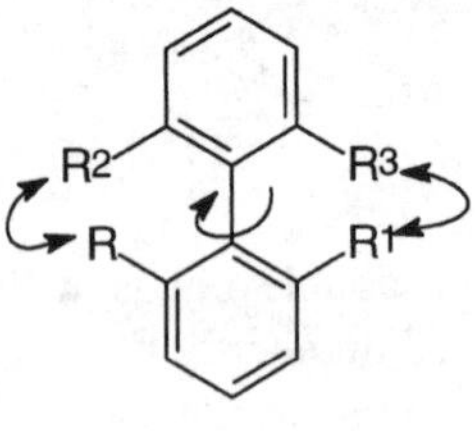

If the substituents R–R^3 are all different, then there is a third conformation for biphenyls, **3.50c**, which is the non-superimposable mirror image of **3.50b**. Thus structures **3.50b** and **3.50c** are enantiomers of one another, and suitably substituted biphenyls possess a stereogenic axis which runs along the bond joining the two aromatic rings. However, structures **3.50b** and **3.50c** are conformational rather than configurational isomers of one another since they can be interconverted simply by rotation around the carbon–carbon single bond joining the two aromatic rings. For this interconversion to occur the molecule must pass through the planar conformation **3.50a** which, as is shown in **Figure 3.10**, is an energy maximum. Thus the rate of interconversion between structures **3.50b** and **3.50c** will depend upon the height of this barrier (i.e. on the activation energy), which is determined by how strongly the *ortho*-substituents repel one another. Many cases of biphenyls are known in which the height of this energy barrier is

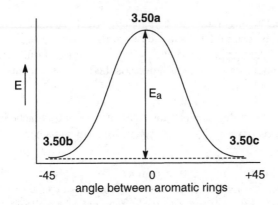

Figure 3.10 Energy diagram for interconversion of biphenyl atropisomers.

such that at room temperature the two enantiomeric structures **3.50b** and **3.50c** are found not to interconvert even over many years. Enantiomeric conformational isomers such as these, which cannot interconvert because of the size of the energy barrier to interconversion, are called **atropisomers**. If a single enantiomer of a chiral biphenyl or any other atropisomer is heated, then eventually there will be enough thermal energy available to overcome the barrier to rotation, and a racemic mixture will be formed.

A biphenyl derivative will exist as a pair of stable enantiomers provided that the *ortho*-substituents are large enough to prevent rotation around the carbon–carbon bond and the symmetry of the aromatic rings is reduced. The latter can be achieved in a number of ways, the simplest being if R and R^1 are different, and R^2 and R^3 are also different to one another as in compound **3.51**. Note that as in allenes it is not necessary for all four substituents to be different, just that two substituents at each end of the stereogenic axis be different. However, the *ortho*-substituents can be identical provided one or more *meta*-substituents are present on the aromatic ring to reduce the symmetry as is the case in the upper ring of compound **3.52**. Usually it is necessary that all four *ortho*-substituents are not hydrogens for a biphenyl to exist as stable atropisomers. However, if the groups on one ring are very large then even a hydrogen atom on the other ring will be big enough to prevent free rotation about the single bond and ensure that a non-planar conformation is adopted, as is the case for biphenyl **3.52**.

The absolute configuration of compounds which contain a stereogenic axis can be defined by use of the R and S descriptors as described in section 3.2.2 for compounds containing a stereocentre. However, an additional rule is needed since for compounds containing a stereoaxis the four substituents may not all be different. This rule is:

Rule 0   View the molecule along the stereoaxis. Groups attached to the front of the axis have a higher priority than groups attached to the rear of the axis.

Note that this new rule is rule number zero, and so is the first rule to be applied in deciding the priorities of the four groups attached to the stereoaxis. The application of the CIP rules to the assignment of the absolute configuration of compounds with a stereogenic axis will be illustrated for compounds **3.42** and **3.51**. In the case of allene **3.42**, we will choose to view the molecule along the stereoaxis so that the two groups on the right side of the allene as shown in structure **3.42** are at the front and will have the highest priorities (rule 0). The CO_2H group will then have priority 1 and the butyl group priority 2 (rules 1–3). The methyl and hydrogen atom attached to the left side of the allene will then have priority 3 and 4 respectively. Since the molecule is being viewed with the group of lowest priority at the back, and the other three groups decrease in order of priority in an anti-clockwise direction, compound **3.42** has the S-configuration. The decision as to which end of the stereoaxis the molecule should be viewed from is not important, because the same result would have been obtained had compound **3.42** been viewed along the axis with the hydrogen and methyl groups at the front.

In the case of biphenyl **3.51**, the molecule is first viewed along the stereogenic axis and, as drawn below, the thick bonds represent groups on the front aromatic

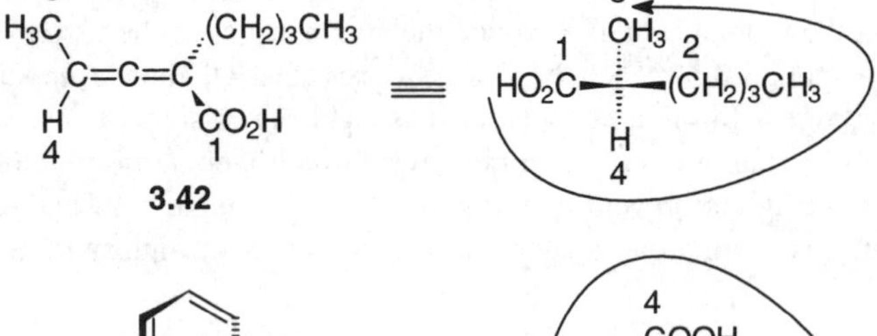

3.42

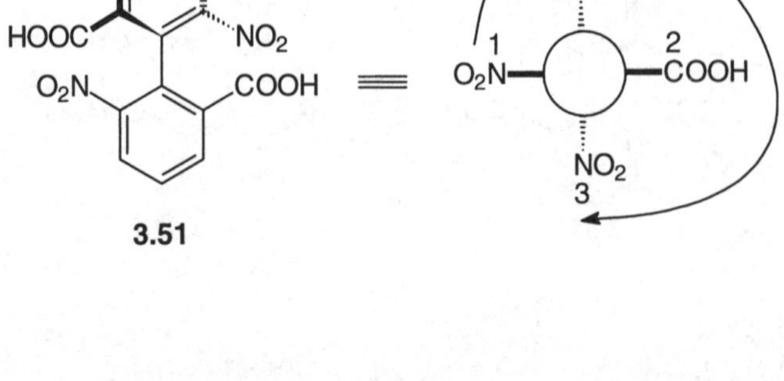

3.51

ring and hashed bonds represent groups on the rear aromatic ring. Application of rule 0 causes the nitro and acid groups attached to the front aromatic ring to have the highest priorities, and the front nitro group will have a higher priority than the front acid (rule 1). Similarly, the nitro group attached to the rear aromatic ring will have priority 3 and the acid group attached to the rear aromatic ring will have priority 4. The three groups of highest priority then decrease in a clockwise direction, which allows the absolute configuration of biphenyl **3.51** to be assigned as *R*.

3.8.2 Stereoplane

Metallocenes are organometallic compounds in which a metal atom is sandwiched between two aromatic rings. Examples include ferrocene **3.53** and (*bis*-benzene)chromium **3.54**. If one or both of the aromatic rings in a metallocene is substituted then these structures can become chiral. The minimum requirement is that on one of the aromatic rings, there should be two different substituents either *ortho* or *meta* to one another. Thus the ferrocene derivative **3.55** is chiral and exists as a pair of enantiomers. Compounds such as this are said to possess a stereoplane, in this case the stereoplane is the plane of the upper aromatic ring. One of the aromatic rings can also be replaced by other groups as in the chromium tricarbonyl derivative **3.56**, and the molecule retains a stereoplane.

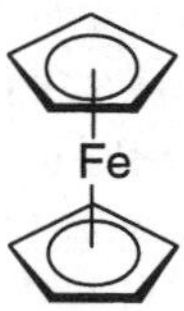

3.53 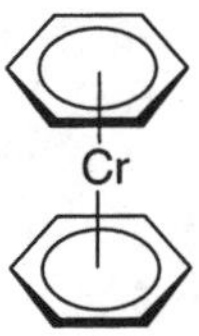   **3.54** 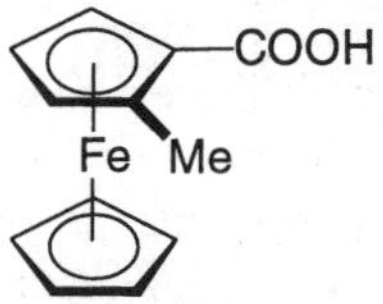   **3.55** 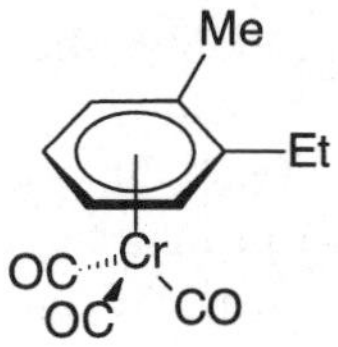   **3.56**

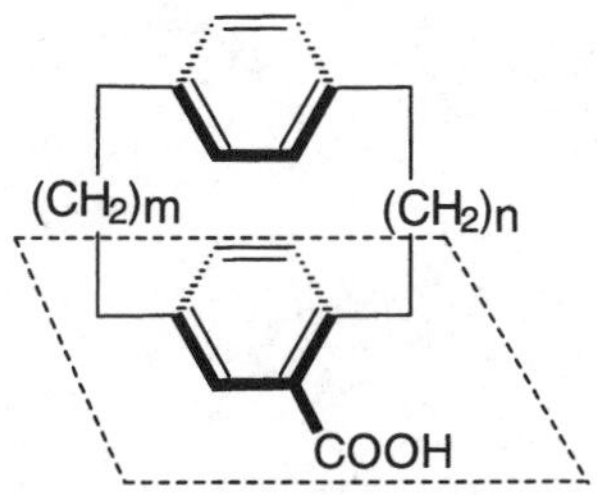

3.57 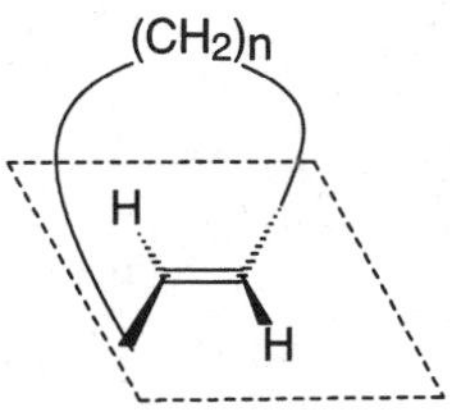   **3.58** 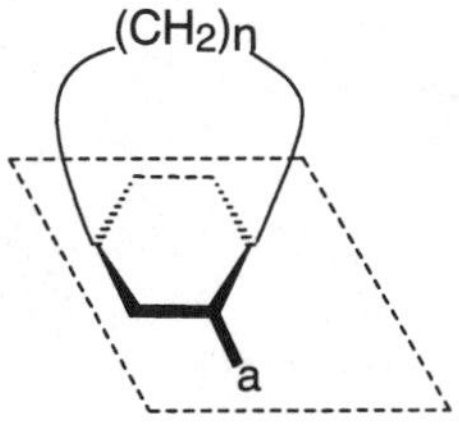   **3.59**

Consider compounds of the general structure **3.57**. If *m* and *n* are small, then there is insufficient room for the two aromatic rings to rotate past one another, and the structure shown is not superimposable on its mirror image and hence is chiral. In practice, it is found that the rotation of the aromatic rings is only prevented if $m < 4$ and $n < 5$. This is another case where the chirality is dependent upon hindered rotation about a single bond, i.e. the enantiomers are atropisomers and are conformers rather than configurations. Other examples of molecules which possess a plane of chirality are small *trans*-cycloalkenes **3.58**, which are chiral provided $n = 5$ or 6, and the cyclohexane derivatives **3.59**. The stereoplane is indicated in each case, and both enantiomers of *trans*-cyclooctene are shown in structures **3.60a** and **3.60b**. There are methods for assigning the absolute configuration of compounds containing a stereoplane using the CIP rules, however, they are beyond the scope of this text.

3.60a   **3.60b**

3.8.3   Helical molecules

A helix, like the thread of a screw, can turn two ways. The two structures are non-superimposable mirror images of one another and are hence enantiomers. Molecules can adopt helical structures and any compound which does so will exhibit chirality. The best known example is DNA, which adopts a right-handed double helical structure (cf. Chapter 8, section 8.10). A simpler example is hexahelicene, the two enantiomers of which are shown in structures **3.61a** and **3.61b**. This molecule cannot be planar as the carbon and hydrogen atoms of the terminal benzene rings would occupy the same space, so the end benzene rings must bend over or under one another giving a chiral, helical molecule.

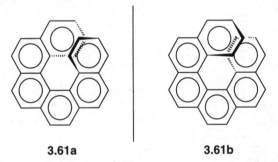

3.61a   **3.61b**

Another class of helical molecules are the propeller shaped molecules such as triarylboranes **3.62**, triarylphosphines **3.63** and triarylmethanes **3.64**. In these compounds, the rotation of the aromatic rings is hindered or prevented (depending upon the size of the *ortho*-substituents) by steric interactions with the other aromatic rings. All of the compounds **3.61**–**3.64** are again atropisomers rather than configurational isomers.

The *R/S* descriptors cannot be used to assign the absolute configuration of helical molecules; instead an alternative system based upon the *M/P* (minus/plus) descriptors is used. To apply these descriptors, the molecule is viewed looking down the helical axis as in structures **3.61**. If the helix turns clockwise from the front to the rear, then it is given the *P* descriptor. However, if the helix turns anti-clockwise from front to rear, then the *M* descriptor is used. Thus **3.61a** represents the (*P*)-enantiomer of hexahelicene whilst **3.61b** represents the (*M*)-enantiomer.

A final class of compounds which can be considered to be propeller shaped, and hence helical, are octahedral metal complexes bearing two or three bidentate ligands such as the chromium complex **3.65**. The *M* and *P* prefixes can also be

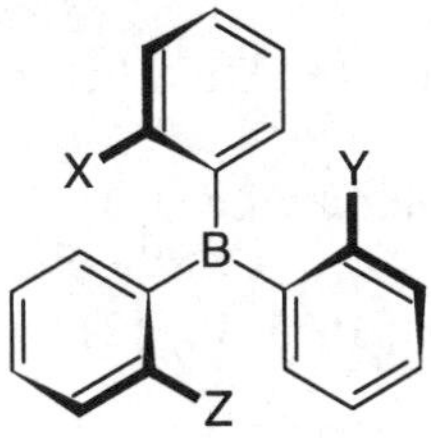

3.62 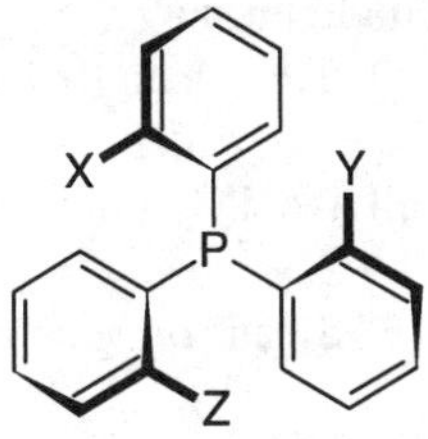   **3.63** 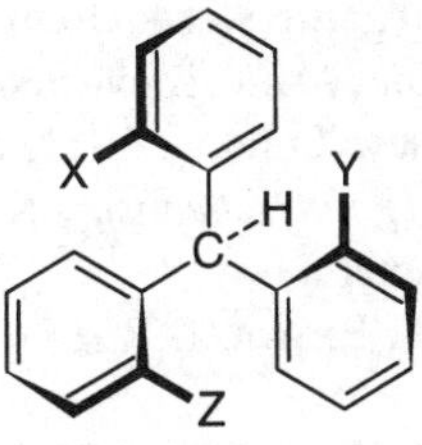   **3.64**

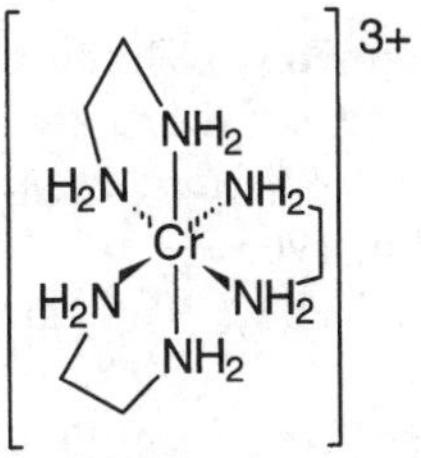

3.65

used for these systems, but the stereochemical descriptors Δ (delta cf. D) and Λ (lambda cf. L) tend to be used instead. If the molecule forms a clockwise helix, the Δ-prefix is used, whilst if the helix turns anti-clockwise, the Λ-prefix is applied.

3.9   Further reading

General
Stereochemistry of Organic Compounds E.L. Eliel and S.H. Wilen. Wiley: London, 1994, chapters 1, 5, 6, 13 and 14.
Stereochemistry, Conformation and Mechanism P.S. Kalsi. Wiley: London, 1990, chapter 1.

Chirality definition (due to Lord Kelvin)
L.D. Barron. *Chem. Eur. J.*, 1996, **2**, 743.
D. Avnir, O. Katzenelson and H.Z. Hel-Or. *Chem. Eur. J.*, 1996, **2**, 744.
And references cited within these two discussion papers.

CIP rules and their application
R.S. Cahn and C.K. Ingold. *J. Chem. Soc.*, 1951, 612.
R.S. Cahn, C.K. Ingold and V. Prelog. *Experientia*, 1956, **12**, 81
R.S. Cahn, C.K. Ingold and V. Prelog. *Angew. Chem., Int. Ed. Engl.*, 1966, **5**, 385.
V. Prelog and G. Helmchem. *Angew. Chem., Int. Ed. Engl.*, 1982, **21**, 567.

Inorganic stereochemistry nomenclature
Topics in Stereochemistry Vol. 12, T.E. Sloan (N.L. Allinger and E.L. Eliel eds). Wiley: London, 1981, chapter 1.
Inorganic Chemistry K.F. Purcell and J.C. Kotz. Holt-Saunders: London, 1977, chapter 11.
M.F. Brown, B.R. Cook and T.E. Sloan. *Inorganic Chem.*, 1975, **14**, 1273.

Biological properties of enantiomers
G.F. Russell and J.I. Hills. *Science*, 1971, **172**, 1043.
A. Richards and R. McCague. *Chem. Ind. (London)*, 1997, 422.

Chiroptical techniques (polarimetry, CD and ORD)
Optical Rotary Dispersion C. Djerassi. McGraw-Hill: London, 1960.
Optical Rotary Dispersion and Circular Dichroism in Organic Chemistry P. Crabbe. Holden-Day: London, 1965.
Optical Circular Dichroism L. Velluz, M. Legrand and M. Grosjean. Verlag Chemie: New York, 1965.
Topics in Stereochemistry Vol. 1, K. Schlogl (N.L. Allinger and E.L. Eliel eds). Wiley: London, 1967, chapter 3.
Asymmetric Synthesis Vol. 1, G.G. Lyle and R.E. Lyle (J.D. Morrison ed.). Academic Press: London, 1983, chapter 2.

Biological origins of optical activity
Origins of Optical Activity in Nature D.C. Walker (ed.). Elsevier: Amsterdam, 1979.
Topics in Stereochemistry Vol. 18, W.A. Bonner (E.L. Eliel and S.H. Wilen eds.). Wiley: London, 1988, chapter 1.
R.A. Hegstrom and D.K. Kondepudi. *Scient. Am.*, 1990, 98.
W.A. Bonner. *Chem. Ind.* (London), 1992, 640.

Optical activity induced by polarized light
Y. Shimizu and S. Kawanishi. *Chem. Commun.*, 1996, 1333 and references cited therein.

Chirality due to isotopic labelling
C.J.M. Stirling. *J. Chem. Soc.*, 1963, 5741.
K.K. Andersen, S. Colonna and C.J.M. Stirling. *J. Chem. Soc., Chem. Commun.*, 1973, 645.
T. Makino, M. Orfanopoulos, T-P. You, B. Wu, C.W. Mosher and H.S. Mosher. *J. Org. Chem.*, 1985, **50**, 5357.

Stereochemistry of Group IV compounds
Topics in Stereochemistry Vol. 15, R.J.P. Corriu, C. Guerin and J.J.E. Moreau (N.L. Allinger, E.L. Eliel and S.H. Wilen eds). Wiley: London, 1984, chapter 2.
Topics in Stereochemistry Vol. 12, M. Gielen (N.L. Allinger and E.L. Eliel eds). Wiley: London, 1981, chapter 5.
Asymmetric Synthesis Vol. 4, C.A. Maryanoff and B.E. Maryanoff (J.D. Morrison and J.W. Scott eds). Academic Press: London, 1984, chapter 3.

Stereochemistry of Group V compounds
Topics in Stereochemistry Vol. 6, J.B. Lambert (N.L. Allinger and E.L. Eliel eds). Wiley: London, 1971, chapter 2.
Topics in Stereochemistry Vol. 6, E.G. Janzen (N.L. Allinger and E.L. Eliel eds). Wiley: London, 1971, chapter 4.
Asymmetric Synthesis Vol. 4, D. Valentine Jr (J.D. Morrison and J.W. Scott eds). Academic Press: London, 1984, chapter 3.
Asymmetric Synthesis Vol. 4, F.A. Davis and R.H. Jenkins Jr (J.D. Morrison and J.W. Scott eds). Academic Press: London, 1984, chapter 4.

Stereochemistry of sulphur compounds
Topics in Stereochemistry Vol. 13, M. Mikolajczyk and J. Drabowicz (N.L. Allinger, E.L. Eliel and S.H. Wilen eds). Wiley: London, 1982, chapter 2.
Asymmetric Synthesis Vol. 4, M.R. Barbachyn and C.R. Johnson (J.D. Morrison and J.W. Scott eds). Academic Press: London, 1984, chapter 2.

Stereochemistry of octahedral systems
Inorganic Chemistry K.F. Purcell and J.C. Kotz. Holt-Saunders: London, 1977, chapter 11.
Topics in Stereochemistry Vol. 10, Y. Saito (N.L. Allinger and E.L. Eliel eds). Wiley: London, 1978, chapter 2.

Six different ligands around a metal
P.K. Baker and D. ap Kendrick. *J. Chem. Soc., Dalton Trans.*, 1993, 1039, and references cited therein.

Stereochemistry of metallocenes
Topics in Stereochemistry Vol. 1, K. Schlogl (N.L. Allinger and E.L. Eliel eds). Wiley: London, 1967, chapter 2.

3.10 Problems

1. Which of the following structures contains a stereocentre?

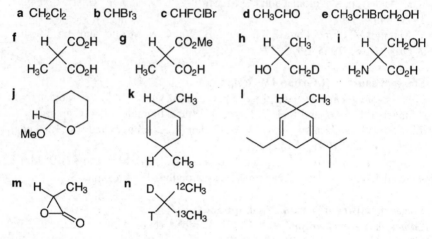

   a CH_2Cl_2 **b** $CHBr_3$ **c** $CHFClBr$ **d** CH_3CHO **e** $CH_3CHBrCH_2OH$

2. Assign the absolute configuration of each stereocentre in each of the following compounds as R or S and in the case of **a–c** also determine whether the structure drawn is the D or L isomer.

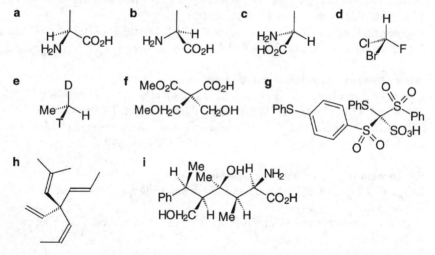

3. Use VSEPR theory (cf. Chapter 1) to determine the shape of each of the following molecules. Hence determine which of the compounds are chiral and identify the stereogenic elements they contain.

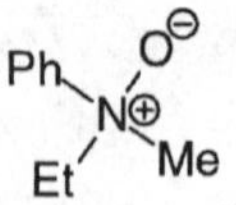

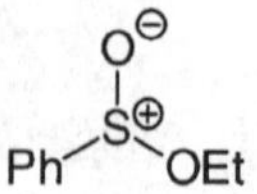

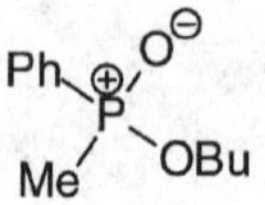

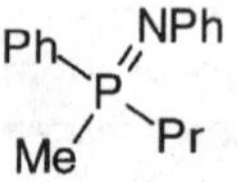

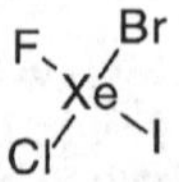

4. Is the cobalt complex shown below chiral? If so, identify the stereogenic element.

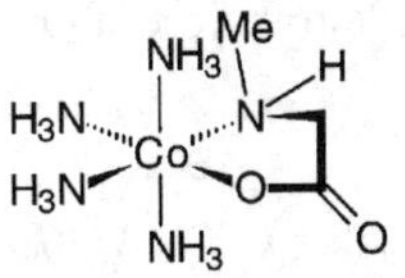

5. Explain why each of the following compounds is chiral and classify the absolute configuration of the enantiomer shown using the CIP rules.

a

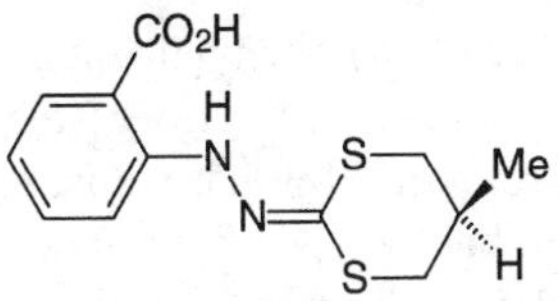

b

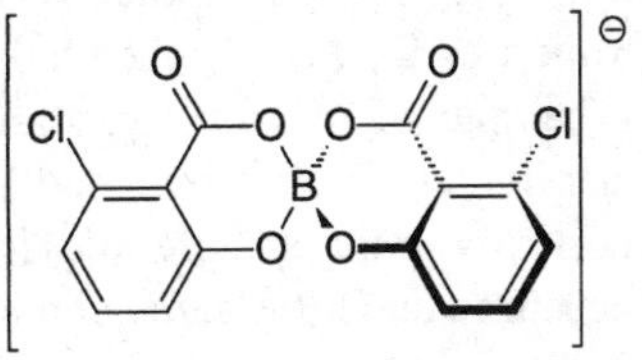

c

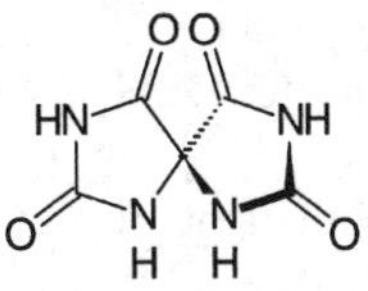

d

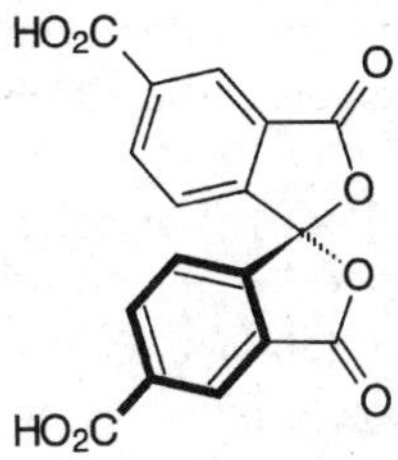

e

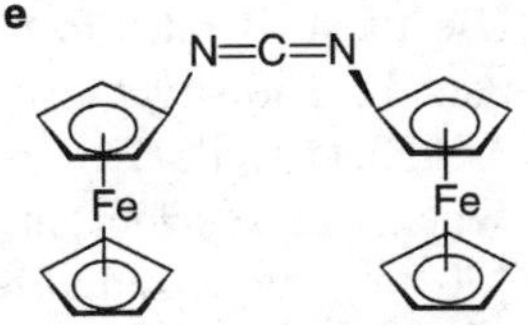

6. The rate constants for the interconversion of the atropisomers of the compounds shown below at 90°C are: (a) 275×10^{-6} s^{-1}; (b) 16×10^{-6} s^{-1}; (c) 6×10^{-6} s^{-1}; (d) too small to measure. Explain why these compounds are chiral and comment on the relative magnitudes of these rate constants. Why is no rate constant quoted for the compound R=H?

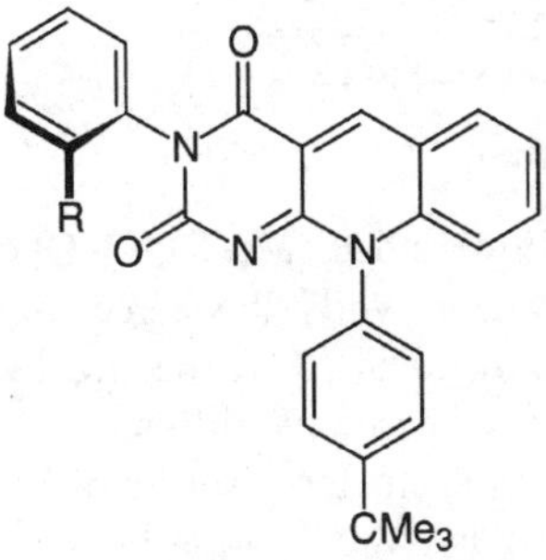

a; R = CH$_2$CH$_3$
b; R = CH(CH$_3$)$_2$
c; R = CF$_3$
d; R = C(CH$_3$)$_3$

7. The cycloheptene derivative shown below is the only known case where both the *cis* and *trans* isomers of an alkene in a seven membered ring are stable. How many stereoisomers of this compound are there? Draw the structure of each stereoisomer and explain their relationship to one another.

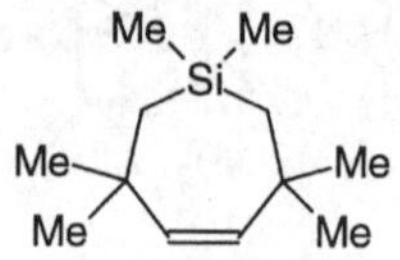

8. (*S*)-Phenylalanine (PhCH$_2$CH(NH$_2$)CO$_2$H) has the following physical properties: $[\alpha]_D^{20} - 34$ (concentration $=2$ g/100 ml in H$_2$O) and melting point 272–274°C. Predict, as far as possible, the corresponding physical properties for (a) (*R*)-phenylalanine and (b) (*RS*)-phenylalanine.

9. A sample of glycidol was found to contain a 4:1 ratio of the *R* and *S* enantiomers. The sample had a specific rotation of +5.5, whilst an enantiomerically pure sample of (*R*)-glycidol had a specific rotation of +12 measured under the same conditions. Calculate both the enantiomeric excess and the optical purity of the sample and comment on their relative values.

10. Use the CIP rules to classify the stereochemistry of the following compounds. The structures are given earlier in this chapter: **3.38**, **3.43**, **3.45**, **3.46**, **3.47** and **3.52**.

11. When *trans*-cyclooctene ($[\alpha] = 0$) is irradiated with light of wavelength 190 nm, isomerization to *cis*-cyclooctene occurs. If this reaction is carried out using circularly polarized light and stopped before the reaction goes to completion, then the reaction mixture is found to be optically active. Explain the reason for this effect and comment on the possible significance of the result.

12. In the molybdenum complex shown below, both metal atoms are stereocentres. Assign the absolute configuration of both stereocentres using the *C*/*A* nomenclature.

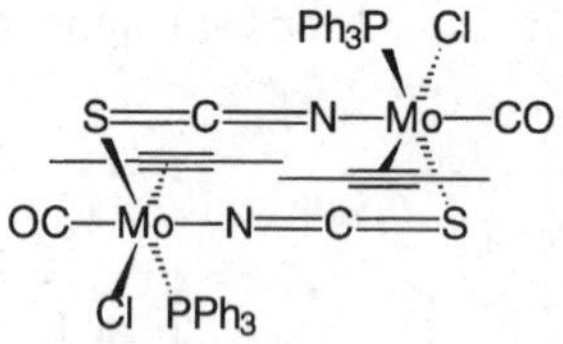

13. Compounds A (MeCH=C=CHMe) and B (MeCH=C=C=CHMe) both exist as a pair of stereoisomers. However, whilst the stereoisomers of compound A are optically active and have identical NMR spectra, the stereoisomers of compound B are optically inactive and have different NMR spectra. Draw clear diagrams of the bonding in each stereoisomer of A and B and hence account for this difference. Predict the relationship between the stereoisomers of the next compound in this series (MeCH=C=C=C=CHMe).

4 Compounds with two or more stereocentres

4.1 Compounds with two stereocentres

In general, a compound with n stereocentres (or stereoaxes, stereoplanes, etc.) can exist in a maximum of 2^n stereoisomers. As an example of this, consider the amino acid threonine which contains two stereocentres (C2 and C3) and so has a maximum of four stereoisomers as shown in structures **4.1**–**4.4**.

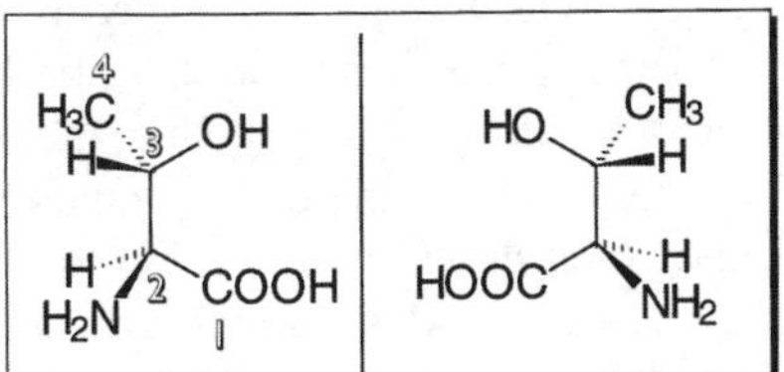

Structures **4.1** and **4.2** are non-superimposable mirror images of one another and hence are a pair of enantiomers. Similarly, structures **4.3** and **4.4** are enantiomers of each other. However, the pairs of structures **4.1** and **4.3**, **4.1** and **4.4**, **4.2** and **4.3**, or **4.2** and **4.4** are not mirror images of one another, but they are stereoisomers of each other as they differ only in the spatial arrangement of their atoms. Stereoisomers which are not mirror images are by definition diastereomers. Unlike enantiomers, diastereomers have different chemical and physical properties and so are, in general, easily separated. Cis–trans isomers, which were discussed in Chapter 2, are also diastereomers since they are also stereoisomers which are not mirror images of one another.

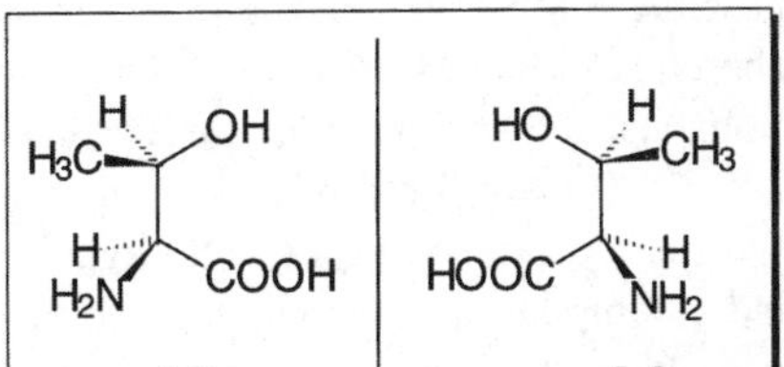

Another way of viewing the relationship between the stereoisomers **4.1**–**4.4** of threonine is to consider a table of the absolute configuration at each stereocentre for each of the isomers as given in **Table 4.1**. For compounds with one stereocentre, enantiomers have opposite absolute configurations at that stereocentre (i.e. one enantiomer has the (R)-configuration and the other enantiomer has the (S)-configuration, cf. Chapter 3). For compounds with more than one stereocentre, enantiomers must have the opposite absolute configuration at every stereocentre. Thus **Table 4.1** shows that the enantiomer of stereoisomer **4.1** is structure **4.2**, whilst **4.3** and **4.4** are not enantiomers of isomer **4.1**. The enantiomeric stereoisomers **4.1** and **4.2** (or **4.3** and **4.4**) are said to have the same

Table 4.1 Absolute configurations of stereoisomers **4.1–4.4**

Stereocentre	Stereoisomer			
	4.1	**4.2**	**4.3**	**4.4**
C2	S	R	S	R
C3	S	R	R	S

relative stereochemistry. That is, in structures **4.1** and **4.2**, the two stereo-centres both have the same absolute configuration, whilst in structures **4.3** and **4.4** the two stereocentres have opposite absolute configurations.

Another molecule which contains two stereocentres is 2,3-dibromobutandioic acid, the stereoisomers of which are shown in structures **4.5–4.8**. Structures **4.5** and **4.6** are again non-superimposable mirror images with the same relative configuration and are therefore enantiomers. Structures **4.7** and **4.8** are also mirror images of one another but they are superimposable since rotation of structure **4.8** by 180° in the plane of the paper as indicated gives structure **4.7**. Therefore, structures **4.7** and **4.8** represent the same stereoisomer and 2,3-dibro-mobutandioic acid has only three stereoisomers **4.5**, **4.6** and **4.7**. Structure **4.7** is called the **meso form** of this compound. In general, any compound containing two stereocentres with identical ligands attached to each will have a meso form and a compound may contain more than one meso form. Meso compounds are achiral, since they are superimposable upon their mirror image and are not optically active, despite containing at least two stereocentres.

An oversimplified, but easily visualized, explanation for the lack of optical activity in meso compounds is that, as a beam of light hits a meso compound, it will first interact with one of the stereocentres and will be rotated by a certain amount. However, the beam of light will then interact with the second stereo-centre, which, being identical to the first but of the opposite absolute configura-tion, will rotate the beam of light by exactly the same amount as the first stereocentre but in the opposite direction (cf. Chapter 3, section 3.4.1). Thus overall, the light ends up in the same plane that it started in and meso compounds are not optically active.

Table 4.2 shows the relationship between structures **4.5–4.8** of 2,3-dibromobu-tandioic acid in terms of their absolute configurations. Structures **4.5** and **4.6** have the opposite absolute configurations at each stereocentre and, as both contain only stereocentres of one absolute configuration (i.e. both R or both S), they are a pair

Table 4.2 Absolute configurations of structures 4.5–4.8

Stereocentre	Structure			
	4.5	4.6	4.7	4.8
C2	R	S	S	R
C3	R	S	R	S

of enantiomers. Structure **4.7** (or structure **4.8** which is identical to **4.7**), however, contains two stereocentres with the same ligands attached, but of opposite absolute configurations (one *R* and one *S*), thus this is the meso isomer.

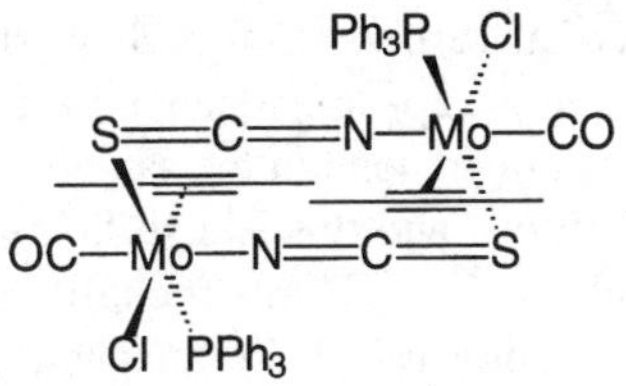

4.9

An interesting example of a meso compound is the dimeric molybdenum complex **4.9**. The structure of this complex was recently determined by X-ray crystallography and it was shown that both molybdenum atoms are stereocentres as they are octahedrally coordinated to six different ligands (cf. Chapter 3, section 3.7.4). Indeed, structure **4.9** represents the only case of an octahedral metal complex in which the metal atom is coordinated to six different groups which, to date, has been stereochemically characterized. In structure **4.9**, the two molybdenum atoms are both coordinated to the same six ligands, but they have opposite absolute configurations, hence the structure is a meso compound. This example shows that meso compounds can be formed by the presence of octahedral as well as tetrahedral stereocentres. The presence of two or more stereoaxes, stereoplanes or helices can also lead to the formation of meso compounds.

4.2 Compounds with more than two stereocentres

As the number of stereocentres in a molecule increases, the stereochemical situation can become more complex. In this section, we shall discuss only one example, trihydroxyglutaric acid. This compound has three stereocentres and hence a maximum of $2^3 = 8$ stereoisomers. These are shown in structures **4.10–4.17** and **Table 4.3** shows the absolute stereochemistry at each stereocentre for each of the eight structures. From a combination of structures **4.10–4.17** and **Table 4.3**, it can be seen that structures **4.10** and **4.11** are identical (they can be

Table 4.3 Absolute configurations of structures **4.10–4.17**

Stereocentre	Compound							
	4.10	**4.11**	**4.12**	**4.13**	**4.14**	**4.15**	**4.16**	**4.17**
2	R	S	R	S	R	S	R	S
3	r	r	/	/	s	s	/	/
4	S	R	R	S	S	R	R	S

/ = Not a stereocentre since two of the ligands are identical.
r, s = Pseudoasymmetric centre.

interconverted by a 180° rotation in the plane of the paper) and that they represent a meso structure. The central carbon atom (C3) in structure **4.10** or **4.11** is a stereocentre as four different groups are attached to it. Although the two groups vertically attached to C3 as drawn contain the same functional groups, they have opposite absolute configurations and this is a sufficient difference to make C3 a stereocentre. A stereocentre such as C3 in structure **4.10** which is a stereocentre only because of the stereochemistry of the attached groups is referred to as a **pseudoasymmetric centre**. The *R/S* nomenclature system can be applied to pseudoasymmetric centres, though lower case letters *r* and *s* are used to denote that the centre is a pseudoasymmetric centre. Note that, despite containing three stereocentres, structure **4.10** is not optically active as it is a meso compound and is thus superimposable upon its mirror image.

Structures **4.12** and **4.13** represent a pair of enantiomers as they differ in the absolute configuration at each of their stereocentres (and so have the same relative stereochemistry). In these two compounds, the central carbon atom C3 is not a stereocentre, as the two groups vertically attached to it are identical both in

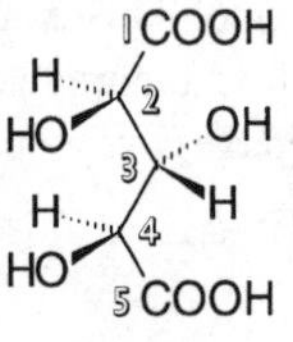

4.10 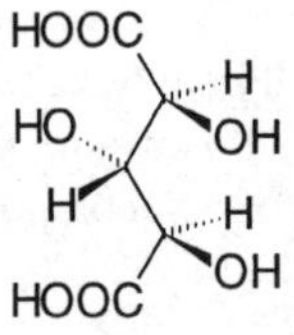   4.11 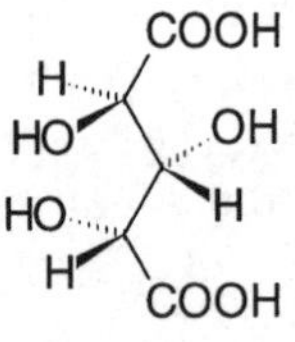   4.12 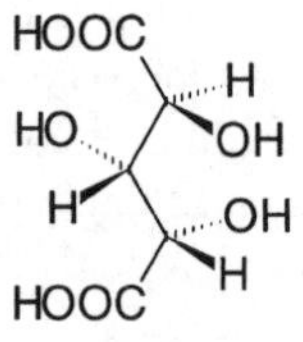   4.13

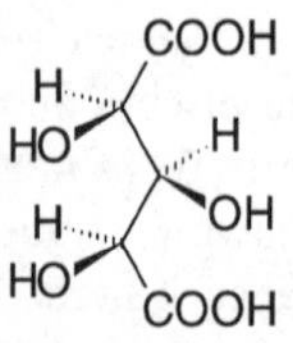

4.14 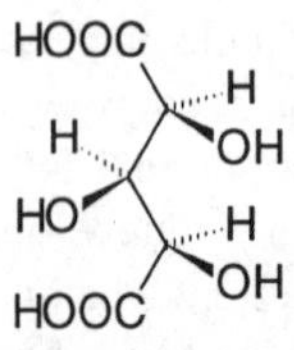   4.15 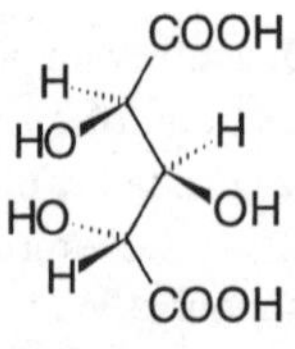   4.16 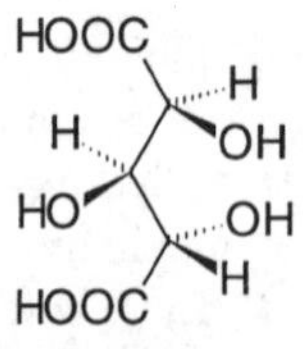   4.17

the functionality contained and in their stereochemistry. Thus, in this pair of stereoisomers, C3 has only three different groups attached to it and the compounds contain only two stereocentres.

Structures **4.14** and **4.15** represent the same meso compound, although this is a different meso compound to that represented by structure **4.10/4.11**. Structures **4.14** and **4.15** can be superimposed by a 180° rotation in the plane of the paper. They do not represent a chiral species as they are superimposable upon their mirror image. As in the case of the meso compound represented by structure **4.10**, the central carbon atom C3 in compound **4.14** or **4.15** is a pseudoasymmetric centre. Structures **4.16** and **4.17** do not represent new species, as they can be superimposed on structures **4.12** and **4.13** respectively by a 180° rotation about a horizontal axis through C3. Thus there are just four stereoisomers of trihydroxyglutaric acid, a pair of enantiomers represented by structures **4.12** and **4.13**, and two meso isomers represented by structures **4.10** (or **4.11**) and **4.14** (or **4.15**). The meso forms are diastereomers of each other and of the pair of enantiomers. Hence, due to the symmetrical nature of the various stereoisomers of trihydroxyglutaric acid, there are considerably fewer than the maximum eight possible stereoisomers.

4.3   Polymer stereochemistry

In sections **4.1** and **4.2**, the stereochemical consequences of a molecule possessing two or three stereocentres were discussed. In this section, we shall consider polymeric molecules that contain large numbers of stereocentres. Only one type of polymer will be considered in this discussion, namely the polymers obtained by the chain growth polymerization of 1-substituted or 1,1-disubstituted alkenes in a head to tail manner as shown in **Scheme 4.1**, although the same principles can be applied to other polymer structures. Many commercially important polymers are prepared by the route shown in **Scheme 4.1**, examples including poly(propylene) **4.18**, poly(styrene) **4.19**, poly(acrylonitrile) **4.20**, and poly-(methyl methacrylate) **4.21**. In each of these cases, X and Y in **Scheme 4.1** are different and the resulting polymer contains a series of stereocentres along the polymer backbone. The polymerization methodologies used to prepare these polymers always result in the formation of polydisperse material, i.e. polymers with a range of different molecular weights (or different values for *n* in **Scheme 4.1**). Thus, a particular polymer sample will consist of molecules of differing molecular weights, each of which contains a different number of stereocentres.

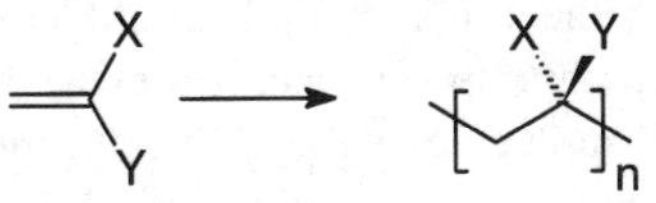

4.18; X= H, Y= Me
4.19; X= H, Y= Ph
4.20; X= H. Y= CN
4.21; X= Me, Y= COOMe

Scheme 4.1

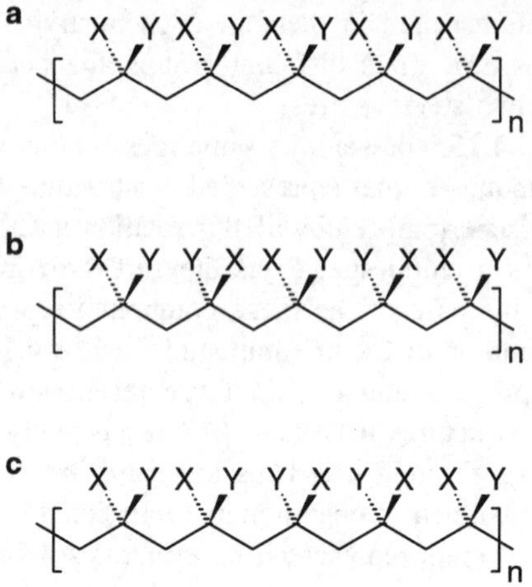

Figure 4.1 **a**, Isotactic; **b**, syndiotactic; and **c**, atactic polymers.

4.3.1 *Stereochemical consequences of backbone stereocentres*

The relative stereochemistry of a polymer is referred to as its **tacticity**, and there are three possible tacticities for polymers **4.18–4.21**. All of the stereocentres may have the same configuration (**Figure 4.1a**), in which case the relative configuration of the polymer is said to be **isotactic**. Alternatively, the stereocentres may alternate in configuration giving a **syndiotactic** polymer (**Figure 4.1b**), or the configuration of the stereocentres may vary randomly along the polymer chain, giving an **atactic** polymer (**Figure 4.1c**). Because the three polymer structures shown in **Figure 4.1a–c** are all diastereomeric with one another, they will have different chemical and physical properties.

When considering the chirality of polymers, it is convenient to consider the polymer chains to be infinitely long. This is, of course, an approximation but it allows the end groups to be neglected and simplifies the stereochemical considerations. Both isotactic and syndiotactic polymers can then be seen to be achiral since they are superimposable on their mirror images as shown in **Figure 4.2a–b** (see also section 4.3.2). In **Figure 4.2**, the substituents attached to stereocentres equidistant from the central atom have been highlighted in bold or italics to aid visualization. Each stereocentre in these polymers is a pseudoasymmetric centre and the polymers are meso compounds.

The situation with the atactic polymer is rather different, since in this case the polymer is not superimposable on its mirror image. However, the configuration at each stereocentre is random, so a large enough sample of the polymer would contain every possible stereoisomer of the polymer and would constitute a racemic mixture of all possible diastereomers. In practice, however, this would require a sample containing at least 2^n molecules, where n is the number of monomer units in the polymer chain, since each monomer unit generates one stereocentre during the polymerization process. For a typical polymer, n might

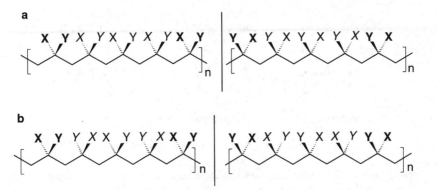

Figure 4.2 **a**, Isotactic; and **b**, syndiotactic polymers and their mirror images.

be equal to 100, giving 2^{100} possible stereoisomers. Since Avogadro's number (the number of molecules in one mole of any substance) is only 6.0×10^{23} (which is approximately 2^{79}), it is clear that a sample containing 2^{21} moles of polymer would be required to allow even one molecule of each stereoisomer to be present. For poly(propylene) **4.18**, which has the lowest molecular weight of polymers **4.18**–**4.21**, this would require a sample of 8,800 tonnes. Thus a realistic sample of an atactic polymer cannot contain all possible stereoisomers of the compound. However, in any randomly generated sample of such an atactic polymer, it is expected that 50% of the molecules will be laevorotatory, whilst the other 50% of the molecules will be dextrorotatory. Further, the average magnitudes of the dextrorotatory and laevorotatory molecules will be expected to be equal and so cancel one another out. Hence a specific rotation of zero will be expected even though the sample may not contain equal amounts of all of the possible stereoisomers of the polymer.

The polymer tacticities discussed so far have been composed of repetitions of two adjacent monomer units. Thus, given that one monomer unit has a specified configuration, in an isotactic polymer the adjacent monomer unit will have the same configuration, whilst in a syndiotactic polymer the adjacent monomer unit will have the opposite configuration and in an atactic polymer the adjacent monomer unit will have a random configuration. These sequences then repeat along the polymer chain. The two adjacent monomer units are referred to as a **diad** and are illustrated in **Figure 4.3**. A polymer whose tacticity is determined entirely by diad units cannot exhibit optical activity for the reasons discussed above. However, the three tacticities discussed so far represent only three of a maximum of 2^n possible stereoisomers of the polymer, where n is the number of monomer units present in the polymer. If three adjacent monomer units (a **triad**) are the repeat unit from which the polymer is composed, then there are four possibilities as illustrated in **Figure 4.4a–d**, although the first of these (**a**) is just the isotactic polymer which has already been discussed. Triad **b** also leads to an achiral polymer, whilst triads **c** and **d** both give chiral polymers. However, triad

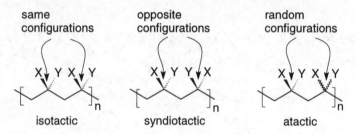

Figure 4.3 Diad units within a polymer.

c gives rise to the same situation discussed above for an atactic diad, so the polymer whilst it is chiral will not be optically active. Triad **d**, however, gives rise to a chiral polymer which can exhibit optical activity.

In summary, polymers derived from 1- or 1,1-disubstituted alkenes can exist in many different diastereomeric forms called tacticities. Only polymers derived from repeat units of triads or higher will be capable of exhibiting optical activity. For triads to pentads some random (atactic) component must be present within the repeat unit to generate an optically active polymer. The hexad **4.22** represents the smallest repeat unit that can give an optically active polymer without containing some random component. The above arguments are based upon a model which assumes the polymer chains are of infinite lengths, so that the effects of end groups can be ignored. However, the arguments can be extended to models of polymers of finite length and incorporating the end groups with only minor modifications. All of the above discussion has been concerned with the stereochemical consequences of the formation of stereo-centres in the polymer chain during a polymerization reaction. Chirality can be caused by other structural features, however, and many polymers adopt helical conformations as will be discussed in the next section.

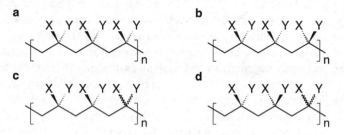

Figure 4.4 Triad units within a polymer.

4.22

4.3.2 Helical polymers

One of the best known examples of a helical polymer is poly(triphenylmethyl methacrylate) **4.23**. This polymer can be prepared by the anionic polymerization of triphenylmethyl methacrylate as shown in **Scheme 4.2**. If the polymerization is initiated with an achiral base (such as BuLi), then the resulting polymer is highly isotactic but achiral. The use of a chiral base (such as the complex formed between (−)-sparteine and BuLi) to initiate the polymerization, however, again gives an isotactic polymer, but in this case the polymer is chiral and possesses an appreciable specific rotation ($[\alpha]_D^{25} + 383$). The optical activity of the polymer is due entirely to the formation of a helical conformation, since the isotactic form of a polymer is an achiral configuration (section 4.3.1). If an achiral base is used to initiate the polymerization then it is equally likely that the left- and right-handed helical conformations of the polymer will be formed and, as these are enantiomers of one another, the polymer will be racemic. The use of a chiral initiator, however, results in the preferential formation of one enantiomer of the helix and so results in an optically active polymer. This is an example of an asymmetric synthesis (a synthesis which produces one enantiomer of the product in excess of the other), a topic which will be explained and discussed in more detail in Chapter 10.

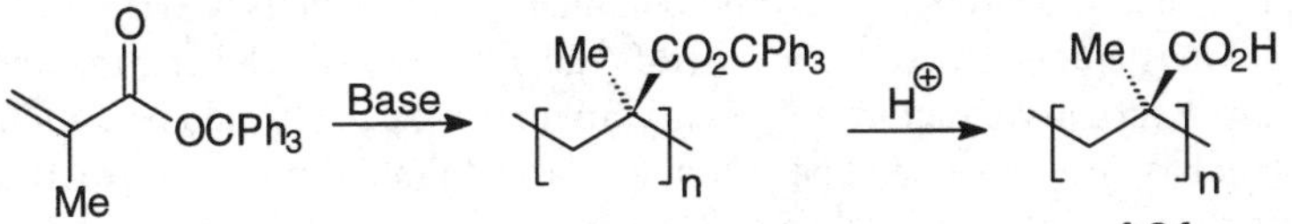

Scheme 4.2

Many isotactic polymers adopt helical conformations; however, unlike poly-(triphenylmethyl methacrylate) most do not show any optical activity. The reason for this is that the two enantiomeric helices can readily interconvert and so the polymer is always a racemic mixture of the two enantiomeric conformations. In the case of poly(triphenylmethyl methacrylate), however, the large triphenylmethyl esters sterically prevent the interconversion of the two enantiomeric helices resulting in the formation of a chiral helix which does not readily racemize. Thus the two enantiomers of poly(triphenylmethyl methacrylate) are atropisomers of one another, as was discussed in Chapter 3, section 3.8 for non-polymeric materials containing stereoaxes, stereoplanes or helices.

That the specific rotation of poly(triphenylmethyl methacrylate) **4.23** is due entirely to its helical conformation can be shown by acidic cleavage of the triphenylmethyl esters from the polymer giving poly(methacrylic acid) **4.24** (**Scheme 4.2**). The poly(methacrylic acid) produced in this way is not optically active since, once the large triphenylmethyl esters have been removed, the two helical conformations of the polymer can rapidly interconvert giving a racemic mixture of the two helical conformations.

Many other sterically hindered esters of methacrylic acid have been polymerized to give optically active polymers due to the formation of chiral helices. Chiral helical conformations have also been obtained from the polymerization of other monomers including isocyanides, isocyanates and chloral. These are discussed in the references cited in the further reading.

4.4   Relative stereochemistry nomenclature

When dealing with compounds containing two or more stereocentres, it is useful to have a nomenclature system to classify the diastereomers since, in general, the diastereomers have different properties and, in a particular sample, each may or may not be composed of a pair of enantiomers which will have identical properties. Consider, for example, the amino acid threonine whose stereoisomers were discussed in section 4.1. A pure sample of isomer **4.1** would have exactly the same chemical properties as a pure sample of isomer **4.2** (unless the reaction was with another chiral species, cf. Chapter 3, section 3.3) and a mixture of isomers **4.1** and **4.2** in any proportion (including a racemic mixture) would also have exactly the same chemical reactivity. Exactly the same is true of stereoisomers **4.3** and **4.4**; however, any combination of isomers **4.3** and **4.4** will have different chemical properties to any combination of isomers **4.1** and **4.2**, since they are diastereomeric to one another. So an easy way of referring to either just structure **4.1**, or just structure **4.2**, or to a mixture of the two would be useful. Such a system has been developed based on the Cahn, Ingold, Prelog priority rules (cf. Chapter 2, section 2.1 and Chapter 3, section 3.2.2). The way in which the rules are used is as follows:

1. Assign the absolute configuration (R or S) at both stereocentres.
2. If both stereocentres have the (R)-configuration, or both have the (S)-configuration, use the prefix l (like).
3. If one stereocentre has the (R)-configuration whilst the other has the (S)-configuration, use the prefix u (unlike).

As an example of this, consider structure **4.25**, which is one of the four possible stereoisomers of the amino acid isoleucine. The absolute configuration at C2 is S and the absolute configuration at C3 is also S. So, as both stereocentres have the same absolute stereochemistry, this structure represents the (l)-diastereomer of the amino acid isoleucine. Note that the enantiomer of structure **4.25** would also be the (l)-diastereomer, since both stereocentres would have the (R)-configuration.

A number of other nomenclature systems are also in widespread use to describe relative stereochemistry. Chemical abstracts uses a system based upon the prefixes R^* and S^*. In this system, the lowest numbered stereocentre in a compound of known relative, but unknown absolute configuration is arbitrarily assigned the (R^*)-configuration. All other stereocentres that have the same

4.25

absolute configuration as the lowest numbered stereocentre are also designated
R^*, whilst stereocentres with the opposite absolute configuration to the lowest
numbered stereocentre are designated S^*. Thus a mixture of compound **4.25** and
its enantiomer, or either enantiomer if the absolute configuration were unknown,
would be described as $(2R^*, 3R^*)$-isoleucine.

Another widely used nomenclature system for describing the relative configuration of two groups is based upon the **syn** and **anti** descriptors. The molecule
whose stereochemistry is being described is drawn in a flying wedge projection
and the main chain identified. The substituent on the lowest numbered stereocentre is used as a reference, with substituents on other stereocentres being
described as syn if they are on the same side of the main chain as the reference
substituent, and anti if they are on the opposite side of the main chain. Thus in
compound **4.26**, the amino group on C2 is the reference substituent. The OH
group on C3 is on the same side of the main chain as the reference group, so the
relative stereochemistry of C2 and C3 is syn. Similarly, the fluorines on C5 and
C6 are both on the opposite side of the main chain to the reference group so the
relative stereochemistry between C2 and C5 is anti as is the stereochemistry
between C2 and C6.

2,3-syn, 2,5-anti, 2,6-anti

4.26

The syn/anti nomenclature is very simple to use and is widely employed;
however, it does suffer from two limitations. The first of these is that it must be
possible to define a main chain to the molecule. The more limiting problem is
that there can be only one non-hydrogen substituent on each stereocentre,
otherwise the nomenclature system becomes ambiguous.

The final nomenclature system for relative configuration which is in common
usage was initially developed for use with sugars, but has subsequently been
extended for use with other systems. The molecule whose stereochemistry is being
defined is drawn in a Fischer projection (cf. Chapter 1, section 1.2.1). If the two
non-hydrogen substituents on the stereocentres are adjacent to one another in the
Fischer projection (as in compound **4.27**), then the compound is named as the
erythro-isomer. If, however, the two substituents are on opposite sides of the
Fischer projection (as in compound **4.28**), then the compound is the threo-isomer.

The threo/erythro nomenclature is fraught with difficulties. It is necessary to be able to define a main chain for the molecule, the molecule has to be drawn in a Fischer projection which are no longer much used, and there must be a single non-hydrogen substituent on each stereocentre. Despite these limitations, the nomenclature is still occasionally used in the chemical literature.

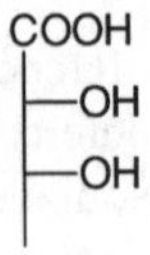

4.27
erythro-2,3-dihydroxy
butanoic acid

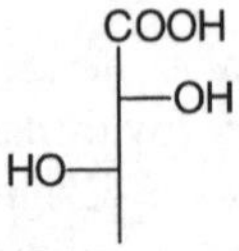

4.28
threo-2,3-dihydroxy
butanoic acid

4.5 Diastereomeric excess

When dealing with mixtures of diastereomers, the term **diastereomeric excess (de)** is defined analogously to enantiomeric excess (Chapter 3, section 3.5). Thus:

$$\textbf{de} = \frac{(\textbf{Major diastereomer} - \textbf{Minor diastereomer}) \times \textbf{100}}{\textbf{Major diastereomer} + \textbf{Minor diastereomer}}$$

or alternatively:

$$\textbf{de} = \textbf{\%Major diastereomer} - \textbf{\%Minor diastereomer}$$

As an example, consider the amino acid **4.29**, which contains two stereocentres. If a given sample of compound **4.29** possesses only the (S)-configuration at C2, but is a 2 : 1 mixture of stereoisomers about C3, then the sample will be a 2 : 1 mixture of diastereomers. Thus the diastereomeric excess will be:

$$\text{de} = \frac{(2-1) \times 100}{2+1} = \frac{100}{3} = 33.3\%$$

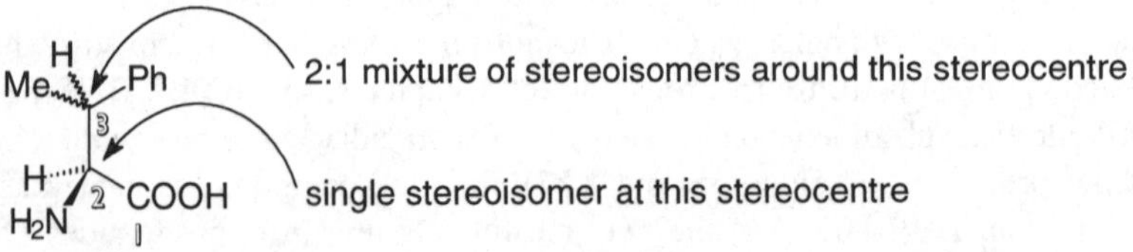

2:1 mixture of stereoisomers around this stereocentre

single stereoisomer at this stereocentre

4.29

4.6 Further reading

General
Stereochemistry of Organic Compounds E.L. Eliel and S.H. Wilen. Wiley: London, 1994, chapters 3 and 5.

***l*/*u* Nomenclature**
D. Seebach and V. Prelog. *Angew. Chem. Int. Ed. Engl.*, 1982, **21**, 654.

Molybdenum complex 4.9
P.K. Baker, M.E. Harman, D. ap Kendrick and M.B. Hursthouse. *Inorg. Chem.*, 1993, **32**, 3395.

Polymer stereochemistry
Topics in Stereochemistry Vol. 2, M. Goodman (N.L. Allinger and E.L. Eliel eds). Wiley: London, 1967, chapter 2.
Topics in Stereochemistry Vol. 17, M. Farina (E.L. Eliel and S.H. Wilen eds). Wiley: London, 1967, chapter 1.
Polymer Spectroscopy F. Ciardelli, O. Pieroni and A. Fissi (A.H. Fawcett ed.). Wiley: London, 1996, chapter 14.
Y. Okamoto and T. Nakano. *Chem. Rev.*, 1994, **94**, 349.
G. Wulff. *Angew. Chem., Int. Ed. Engl.*, 1989, **28**, 21.
M.M. Green and B.A. Garetz. *Tetrahedron Lett.*, 1984, **25**, 2831.

4.7 Problems

1. Apply the *l*/*u* and syn/anti nomenclature systems to each of the stereo-isomers of threonine **4.1–4.4**, 2,3-dibromobutandioic acid **4.5–4.7** and tri-hydroxyglutaric acid **4.10**, **4.12**, **4.13**, **4.14**.

2. The *l*/*u* nomenclature system can be easily extended to species for which the *C*/*A* or *M*/*P* stereochemical descriptors are used instead of *R* or *S*. Is compound **4.9** the *l* or the *u* diastereomer?

3. How many stereoisomers are there of the two compounds shown below? For each stereoisomer determine whether it will be an enantiomer or a diastereomer of each other stereoisomer and whether or not it will be optically active. You may find it helpful to build a model of these compounds and to review Chapter 3.

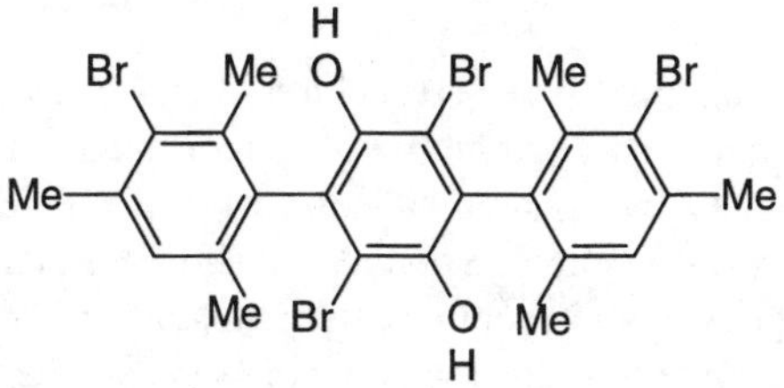

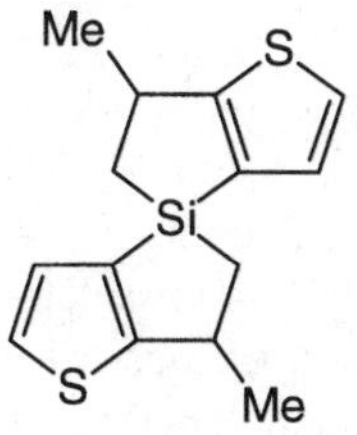

4. Consider the reaction shown below. How many stereoisomers are there of both the starting material and the product? Draw wedge/hash structures of each stereoisomer.

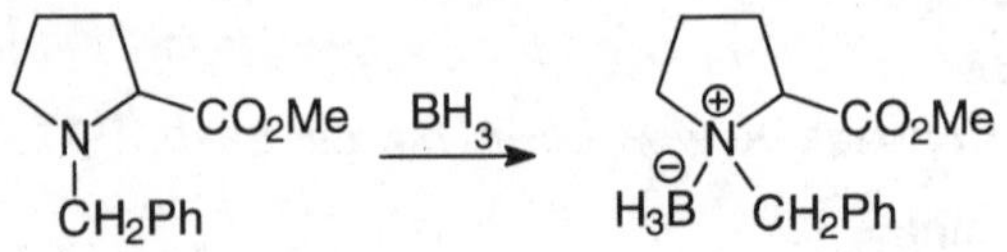

5. The structure shown below represents one of the stereoisomers of tartaric acid and has a melting point of 170–172°C and a specific rotation of +12.4 (in water). How many other stereoisomers of tartaric acid are there? Draw a wedge/hash structure of each stereoisomer, construct a table showing the stereochemistry at each stereocentre of each stereoisomer and specify the stereochemical relationship between the stereoisomers. Predict, as far as possible, the melting point and specific rotation of each stereoisomer and use the l/u, syn/anti and erythro/threo nomenclature systems to specify the stereochemistry of each stereoisomer.

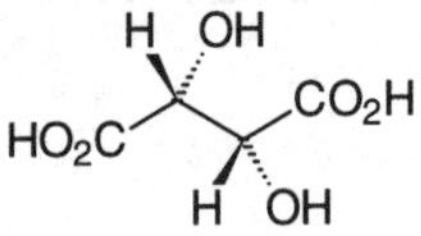

6. For the reaction shown below determine how many stereoisomers there are of both the starting material and the product. Which stereoisomer(s) of the product will be produced from each stereoisomer of the starting material?

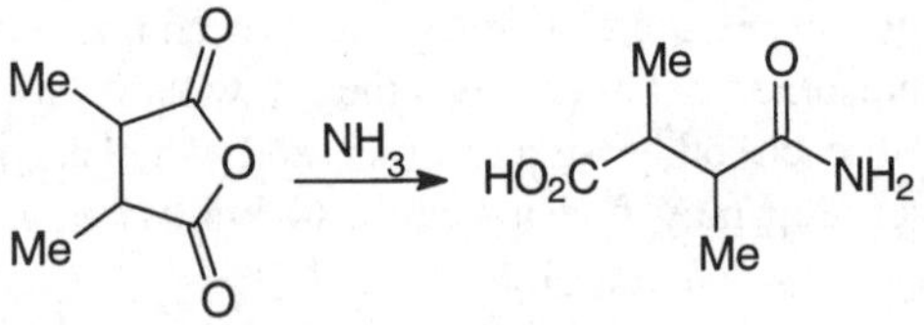

7. The anhydride shown below contains four stereocentres. Identify each of them and determine the maximum possible number of stereoisomers of this compound. In fact, there are only two known stereoisomers of this anhydride; construct a molecular model of the compound and show that formation of most of the stereoisomers would require extreme distortions of the bond lengths or bond angles. What is the relationship between the two

stereoisomers that can be formed? Will either stereoisomer be optically active? This problem is important, since it illustrates a second way in which the number of stereoisomers of a compound may be less than the theoretically maximum number.

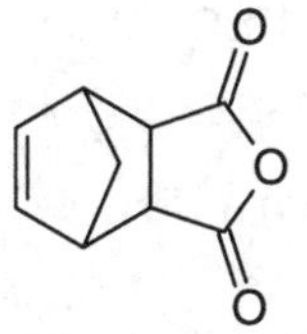

8. Reaction of lead tetraacetate with a 1,2-diol results in oxidation of the diol to two aldehydes as shown below. An unknown compound (A) when treated with lead tetraacetate gave the enantiomerically pure aldehyde (B) as the only product. Determine, as fully as possible, the structure of (A). How many of the stereoisomers of (A) would give (B) as the only product? What product(s) would be obtained from each of the other possible stereoisomers of (A)?

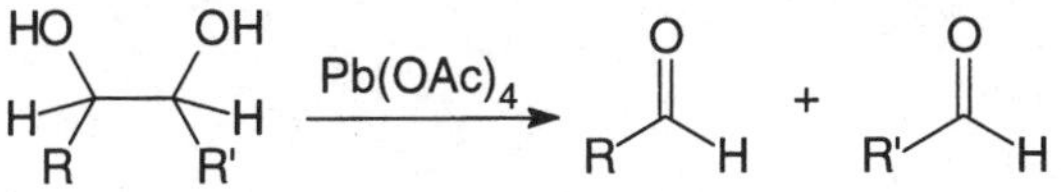

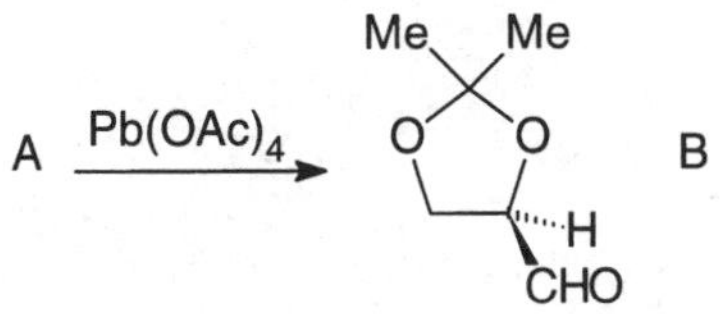

9. Monomers A–C below all undergo anionic polymerization initiated by the complex formed from BuLi and (−)-sparteine. However, monomer A gives an optically active polymer whilst monomers B and C give polymers with no optical activity. Suggest an explanation for these results.

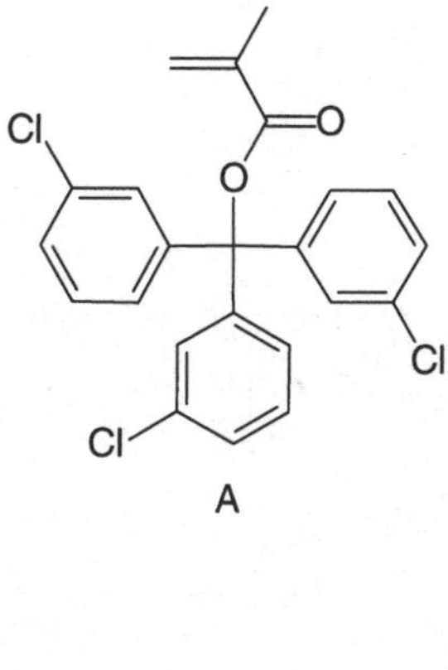

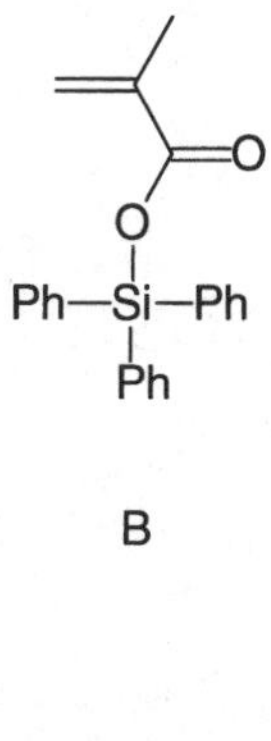

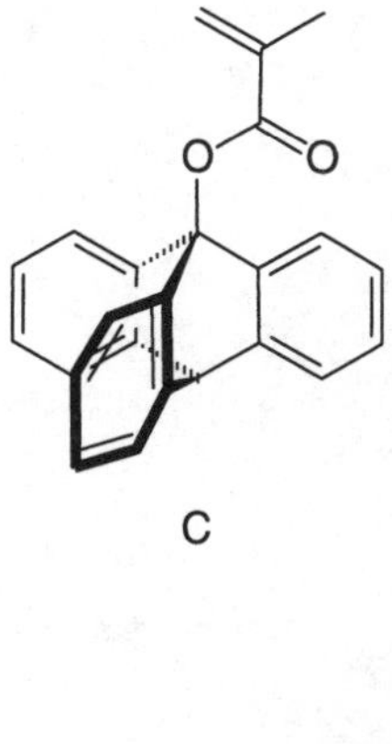

10. Polymers with the structure shown below can be prepared in which the stereochemistry of the stereocentres adjacent to the acids is fixed as shown but the stereochemistry of the stereocentre adjacent to the phenyl ring is not controlled. Will this polymer be optically active?

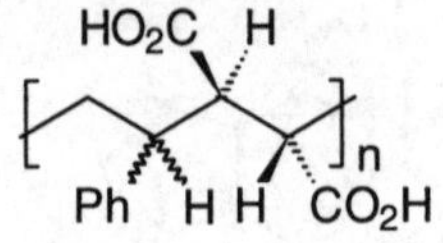

5 Interconversion and analysis of stereoisomers

In this chapter, the processes that can change the enantiomeric or diastereomeric excess of a compound will be discussed, then the various techniques which can be used to determine the enantiomeric excess, absolute configuration and relative configuration of a sample will be surveyed.

5.1 Racemization

Racemization is the process whereby one enantiomer of a compound is converted into a 1 : 1 mixture of the two possible enantiomers, i.e. into a racemic mixture (cf. Chapter 3, section 3.4.1). A racemic mixture may have different physical properties to those of the pure enantiomer and the chemical properties of a racemate will reflect the fact that both enantiomers of a compound are present. Thus when reacting with chiral compounds, the two enantiomers present in the racemic mixture may react differently, whilst in reactions with achiral species the two enantiomers will react identically.

Often, a chemist is concerned with preventing racemization since it is usually desired to produce a single enantiomer of a product. However, when combined with the process which is the reverse of racemization (resolution cf. section 5.3), racemization can be a useful tool in increasing the amount of the desired enantiomer of a product which can be obtained. Racemization may occur in a number of ways, three of the more important of which are discussed below. Racemization may also occur during substitution reactions, a topic which will be discussed further in Chapter 9.

5.1.1 Thermal racemization

A number of examples of compounds which racemize upon heating were discussed in Chapter 3 (section 3.8). Such compounds are usually atropisomers (i.e. enantiomeric conformations with an energy barrier preventing interconversion of the conformations).

5.1.2 Base induced racemization

This is a very important process if one of the substituents on the stereocentre is electron withdrawing and a hydrogen atom is also attached to the stereocentre.

The hydrogen atom will be acidified by the electron withdrawing group and, in the presence of a suitable base, will be reversibly removed. The resulting carbanion/enolate will have a trigonal planar geometry and hence be achiral. Reprotonation can then occur equally easily from either face giving a racemic mixture. An example of this process is the racemization of amino acids when treated with hot, aqueous base as is shown in **Scheme 5.1** for the amino acid phenylalanine. It should be noted that racemic phenylalanine **5.2** has a lower melting point than the pure (S)-enantiomer **5.1**, illustrating that a racemic mixture may have different physical properties to those of a single enantiomer.

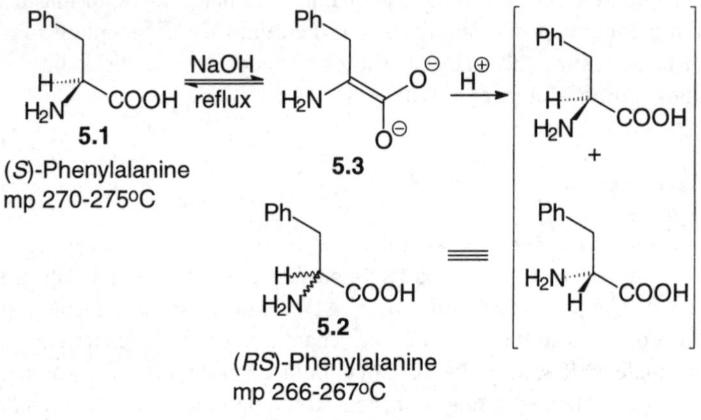

Scheme 5.1

5.1.3   Acid induced racemization

Aldehydes and ketones enolize under acidic as well as basic conditions, and this can cause racemization of a stereocentre adjacent to a carbonyl group if a hydrogen atom is also present on the stereocentre. In 1,3-dicarbonyl compounds such as **5.4**, the enol form is particularly stable due to both the formation of an intramolecular hydrogen bond and conjugation between the alkene and remaining carbonyl bond. As a result of this, 1,3-dicarbonyl compounds are particularly prone to racemization as shown in **Scheme 5.2**.

5.2   Epimerization

If a compound contains two or more stereocentres, then a change in the stereochemistry of just one of these centres will lead not to the enantiomer of the starting compound but to a diastereomer of it. Diastereomers which differ in their configuration at a single stereocentre are called **epimers**, and any process which converts a compound into one of its epimers is called **epimerization**. An example of this process is shown in **Scheme 5.3**: the epimerization of the naturally occurring amino acid (2S, 3S)-isoleucine **5.5** on treatment with base.

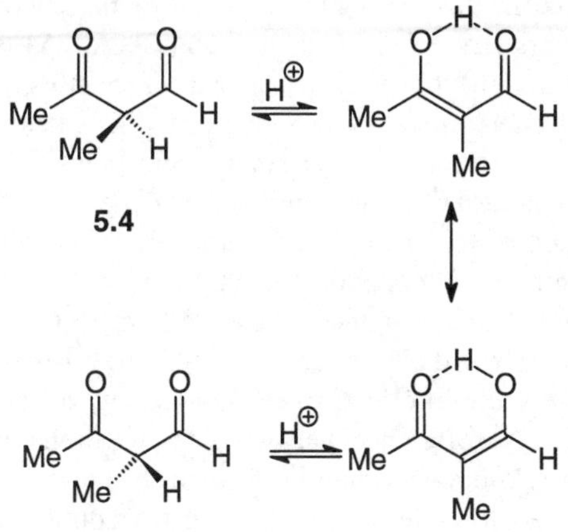

Scheme 5.2

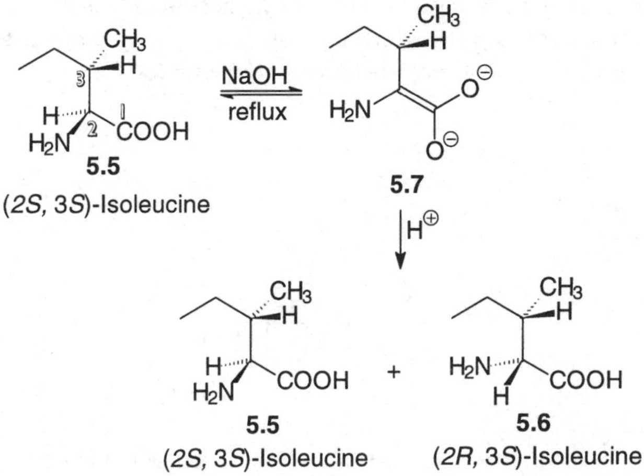

Scheme 5.3

It is instructive to compare **Scheme 5.3** with the almost identical chemistry illustrated in **Scheme 5.1** for the amino acid (*S*)-phenylalanine **5.1**. In both cases, the sodium hydroxide acts as a base and abstracts the two most acidic hydrogens from the amino acid forming an enediolate (**5.3** and **5.7**) with trigonal planar geometry. This process destroys the stereocentre which was present adjacent to the carboxylic acid group. In the case of phenylalanine (**Scheme 5.1**), this is the only stereocentre present in the starting material, so the enediolate **5.3** is achiral and reprotonation can occur equally easily on either face giving a 1:1 mixture of the two enantiomers of phenylalanine **5.2**. In the case of isoleucine, however, a second stereocentre is also present and is unaffected by the sodium hydroxide since no electron withdrawing groups are attached to it. Hence in this case, reprotonation of the enediolate **5.7** gives a mixture of the epimers (2*S*, 3*S*)-isoleucine **5.5** and (2*R*, 3*S*)-isoleucine **5.6**.

Unlike racemization, which produces a pair of enantiomers, epimerization produces a pair of diastereomers. Diastereomers, unlike enantiomers have different physical properties including bond lengths and bond angles which lead to them having different heats of formation (cf. Chapter 1, section 1.5.1). Thus whereas reprotonation of the enediolate **5.3** of phenylalanine (**Scheme 5.1**) will always give a 1 : 1 mixture of enantiomers, reprotonation of the enediolate **5.7** of isoleucine (**Scheme 5.3**) may produce epimers **5.5** and **5.6** in unequal amounts. A reaction pathway/energy diagram representing both cases is shown in **Figure 5.1**. This can be a very useful effect, since if the energy difference between two epimers (or the energy difference between the two transition states) is significant, one of the two epimers may be formed to the exclusion of the other. This provides a means for selectively inverting the configuration of a specific stereocentre, or for converting a mixture of stereoisomers into a single diastereomer. As an example of this, treatment of compound **5.8** with a catalytic amount of base (DBU; 1,8-diazabicyclo[5.4.0]undec-7-ene) results in epimerization of the acidic hydrogen (H_a which is acidic since it is vinylogous to the ester carbonyl) to give compound **5.9** as shown in **Scheme 5.4**.

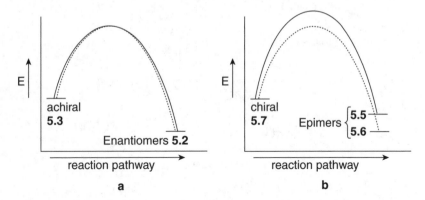

Figure 5.1 Reaction pathway/energy diagrams for the reprotonation of: **a**, enediolate **5.3**; and **b**, enediolate **5.7**.

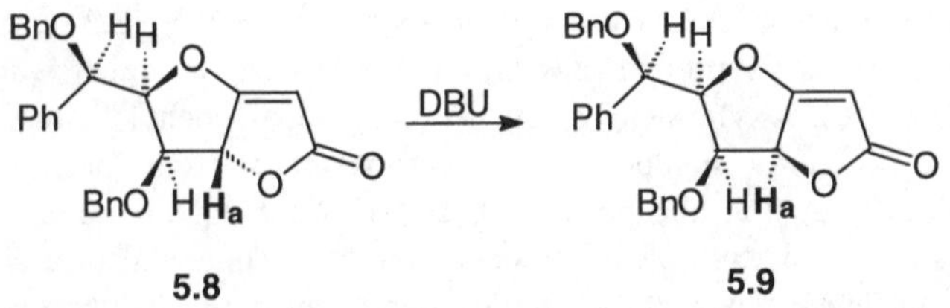

Scheme 5.4

5.3 Resolution

Resolution is the opposite of racemization, i.e. it is the process whereby a mixture of two enantiomers is separated into its constituent isomers. Resolution is an extremely important process, since chemical reactions which use only achiral starting materials and reagents, but which produce chiral products will always produce the product as a racemic mixture (cf. Chapter 3, section 3.6.1). Thus if only one enantiomer is wanted, it has to be separated, i.e. resolved. There are a number of methods available that can be utilized to resolve a racemic mixture.

5.3.1 Resolution by crystallization

When a solution of a racemate is allowed to crystallize, there are three different ways in which the crystals may form:

1. Each crystal contains a 1 : 1 mixture of the two enantiomers.
2. Each crystal contains only one enantiomer of the compound but the crystals are indistinguishable.
3. Each crystal contains only one enantiomer of the compound and the crystals can be distinguished.

In the first case, the racemate is said to form a **racemic compound**, whilst in the second and third cases the crystals are said to form a **conglomerate**. Only 5–10% of racemates crystallize as conglomerates, and only conglomerates can be resolved by crystallization. The third possibility represents the ideal situation for a resolution, since the two types of crystals (which will be enantiomers of one another) can be distinguished by visual inspection and separated by hand. This is a tedious process which can be used to separate a few grams of material but which is impractical for larger scale resolutions.

Resolution by crystallization is of significant historical importance, since the first resolution, that of racemic sodium ammonium tartrate **5.10**, was achieved in this way by Louis Pasteur. It is notable that racemic tartaric acid itself crystal-lizes as a racemic compound and so cannot be resolved by crystallization. This illustrates a general principle, that it is not possible to predict how a racemate will crystallize, and that if a given compound fails to form a conglomerate then it may be possible to convert it into a derivative that does crystallize as a conglomerate. Resolution by crystallization is still used occasionally, partic-ularly in chemical research laboratories where only a few milligrams of a compound may be needed to carry out tests for biological activity.

If a racemate crystallizes as a conglomerate, but the crystals cannot be visually distinguished, then the situation becomes even more tedious as each crystal must individually be dissolved in a suitable solvent and its specific rotation recorded. Half the crystals will give a laevorotatory solution, whilst the other half will give a dextrorotatory solution. The solutions which give specific

5.10

rotations of the same sign can then be combined and the solvent evaporated to accomplish the resolution. This is feasible only if the crystals are of sufficient size as to give a measurable optical rotation.

Before attempting a resolution by crystallization, it is important to know whether the crystals are of a conglomerate or a racemic mixture. This can also be determined by specific rotation since a single crystal of a racemic mixture will always have a specific rotation of zero, whilst a single crystal of a conglomerate will have a non-zero specific rotation provided the crystal contains sufficient compound to give a measurable rotation. Other methods that can be used for distinguishing between racemic mixtures and conglomerates are discussed in the further reading.

The crystallizations discussed above are carried out under equilibrium conditions, that is the crystallizing solution starts as a racemic mixture and remains racemic throughout the crystallization process as crystals of both enantiomers form at the same rate. However, it is also possible to carry out crystallizations under non-equilibrium conditions. If a hot, saturated solution of a conglomerate forming racemate is seeded with a few crystals of one enantiomer of the racemate and allowed to cool, then these crystals will induce the crystallization of only that enantiomer from the solution. This causes the solution to cease to be racemic since there will be an excess of the other enantiomer present in the residual solution. The enantiomerically pure crystals can be filtered, the solution reheated, and further racemic solid added to generate again a saturated solution. This solution is not racemic but is enriched in the enantiomer that did not crystallize during the earlier crystallization. Hence, if a few crystals of the enantiomer now present in excess are added, then this enantiomer will selectively crystallize from the solution. This process is called **entrainment**, and can be repeated indefinitely giving alternating crops of crystals of each enantiomer of the compound.

Unlike equilibrium crystallization, entrainment has significant industrial importance, since the crystals of each enantiomer are produced during alternating crystallization cycles and thus do not require separation by hand. In addition, no chemicals (other than the solvent) are required for the resolution so this can be a very cost effective process. Examples of racemates that can be resolved in this way include glutamic acid **5.11**, 1,1′-binaphthalene-2,2′-diol dimethyl ether **5.12**, and epoxide **5.13** which is an intermediate in the synthesis of a pharmaceutically important class of drugs called β-blockers. Compound **5.12** demonstrates that the process is applicable to all chiral compounds which crystallize as a conglomerate, not just those which contain a stereocentre.

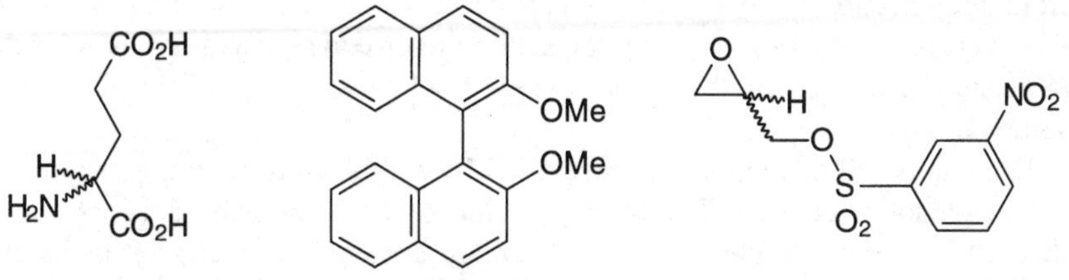

5.11   **5.12**   **5.13**

If a single enantiomer of a compound is desired, then the maximum yield which can be obtained from the above resolutions is 50%, since the other 50% of the starting material constitutes the unwanted enantiomer. In favourable cases, it may be possible to racemize and recycle the unwanted enantiomer. Thus in the case of glutamic acid **5.11**, the commercially important enantiomer is the (S)-enantiomer, the monosodium salt of which (monosodium glutamate) is an important flavour enhancer. The stereocentre present in **5.11** has substituents which include a hydrogen atom and an electron withdrawing carboxylic acid function. Hence, this stereocentre is susceptible to racemization under basic conditions (section 5.1.2). Thus after resolution by entrainment, the unwanted (R)-enantiomer is treated with aqueous sodium hydroxide to convert it back into racemic **5.11** and is recycled as shown in **Scheme 5.5**.

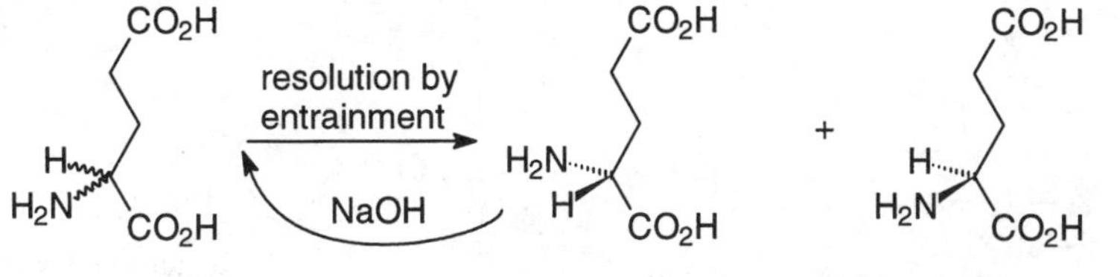

5.11   (R)-glutamic acid   (S)-glutamic acid

Scheme 5.5

Probably the ultimate resolution by crystallization technology, however, occurs when the racemization of the unwanted enantiomer can be carried out *in situ* with the resolution by entrainment. A good example of this is the resolution of ketone **5.14**, an intermediate in the synthesis of (2S, 3S)-paclobutrazol **5.15**, a commercially important plant growth regulator. Ketone **5.14** crystallizes as a conglomerate and so can be crystallized by entrainment (in this case from a 3:1 methanol/water mixture). However, the hydrogen atom attached to the stereocentre in ketone **5.14** is highly acidic due to the effects of the ketone and triazole groups. Thus, addition of 1% sodium hydroxide solution to enantiomerically pure samples of ketone **5.14** results in rapid racemization. These two factors allow the resolution of racemic ketone **5.14** to be carried out by entrainment from a 3:1 methanol/water solvent containing 1% sodium hydroxide. The solution is seeded with a few crystals of enantiomerically pure (S)-ketone **5.16**, inducing the crystallization of this enantiomer from the solution. The (*R*)-enantiomer **5.17** which remains in the solution, however, is rapidly racemized by the sodium hydroxide, generating more of the (*S*)-enantiomer **5.16** of the ketone which continues to crystallize out of the solution.

Eventually, all of the initially racemic **5.14** will be converted into crystals of the (*S*)-enantiomer **5.16**. The process is illustrated in **Scheme 5.6**, and has a theoretical maximum yield of 100%. Sodium borohydride reduction of ketone **5.16** then gives enantiomerically pure (2*S*, 3*S*)-paclobutrazol **5.15**. The stereochemistry of this step of the synthesis will be discussed in Chapter 9.

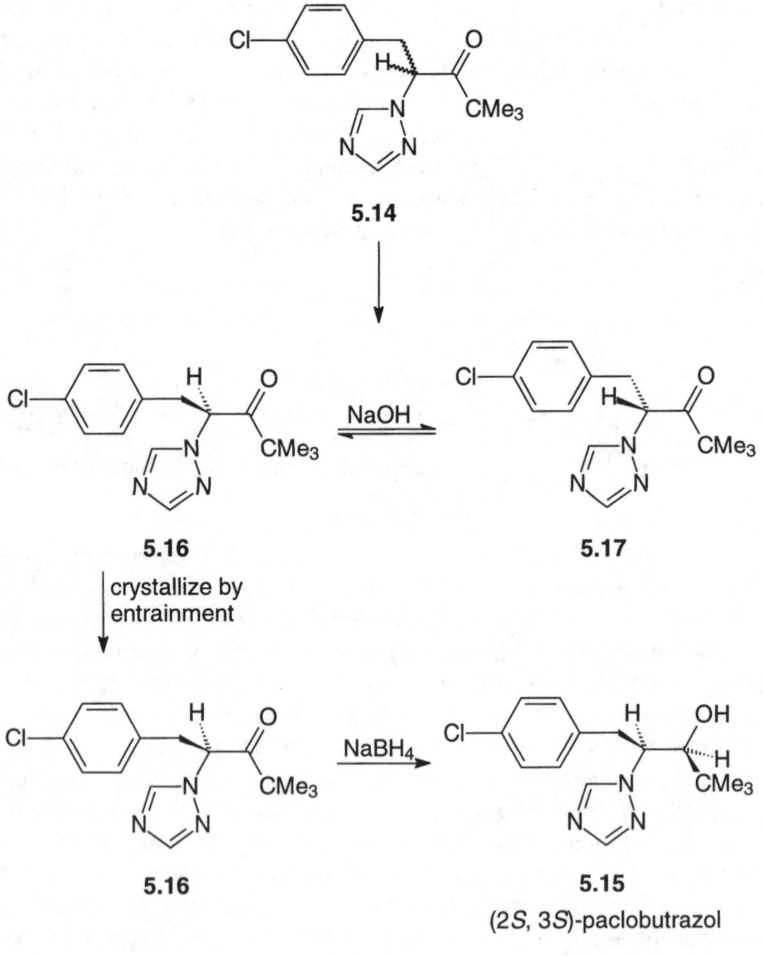

Scheme 5.6

5.3.2 Resolution by formation of diastereomers

In section 5.3.1, the direct crystallization of enantiomers was discussed as a method for their resolution. Unfortunately, only about 10% of all racemates crystallize as a conglomerate and so can be resolved in this way; for the other 90% of racemates some other method must be found for their resolution. Unlike enantiomers, diastereomers have different physical properties and thus can usually be easily separated. Thus, if a pair of enantiomers can be converted into diastereomers, they can be separated. If the diastereomers can then be converted back into enantiomers a resolution will have been achieved. This is most commonly carried out by salt formation.

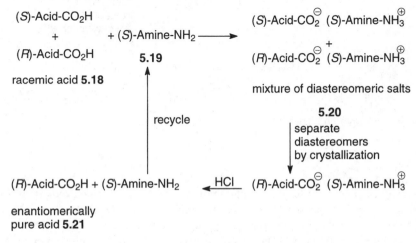

Scheme 5.7

Suppose a racemic carboxylic acid **5.18** needs to be resolved. If a single enantiomer of a chiral amine **5.19** is added to it, a salt **5.20** will be formed. This salt is a 1 : 1 mixture of two diastereomers which can be separated by any convenient physical technique (usually crystallization although chromatography is also sometimes used). The enantiomerically pure carboxylic acid **5.21** can then be obtained by adding dilute acid, a process which also regenerates the chiral amine **5.19**. The process is shown in **Scheme 5.7**, using the (S)-enantiomer of an amine to accomplish the resolution, although the (R)-enantiomer could have been used instead. Only one of the two diastereomeric salts is shown as being converted into the free acid **5.21**, although, in practice, one salt would crystallize out during the recrystallization and the other could be obtained from the mother liquor, thus allowing both enantiomers of the acid to be obtained. An example of resolution by diastereomeric salt formation is shown in **Scheme 5.8**. As this example shows, the method is not restricted to compounds containing a stereocentre; any chiral acid can be resolved in this way. This methodology can also be used to resolve a racemic amine by using a single enantiomer of an acid to form the diastereomeric salts.

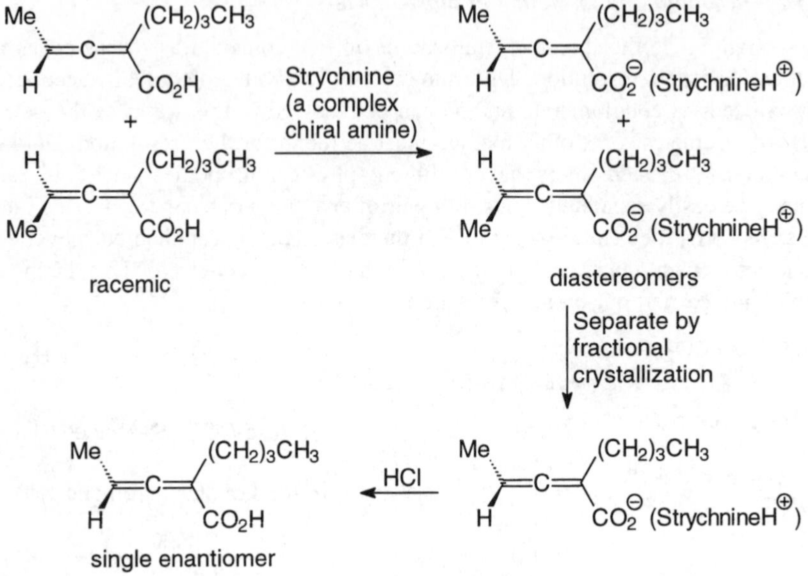

Scheme 5.8

A more sophisticated example of the resolution of an amine in this way is shown in **Scheme 5.9** for the resolution of an intermediate in the synthesis of a cholecystokinin antagonist. In this example, the amine is resolved with (S)-camphor sulphonic acid in the presence of a catalytic amount of 2-hydroxy-3,5-dichlorobenzaldehyde. The aldehyde reversibly forms an imine with the amine, which increases the acidity of the hydrogen atom attached to the stereocentre by a factor of about 10^8. This causes the amine dissolved in the recrystallization solvent to undergo rapid racemization during the recrystallization but has no effect upon the solid amine salt once it has crystallized from the solution. In this way, a greater than 50% yield of the desired amine can be obtained. The process is closely analogous to the resolution of ketone **5.16** shown in **Scheme 5.6**.

It is because acids and amines are so easy to resolve in this way that many of the examples of chiral compounds discussed in Chapter 3 contain acid or amine functional groups. A fundamental requirement for this sort of resolution, however, is the availability of enantiomerically pure amines or acids. Fortunately, as was discussed in Chapter 3, section 3.6, many naturally occurring compounds can be isolated from biological sources in enantiomerically pure form. Examples of enantiomerically pure acids and bases which are naturally occurring or readily derived from natural products are given in **Table 5.1**. In addition to being available in enantiomerically pure form, acids and bases which are to be used as resolving agents should have a number of other properties including: low cost, ease of recycling, stability of supply, chemical and chiral

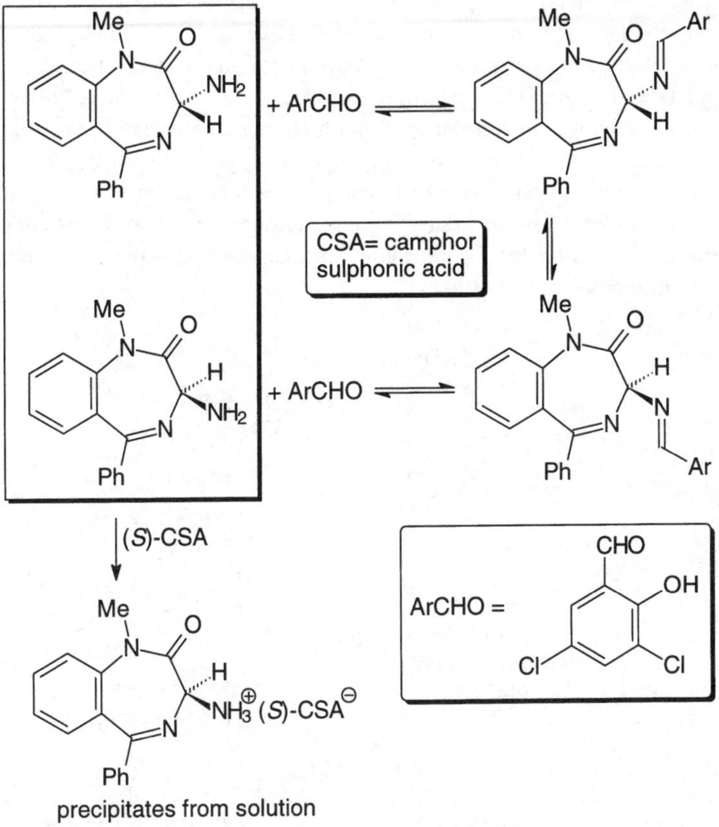

Scheme 5.9

stability (the compound should not decompose or racemize under the resolution conditions), and low toxicity, which is not the case for some of the naturally occurring bases such as strychnine.

Chiral alcohols can also be resolved by crystallization of a diastereomeric salt. The alcohol is first reacted with phthalic anhydride to convert the alcohol into an acid as shown in **Scheme 5.10**. The phthalic acid derivative is then resolved

Table 5.1 Examples of readily available enantiomerically pure acids and bases

Acids	Bases
Tartaric acid	$PhCHMeNH_2$
Lactic acid	Brucine
Camphor sulphonic acid	Morphine
Amino acids	Strychnine
	Ephedrine
	Amino acids

using a suitable enantiomerically pure amine and the alcohol is reformed by treatment of the resolved phthalic acid derivative with sodium hydroxide. In **Scheme 5.10**, the R* represents an R group containing at least one stereocentre (or stereoaxis etc.) somewhere within its structure. An alternative process for resolving an alcohol is to form an ester with an enantiomerically pure acid. The diastereomeric esters can then be separated by crystallization or chromatography, and acidic or basic hydrolysis used to regenerate the now enantiomerically pure alcohol. Chiral aldehydes and ketones can also be resolved by diastereomer formation by forming chiral imines or hydrazones with enantiomerically pure amines or hydrazines.

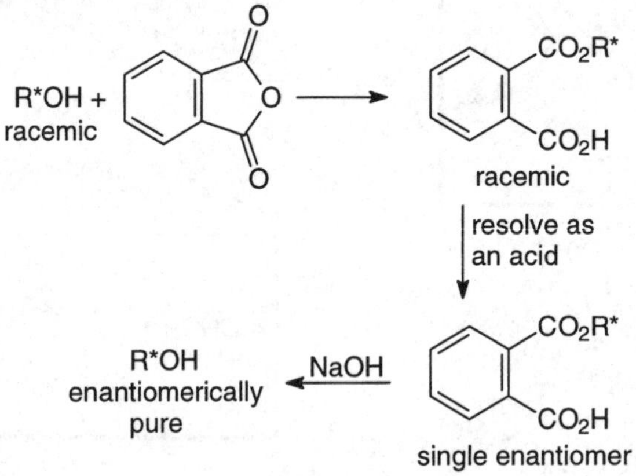

Scheme 5.10

5.3.3   Kinetic resolution

The chemical properties of enantiomers were discussed in Chapter 3, section 3.3. It was shown that enantiomers would react identically with achiral species but that they could react differently with other chiral species. The cause of this difference in reactivity with other chiral species is the diasteomeric nature of the transition states for the reaction and this is the basis of a process called **kinetic resolution**. Kinetic resolution has much in common with asymmetric synthesis (the preparation of enantiomerically pure products from achiral starting materials) which will be discussed in Chapter 10.

If a racemic mixture of two enantiomers ((R)-SM and (S)-SM in **Scheme 5.11**) is allowed to react with a single enantiomer of a chiral reagent ((R)-R), then the reagent may react with the two enantiomers at different rates due to the difference in energies of the diastereomeric transition states as shown in **Figure 5.2**. In the extreme case, the energy difference in the transition states will be so large that one of the enantiomers of the starting material will have completely reacted, whilst the other enantiomer remains completely unreacted and can be recovered from the reaction mixture thus constituting a resolution. Note that, in

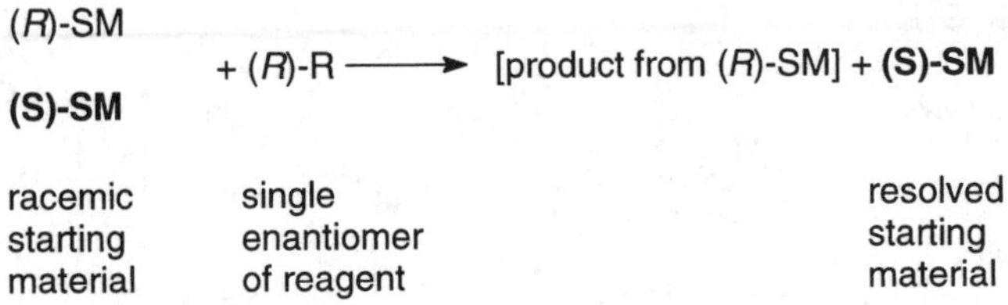

(R)-SM
                 + (R)-R ⟶ [product from (R)-SM] + **(S)-SM**
(S)-SM

racemic   single   resolved
starting   enantiomer   starting
material   of reagent   material

Scheme 5.11

this process, the nature of the products which are obtained from the two enantiomers is completely irrelevant and so is not shown in **Figure 5.2**. The enantiomeric starting materials may be converted into enantiomeric, diastereomeric or the same achiral product. All that matters is that the two transition states for the reaction should have different energies or, equivalently, that the two processes should have different activation energies.

A simple example of a kinetic resolution is the Sharpless epoxidation of racemic allylic alcohols **5.22** as shown in **Scheme 5.12**. The chiral reagent in this case is the complex formed between the titanium(IV) isopropoxide, diisopropyl (R,R)-tartrate and tertiary-butyl hydroperoxide, and this oxidizes allylic alcohols to epoxyalcohols. Due to the diisopropyl (R,R)-tartrate present in the oxidizing agent, the two enantiomers of the allylic alcohol **5.22** are oxidized at different rates. If the reaction is stopped at about 60% conversion, then it is found that the (S)-enantiomer of the allylic alcohol has been almost completely converted into the epoxide **5.23**, whilst the resolved (R)-enantiomer of the allylic alcohol can be recovered usually with an enantiomeric excess >96%. The Sharpless epoxidation is a very general process that can be used on a wide range of allylic alcohols.

The main problem with a simple kinetic resolution, such as the Sharpless epoxidation shown in **Scheme 5.12**, is that a full equivalent of the chiral reagent is required which can make the process expensive. However, it is possible to carry out the kinetic resolution of a racemic mixture using only achiral reagents,

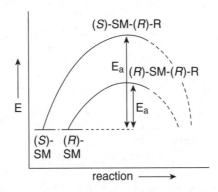

Figure 5.2 Energy/reaction pathway diagram for kinetic resolution using a chiral reagent.

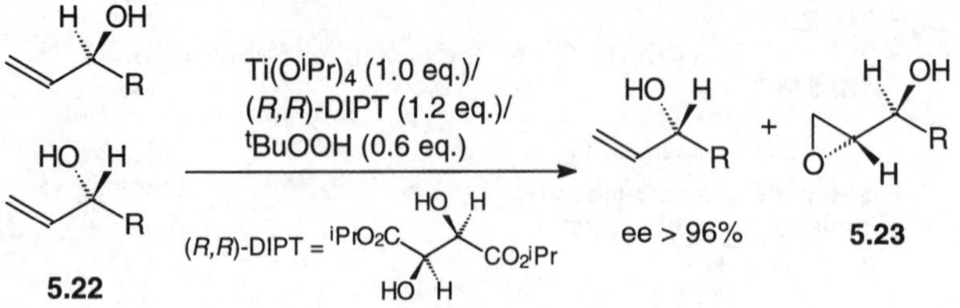

Scheme 5.12

provided that a chiral catalyst is also employed. Modification of the general reaction scheme, shown in **Scheme 5.11**, and the reaction pathway/energy diagram, shown in **Figure 5.2**, to incorporate the effect of a chiral catalyst and achiral reagent gives **Scheme 5.13** and **Figure 5.3** respectively. Provided that the chiral catalyst is bound to the substrates during the transition state, the transition states remain diastereomeric and hence at different energies.

The Sharpless epoxidation of racemic allylic alcohols can be made catalytic (in both titanium(IV) isopropoxide and diisopropyl (R,R)-tartrate) simply by the addition of 4 Å molecular sieves to the reaction mixture. The molecular sieves absorb any water present and, under these conditions, only 5–10 mol% of the catalyst is needed. Another example of a kinetic resolution using a chiral catalyst is the hydrogenation of racemic allylic alcohol **5.24** using just 0.2 mol% of the enantiomerically pure rhodium complex **5.25** as shown in **Scheme 5.14**. If the reaction is stopped after 65% of the allylic alcohol has been hydrogenated, then the (S)-enantiomer of alcohol **5.24** is recovered in 35% yield and with 98% enatiomeric excess.

Enzymes are naturally occurring catalysts that are used by living organisms to catalyse the chemical reactions which are necessary for life. Enzymes are proteins (cf. Chapter 8, section 8.10.1), that is they are composed of long chains of amino acids, and all but one of the amino acids found in proteins contains at

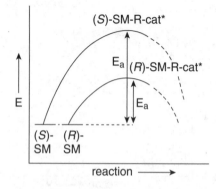

Figure 5.3 Energy/reaction pathway diagram for kinetic resolution using an achiral reagent and a chiral catalyst.

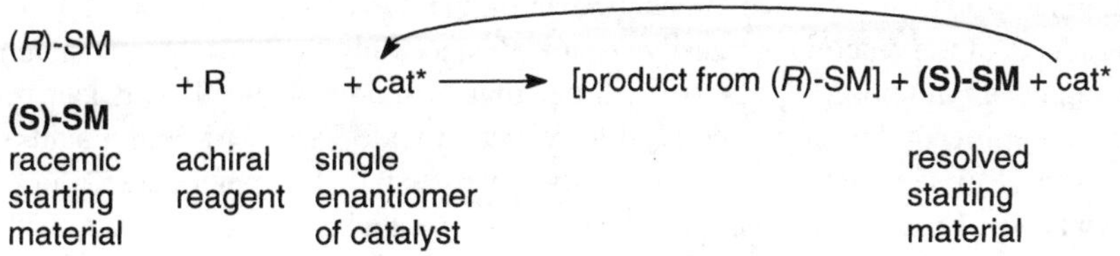

(R)-SM

(S)-SM
racemic   achiral   single   resolved
starting   reagent   enantiomer   starting
material   of catalyst   material

$+R$ $+ cat^* \longrightarrow$ [product from (R)-SM] + **(S)-SM** + cat*

Scheme 5.13

5.24

5.25

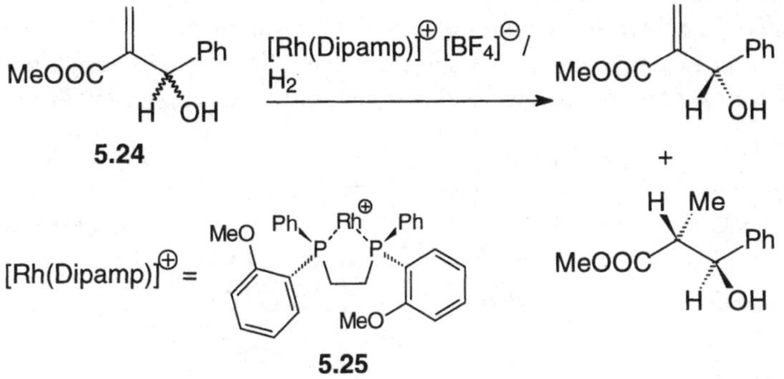

Scheme 5.14

least one stereocentre and is always found as the L-enantiomer. Hence enzymes are enantiomerically pure catalysts. It is possible to isolate enzymes from their natural host or, if the host is a yeast or bacterium, to use them *in situ* to carry out chemical transformations. Many enzymes will carry out the same chemical reaction on a wide range of substrates and are highly enantioselective (that is they react with only one enantiomer of a substrate). The reason for the enantio- selectivity exhibited by enzymes is the same reason why the two enantiomers of a compound often exhibit different biological properties (cf. Chapter 3, section 3.6). These properties make enzymes ideal for carrying out kinetic resolutions. Two examples are shown in **Scheme 5.15** and **Scheme 5.16**.

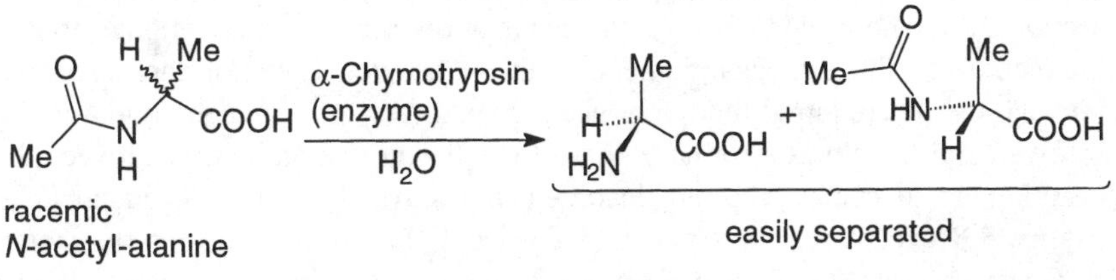

racemic
N-acetyl-alanine

easily separated

Scheme 5.15

In the first example (**Scheme 5.15**), the enzyme α-chymotrypsin is used to resolve N-acetyl alanine. α-Chymotrypsin is an enzyme which hydrolyses amide bonds and does so selectively from the L-enantiomer of an amino acid substrate. Thus, the enzyme cleaves the acetyl group from the L-enantiomer of N-acetyl alanine whilst leaving the D-enantiomer of the substrate untouched. The two products of the reaction are easily separated, since only the hydrolysed L-alanine contains an amino group and so will be soluble in aqueous acid. This is a very general procedure that can be used to resolve a wide variety of amino acids.

The (R)-enantiomer of glycidyl butyrate 5.26 is a key intermediate in the synthesis of a class of pharmaceuticals called β-blockers. The racemate of this compound is readily prepared and can be resolved using the enzyme porcine pancreatic lipase which selectively hydrolyses the ester bond in the (S)-enantiomer of the substrate whilst leaving the desired (R)-enantiomer unreacted, as shown in **Scheme 5.16**. Kinetic resolutions can also be combined with *in situ* racemization of the starting material to provide a highly effective means for asymmetric synthesis. This topic will be discussed in Chapter 10.

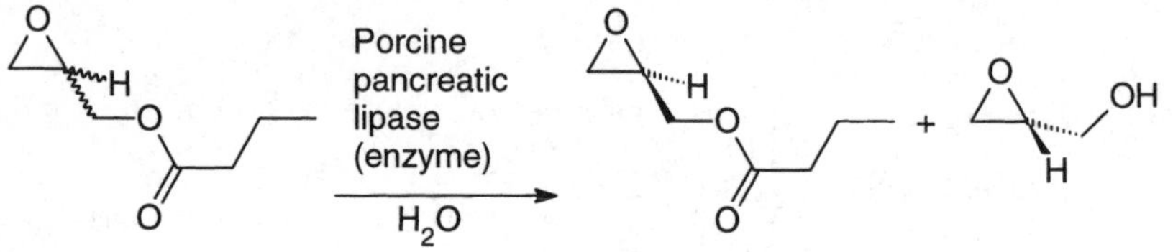

5.26

Scheme 5.16

5.3.4   Resolution by preferential absorption

Enantiomers cannot be separated by chromatography on a 'normal' achiral stationary phase such as silica or alumina since the enantiomers have identical physical properties, including retention time, on the chromatography column. However, if the stationary phase from which the chromatography column is constructed is itself composed of an enantiomerically pure chiral compound, then the two enantiomers of a compound will interact differently with this stationary phase and will have different retention times on the chromatography column, allowing them to be separated. The origin of this effect is identical to the cause of enantiomers having different biological properties (cf. Chapter 3, section 3.6). The biological receptor or enantiomerically pure stationary phase interacts with the two enantiomers of the substrate. Three or more sites of interaction are required between each enantiomer of the compound to be resolved and the chiral stationary phase for efficient discrimination between the enantiomers to occur. This is illustrated in **Figure 5.4**, which is identical to **Figure 3.8** except for the change of the label 'biological receptor' to 'chiral stationary phase'. In **Figure 5.4b**, the enantiomer whose interactions match those of the chiral stationary phase will be more tightly bound and hence retained longer on the chromatography column than the other enantiomer.

One of the first chiral stationary phases to be utilized was paper, which is composed mostly of cellulose, a polymer of the naturally occurring, enantiomerically pure carbohydrate D-glucose (cf. Chapter 8, section 8.10.2). This can be

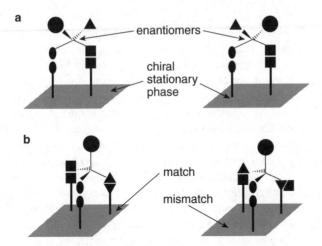

Figure 5.4 Interaction of enantiomers with a chiral stationary phase containing: **a**, two; and **b**, three binding sites.

used for thin layer chromatography and to separate a few milligrams of enantiomers but is not practicable for larger scale separations. Other chiral stationary phases are available composed of chiral molecules, often amino acid derivatives bound to silica and these can be used for HPLC, but again only a few hundred milligrams or at most a few grams can be separated in a single injection. Hence, chiral chromatography is best used for resolutions when only a small amount of material is needed (e.g. for initial biological screening of a potential pharmaceutical) or to provide the seed crystals that will be used for a subsequent large scale resolution by crystallization.

5.4   Methods for determining enantiomeric excess

In section 5.3, the methods that are available for the separation of enantiomers were discussed. In this section, the ways in which the enantiomeric composition of a sample can be determined will be surveyed. There is considerable overlap between the two topics, since any method which separates enantiomers will allow the enantiomeric excess to be determined simply by weighing the separated enantiomers.

5.4.1   Chromatography using a chiral stationary phase

It was shown in section 5.3.4 that an enantiomerically pure, chiral stationary phase could be used to separate enantiomers. The same methodology can be used to determine the enantiomeric excess of a sample simply by incorporating an electronic integration unit into the detector used to determine when each component elutes from the column. The enantiomeric excess of the sample is then obtained directly from the relative integrals of the detector peaks corresponding to the two enantiomers. This is a very common method for determining the enantiomeric excess of a sample, as it is not subject to any of the limitations of the methods that will be discussed in the following sections. A large number of chiral stationary phases are available for high pressure liquid chromatography (HPLC) or gas chromatography (GC) systems, allowing the enantiomeric excess of virtually any sample to be determined in this way.

One precaution needs to be taken in using chromatographic techniques to determine the enantiomeric excess of a sample. That is to prove that under the conditions being used (the chiral stationary phase, solvent, temperature, etc.) the enantiomers are actually being separated. This is most conveniently done by analysing an authentic, racemic mixture of the compound under investigation, or alternatively, by separately analysing authentic samples of both enantiomers. In the first case, two peaks of equal intensity should be observed and, in the latter case, the two samples should be shown to have different retention times. It should also be proven that the two peaks detected as eluting from the chromatography column really do correspond to the two enantiomers of the compound in question and not to the unseparated enantiomers and some impurity. For GC, this is often conveniently achieved by coupling the gas chromatograph to a mass spectrometer and using the latter to show that the two components have the same molecular weight and fragmentation pattern. For HPLC, a diode array UV detector can be used to show that the two components have the same UV spectrum, or the two components can be isolated as they elute from the column and analysed by NMR or spectrometry.

For accurate determination of enantiomeric excesses, there should be baseline separation between the peaks corresponding to the two enantiomers of a compound. For GC, the temperature and/or temperature gradient can be varied to achieve this, whilst for HPLC, the solvent system and/or solvent gradient can be varied. Assuming that baseline separation of the peaks corresponding to each enantiomer can be achieved, the limiting factor in the accuracy of the enantiomeric excess determined by HPLC or GC is then the signal-to-noise ratio. This is particularly problematic when samples which are nearly enantiomerically pure are being analysed, as the peak corresponding to the minor enantiomer may be lost in the noise. Thus a nearly enantiomerically pure sample analysed in this way is often reported as having an enantiomeric excess >98% (or greater than some other percentage), meaning that only one peak was detected and that the minor enantiomer constitutes not more than 1% of the sample or it would have been detected.

An example of the determination of the enantiomeric excess of a sample by HPLC using a chiral stationary phase is shown in **Figure 5.5**. In this case, the compound being analysed, **5.27**, is an N-protected allylic amine, and a chiral stationary phase composed of N-(2,4-dinitrophenyl)-(S)-phenylglycine ionically bound to silica was found to separate the two enantiomers. **Figure 5.5a** shows the chromatogram obtained by the analysis of a racemic sample of compound

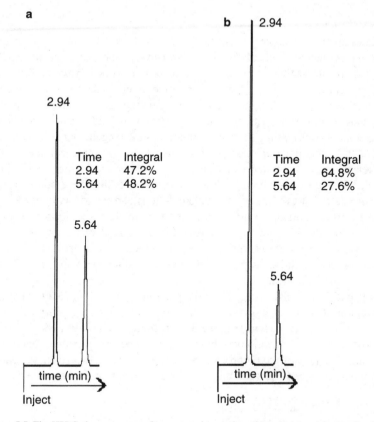

Figure 5.5 The HPLC chromatogram of: **a**, a racemic sample and **b**, a non-racemic sample of amine **5.27** recorded using a chiral stationary phase.

5.27

5.27, whilst **Figure 5.5b** shows the chromatogram of a sample of compound **5.27** which was determined to have an enantiomeric excess of 40% ($\pm 1\%$). The two chromatograms were obtained under identical conditions. Note that in **Figure 5.5a** the two peaks are not of equal height but they are of equal area (within a 1% error) as determined by the peak integration. This is not uncommon, as the longer a compound is absorbed on a chromatography column, the broader the band in which it elutes.

5.4.2 NMR

It was shown in Chapter 1, section 1.5.1 that in an achiral environment enantiomers have identical physical properties, including NMR spectra. However, if the NMR spectrum of a racemate is recorded in the presence of a chiral solvent or some other enantiomerically pure chiral species with which the racemate can interact, then it is possible to distinguish enantiomers by NMR since their interactions with the chiral species will lead to the formation of diastereomers. Both chiral solvents and chiral shift reagents have been used to achieve this, although for reasons of cost the latter are more common. Chiral shift reagents are enantiomerically pure, camphor derived complexes of lanthanide metals, an example being shown in structure **5.28**. These complexes are six-coordinate, but the size of the lanthanide ions and the availability of f-shell electrons allow lanthanides to coordinate to more than six ligands. The best ligands for lanthanides are oxygen based, for example alcohols, ketones or acid derivatives, so if a chiral shift reagent is added to a solution of a chiral compound that contains ligands capable of coordinating to the metal then a complex will be formed.

As an example, consider the ^{1}H NMR spectrum of compound **5.29**, the methyl ester region of which is shown in **Figure 5.6**. For the racemic compound in the absence of any shift reagent (**Figure 5.6a** bottom trace), only a single peak is observed as the two enantiomers have identical spectra. However, addition of increasing amounts of the chiral shift reagent **5.28** causes two separate peaks of equal intensity to be observed, one peak for each enantiomer (**Figure 5.6a**). The

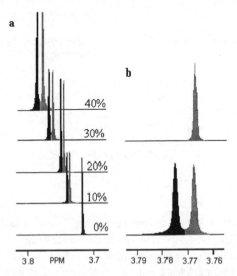

Figure 5.6 a, The methyl ester region of the ^{1}H NMR spectra of racemic **5.29** in the presence of increasing amounts of chiral shift reagent **5.28**. **b**, the methyl ester region of the NMR spectra of racemic (bottom) and enantiomerically pure (top) samples of **5.29** in the presence of 40 mol% of shift reagent **5.28**.

two enantiomers of compound **5.29** reversibly coordinate to the shift reagent **5.28** as shown in **Scheme 5.17**, to give diastereomers **5.30a** and **5.30b**. As less than one equivalent of the chiral shift reagent is used, an equilibrium is established between coordinated and uncoordinated **5.29**, and exchange between the coordinated and uncoordinated molecules is rapid on the NMR time scale so that only an average species is detected. Since **5.30a** and **5.30b** are diastereomers, they have different NMR spectra and hence produce two peaks in the NMR spectra.

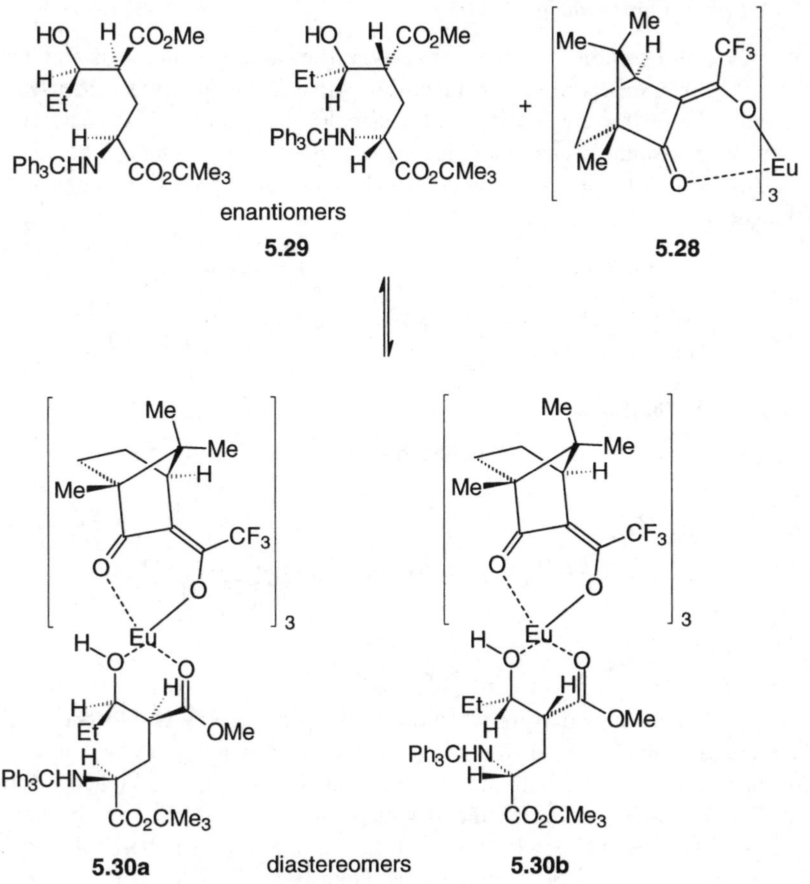

enantiomers

5.29

5.28

5.30a   diastereomers   **5.30b**

Scheme 5.17

As increasing amounts of shift reagent **5.28** are added, the chemical shift difference between the two peaks increases. However, there is an upper limit to the amount of shift reagent that can be added since compound **5.28** is paramagnetic and so causes broadening of the lines in the NMR spectrum and loss of resolution. In the present case, 40 mol% of shift reagent **5.28** was found to give optimal separation. **Figure 5.6b** then compares the spectra obtained for racemic (bottom) and enantiomerically pure (top) samples of **5.29** in the presence of 40 mol% of shift reagent **5.28**. In the latter case, only a single peak is observed and the enantiomeric excess was calculated as >98% based upon the signal-to-noise ratio. In this case, it was not possible to obtain baseline separation between the peaks for the two enantiomers, which also limits the accuracy of enantiomeric excess determinations.

5.4.3 Conversion to diastereomers

It was shown in section 5.3.2 that conversion of enantiomers into diastereomers allows the enantiomers to be separated. The same methodology can be used to determine the enantiomeric excess of a sample: the sample to be analysed is reacted with a suitable enantiomerically pure compound (a chiral derivatizing agent) to form a pair of diastereomers which can be analysed by any convenient technique.

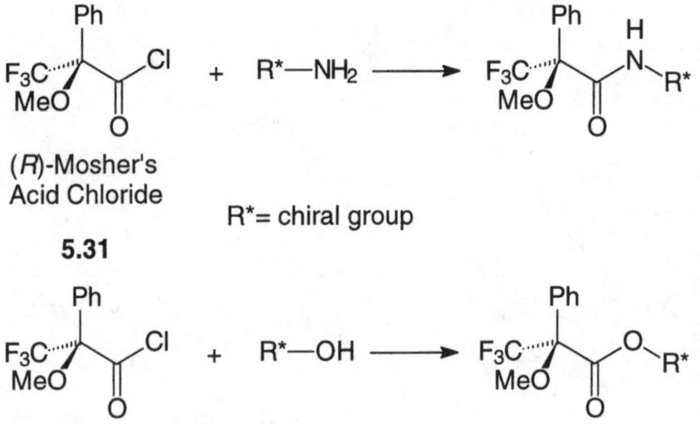

(R)-Mosher's Acid Chloride

R* = chiral group

5.31

Scheme 5.18

One of the most widely used chiral derivatizing agents is Mosher's acid chloride **5.31**, which is readily available in enantiomerically pure form and can form diastereomers with both amines and alcohols as shown in **Scheme 5.18**. The diastereomeric excess of the resulting amides or esters (which should correspond to the enantiomeric excess of the starting amine or alcohol) can be determined by a variety of techniques including ^{1}H or ^{19}F NMR, GC or HPLC. It is the fact that Mosher's acid chloride will react with two of the most common functional groups, and gives diastereomers that can be analysed by so many different techniques, that makes this compound so widely used as a chiral derivatizing agent. Some care is needed in the use of the R and S-descriptors to describe the stereochemistry of Mosher's acid derivatives, since the priorities of the groups around the stereocentre change during the reactions of Mosher's acid chloride, a point that was discussed in detail in Chapter 3, section 3.3.

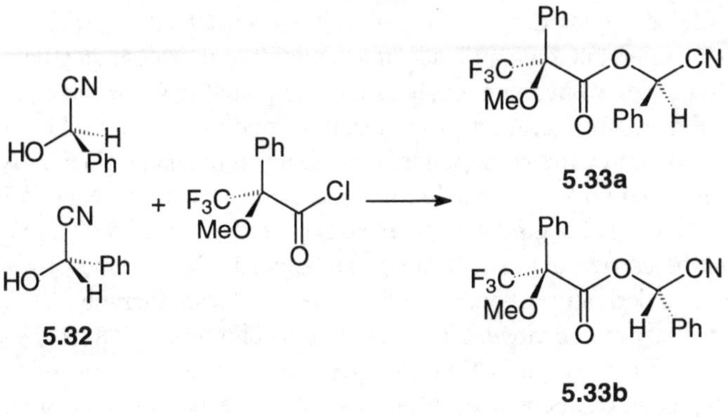

Scheme 5.19

An example of the use of Mosher's acid chloride to determine the enantiomeric excess of a compound is shown in **Scheme 5.19** and **Figure 5.7**. In this case, the compound whose enantiomeric excess was to be determined was the cyanohydrin **5.32**. Reaction of a racemic sample of cyanohydrin **5.32** with (R)-Mosher's acid chloride in the presence of pyridine gave α-cyano esters **5.33a,b** as a 1 : 1 mixture of diastereomers. Compound **5.33** was analysed by ^{1}H NMR, and an extract of the spectrum is shown in **Figure 5.7a**. Two singlets are observed at 3.5 ppm, these are the resonances for the OCH$_3$ groups of the two diastereomers. The integration curve shows that, as expected, the two peaks have equal intensities. The peak at 6.6 ppm corresponds to the H–C–CN resonance and again this is split into two singlets, although in this case baseline separation between the two resonances was not achieved. Hence, the OCH$_3$ resonances were used for enantiomeric excess determination.

Figure 5.7b shows an extract of the ^{1}H NMR spectrum obtained for compound **5.33**, starting from a non-racemic sample of cyanohydrin **5.32**. In this case, two OCH$_3$ signals were again observed though with different intensities. From the integrals of the two peaks, the enantiomeric excess was calculated as 59%. The formula for calculating the enantiomeric excess of a sample given the amount of each enantiomer present was presented in Chapter 3, section 3.5. Many other chiral derivatizing agents are available and, in particular, chiral alcohols such as methyl mandelate [PhCH(OH)CO$_2$Me] can be used to convert enantiomeric carboxylic acids into diastereomeric esters. Additional examples will be found in the further reading.

Whichever chiral derivatizing agent is used, a number of important precautions must be taken. The first of these is to prove that no kinetic resolution (cf. section 5.3.3) is taking place. Enantiomers may have different chemical properties when reacting with other chiral species so it is possible that, under a given set of reaction conditions, one enantiomer of the compound whose enantiomeric excess is being determined will react faster with the chiral derivatizing agent than the other. If the reaction is then stopped before it has gone to completion, or if insufficient chiral derivatizing agent has been added to react completely with both enantiomers, then an inaccurate enantiomeric excess may be obtained. In the extreme case, one enantiomer of the compound may not react at all with the chiral derivatizing agent under the reaction conditions and, in this case, even a racemic mixture of the compound would seem to be enantiomerically pure.

To confirm that no kinetic resolution is occurring, a control reaction in which the racemate of the compound is reacted with the chiral derivatizing agent should first be carried out. This should give signals of equal intensity when the diastereomeric derivative is analysed by NMR or chromatography. If signals of unequal intensity are obtained, then the reaction conditions for the derivatization must be varied (the reaction time, temperature or concentration of the chiral derivatizing agent can be increased) until signals of equal intensity are obtained.

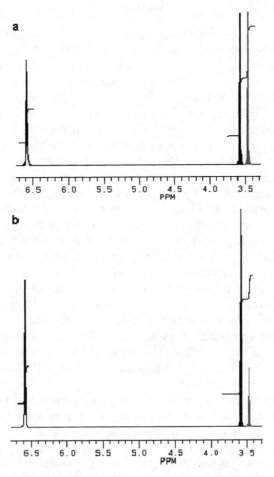

Figure 5.7 Extract of the ^{1}H NMR spectrum of compound **5.33** obtained from: **a**, racemic **5.32** and **b**, **5.32** with an enantiomeric excess of 59%.

Alternatively, it may be necessary to use a different chiral derivatizing agent. If the racemate of the compound whose enantiomeric excess is being determined is not available, then it is possible to use the racemate of the chiral derivatizing agent with an enantiomerically pure sample of the compound of interest instead, as this will again result in the formation of a $1:1$ mixture of the two diastereomers.

Once suitable reaction conditions have been found, care should be taken when analysing the sample of interest to ensure that the chiral derivatizing agent is present in excess. Finally, it is often necessary to purify the diastereomeric derivatives prior to analysis to remove the excess of derivatizing agent and other reagents (such as bases) which may have been added to the reaction mixture. Under no circumstances should the diastereomeric derivatives ever be purified by crystallization, as this is likely to result in the selective crystallization of one diastereomer. Similarly, chromatography should be avoided if possible since the diastereomers may have different retention times and so be completely or partially separated. Ideally, the sample should be analysed without purification, but if this is not possible then the only purifications used should be to wash an organic solution of the diastereomeric derivative with aqueous acid or base, as these are unlikely to result in any change in the ratio of the diastereomers present.

5.4.4   Polarimetry

Polarimetry can be used to determine the optical purity of a compound as was discussed in Chapter 3, section 3.5. In many cases, the optical purity is numerically equal to the enantiomeric excess and so can be used to determine the latter. However, before polarimetry is used to determine the enantiomeric excess of a compound, it should first be proved that for that compound the specific rotation varies in direct proportion to the enantiomeric excess. This entails preparing a series of samples of known enantiomeric excess, and showing that a plot of the specific rotation of these samples against their enantiomeric excess is linear. In order to determine either the optical purity or the enantiomeric excess of a sample using polarimetry, it is necessary to know the specific rotation of an enantiomerically pure sample. Hence, polarimetry is only useful for determining the enantiomeric excess of known compounds since there is no method available to predict the specific rotation of a compound which has never previously been studied by polarimetry.

If the above requirements have been met, and polarimetry is to be used to determine the enantiomeric excess of a sample, then the specific rotation must be measured under identical conditions to those reported for the enantiomerically pure sample. Thus the solvent, concentration, wavelength and temperature must all be kept constant, since they can all affect the magnitude (and even the sign) of the measured specific rotation. Finally, it is essential that the sample whose specific rotation is being measured is analytically pure, since even small amounts of an impurity with a larger specific rotation than the substance being investigated will lead to a large error in the specific rotation. In purifying the compound for polarimetric investigation, however, recrystallization should be avoided as this is likely to result in preferential crystallization of enantiomerically pure material whilst leaving the minor enantiomer in the mother liquor. Chromatographic purification or distillation are, however, not likely to change the enantiomeric excess of the sample, although if chromatographic purification is employed, then care should be taken to collect all of the fractions corresponding to the desired compound as cases have been reported (see further reading) where the enantiomeric excess of a sample varies as the compound elutes from the chromatographic support.

As an example of the importance of sample purity on polarimetric measurements, consider cyanohydrin **5.32**. The specific rotation of enantiomerically pure (S)-**5.32** is known to be −45.5 (concentration = 1 g/100 ml in chloroform at 25°C and the sodium D-line). However, when samples of (S)-**5.32** were prepared, using *bis*-imine **5.34** as a chiral catalyst (see Chapter 10 for more details on chiral catalysts), the enantiomeric excess of the samples determined by polarimetry were found to differ from the values determined by conversion to the diastereomeric Mosher esters (**Scheme 5.19**). The inconsistency was eventually determined to be due to contamination of the cyanohydrin samples with 1–2% of catalyst **5.34**, the catalyst having a very large specific rotation ($[\alpha]_D^{25} = -318$ ($c = 1$ in chloroform)). Thus the presence of just 2% of catalyst **5.34** in the samples of the cyanohydrin could introduce an error of −6.4 in the measured specific rotation, which corresponds to an error of about 20% in the measured value. Two percent contamination is difficult to detect by common analytical techniques such as NMR, and in the case of cyanohydrin **5.32** further purification was not feasible due to the instability and ease of racemization of cyanohydrins.

(S)-**5.32**

$[\alpha]_D^{25}$ -45.5

5.34

$[\alpha]_D^{25}$ -318

5.5 Determination of absolute configuration

In section 5.4, the various methods that can be used to determine the enantiomeric excess of a sample were discussed. For compounds where an authentic, enantiomerically pure sample of known absolute configuration is available, these methods will also provide information as to the absolute configuration of the major enantiomer of the sample. For example, the sign of the specific rotation of the sample will correspond to the sign of the specific rotation of the major enantiomer. Alternatively, analysis of the enantiomerically pure sample by chiral chromatography or of a diastereomeric derivative by any suitable technique will, by comparison with the results obtained for the unknown sample, allow the absolute configuration of the unknown to be determined. Often, however, no enantiomerically pure sample of known absolute configuration is available and, in these cases, it is necessary to resort to other methods to determine the absolute configuration of a sample.

5.5.1 Chiroptical methods

Whilst the magnitude and sign of the specific rotation of a compound cannot currently be predicted, it is found that closely related compounds give ORD or CD curves (cf. Chapter 3, section 3.4.2) of the same shape. Hence, the CD or ORD curve obtained from a compound of unknown absolute configuration can be compared with that of a similar compound of known absolute configuration. The two compounds being compared must have the same conformation and, for this reason, the technique is most useful for cyclic compounds, the conformations of which are relatively easily predicted (Chapter 8). The main problem with this technique is knowing how similar the two compounds have to be for the method to be reliable and therefore the methodology is not widely used.

An example of the use of ORD to determine the absolute configuration of a compound is shown in **Figure 5.8**. Two of the curves in **Figure 5.8** show the

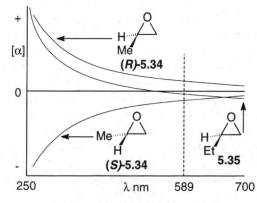

Figure 5.8 The ORD curves obtained for epoxides **5.34** and **5.35**.

ORD spectra of the enantiomers of compound **5.34**, whilst the third curve is the ORD spectrum of one enantiomer of compound **5.35**. It is apparent that this curve has the same shape as the ORD curve corresponding to the (*R*)-enantiomer of epoxide **5.34** and hence that the enantiomer of compound **5.35** giving rise to the ORD spectrum also has the (*R*)-configuration. It should be noted that at 589 nm (*R*)-**5.34** and **5.35** have specific rotations of opposite sign showing that, unlike ORD or CD, the specific rotation measured at a single wavelength cannot be used to determine the absolute configuration of an unknown compound.

5.5.2   X-ray crystallography

For a brief introduction to X-ray crystallography, see Chapter 2, section 2.6.2. In favourable cases, it is possible to obtain the absolute configuration of a molecule by X-ray crystallography. It is often much easier, however, to form a diastereomeric derivative between the compound under investigation and one of known absolute configuration, and then to use X-ray crystallography to determine the relative stereochemistry between the two stereocentres of the derivative. The absolute stereochemistry of the stereocentre in the original compound can then be derived.

5.5.3   Conversion to a compound of known absolute configuration

If the above physical methods are not appropriate for a particular case then the only remaining option is to transform chemically the compound whose absolute configuration is to be determined into a known compound whose absolute configuration has previously been determined. This can be a very difficult and time consuming process which requires a substantial amount of material but it is often the only way to determine the absolute configuration of an unknown compound. It is important that, during the conversion process, no reaction be carried out which might alter the stereochemistry at the stereocentre of the unknown compound.

   As an example, consider the chemical reactions shown in **Scheme 5.20**. Addition of methyl lithium to imine **5.36** in the presence of a chiral catalyst gives an excess of one enantiomer of allylic amine **5.37** over the other enantiomer. However, the absolute configuration of the major enantiomer was unknown. The reason why this process does not give a racemic mixture of the enantiomers of amine **5.37** will be discussed in Chapter 10. Subsequent oxidation of the alkene to the corresponding carboxylic acid, however, gave a well known derivative **5.38** of the naturally occurring amino acid alanine. The specific rotation of a sample of compound **5.38** prepared by the route shown in **Scheme 5.20** was found to have the same sign as that of an authentic sample prepared from (*S*)-alanine. Thus it could be concluded that the major enantiomer

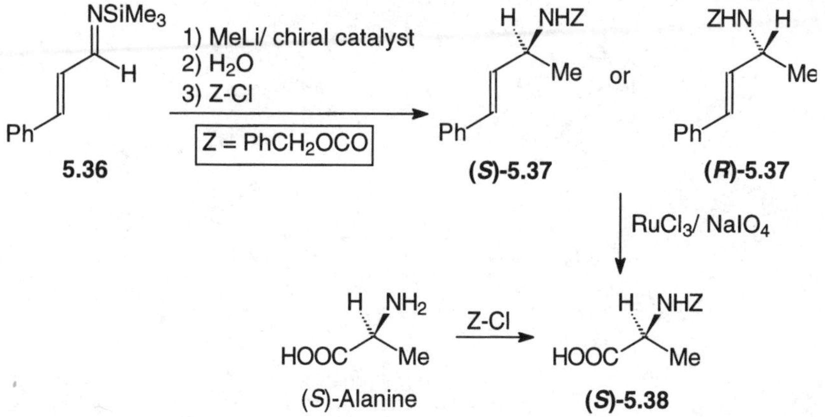

Scheme 5.20

of amine **5.37** formed by the chemistry shown in **Scheme 5.20** was the (*S*)-enantiomer.

5.6   Determination of relative configuration

Diastereomers have different physical properties and so can easily be distinguished by spectroscopic and chromatographic techniques. However, most analytical techniques will determine only how many diastereomers are present in a sample, not which diastereomer(s) are present. If authentic samples of the various diastereomers of a compound are available, then the task of determining which diastereomer(s) are contained within a particular sample of that compound is relatively straightforward as a wide range of chromatographic (e.g. HPLC, GC) and spectroscopic (e.g. ^1H and ^{13}C NMR) techniques can be used to compare the components of the sample with the authentic samples.

If, however, authentic samples are not available, then the task of determining the relative configuration of an unknown diastereomer is more difficult. Probably the most powerful technique is X-ray crystallography since, as was discussed in section 5.5.2, this will readily provide information on the relative configuration of the compound. However, X-ray analysis requires that the sample be a crystalline solid (although not necessarily at room temperature) and this is not always the case. For compounds that do not form suitable crystals, it is often possible to prepare a derivative which is crystalline and analyse this by X-ray crystallography. A certain amount of trial and error and a considerable amount of material will, however, be required to find an appropriate derivative.

^1H NMR will sometimes allow the relative configuration of a compound to be determined by analysis of the coupling constants or nuclear Overhauser effects (nOe's) (see Chapter 2, section 2.6.1 for an introduction to nOe's). NMR analysis requires that the conformation of the compound be known and this is

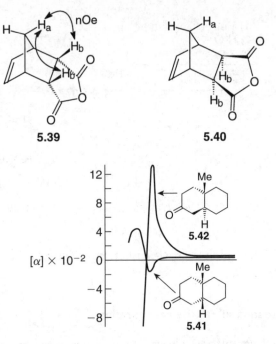

Figure 5.9 ORD traces for the diastereomeric ketones 5.41 and 5.42.

usually only the case for cyclic compounds, which restricts the applicability of the technique. An example of a case where nOe effects are useful in determining the relative configuration of a compound is the diastereomers 5.39 and 5.40. For compound 5.39, an nOe is observed between H_a and the two H_b's, as these hydrogens are close together. In the diastereomer 5.40, however, no such nOe is observed as the H_b's are too far away from H_a.

Diastereomers have different ORD and CD spectra, as shown in **Figure 5.9**, for the ORD traces of diastereomeric ketones 5.41 and 5.42. Comparison of the ORD or CD spectra of a compound of unknown relative configuration with those of closely related compounds of known relative configuration may then allow the relative configuration of the unknown to be determined.

5.7 Further reading

General
Stereochemistry of Organic Compounds E.L. Eliel and S.H. Wilen. Wiley: London, 1994, chapters 6 and 7.

Racemization
E.J. Ebbers, G.J.A. Ariaans, J.P.M. Houbiers, A. Bruggink and B. Zwanenburg. *Tetrahedron*, 1997, **53**, 9417.

Epimerization of 5.8
P.J. Murphy and S.T. Dennison. *Tetrahedron*, 1993, **49** 6695.

Resolution
Chirality in Industry J. Crosby (A.N. Collins, G.N. Sheldrake and J. Crosby eds). Wiley: Chichester, 1992, chapter 1.
S. Caddick and K. Jenkins. *Chem. Soc. Rev.*, 1997, **25**, 447.

Resolution by crystallization
Topics in Stereochemistry Vol. 6, S.H. Wilen (N.L. Allinger and E.L. Eliel eds). Wiley: London, 1971, chapter 3.
A. Collet, M.-J. Brienne and J. Jacques. *Chem. Rev.*, 1980, **80**, 215.

Kinetic resolution
Topics in Stereochemistry Vol. 18, H.B. Kagan and J.C. Fiaud (E.L. Eliel and S.H. Wilen eds). Wiley: Chichester, 1988, chapter 4.
Topics in Stereochemistry Vol. 19, C.J. Sih and S.-H. Wu (E.L. Eliel and S.H. Wilen eds). Wiley: Chichester, 1989, chapter 2.
J.M. Brown. *Angew. Chem., Int. Ed. Engl.*, 1987, **26**, 190.
H. Stecher and K. Faber. *Synthesis*, 1997, 1.

Enantiomeric excess determination
Asymmetric Synthesis D. Parker and R.J. Taylor (R.A. Aitken and S.N. Kilenyi eds). Blackie: London, 1994, chapter 3.
Topics in Stereochemistry Vol. 2, M. Raban and K. Mislow (N.L. Allinger and E.L. Eliel eds). Wiley: London, 1967, chapter 4.
Asymmetric Synthesis Vol. 1 (J.D. Morrison ed.). Academic Press: London, 1983, chapters 2–9.

Chiral shift reagents
K.A. Kime and R.E. Sievers. *Aldrichimica Acta*, 1977, **10**, 54.
Topics in Stereochemistry Vol. 10, G.R. Sullivan (E.L. Eliel and N.L. Allinger eds). Wiley: Chichester, 1978, chapter 4.

Variation of ee during chromatography
R. Matusch and C. Coors. *Angew. Chem., Int. Ed. Engl.*, 1989, **28**, 626.

5.8   Problems

1. Which of the following compounds will be racemized on treatment with acid and which will racemize under basic conditions?

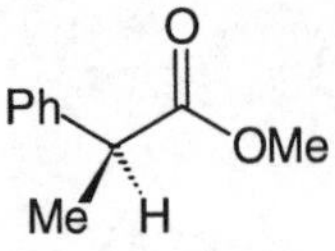

2. The structures of the compounds **3.42**, **3.43**, **3.45**, **3.46**, **3.51**, **3.52**, **3.55–3.59** and **3.61–3.64** are given in Chapter 3, section 3.8. Which of them will undergo thermal racemization?

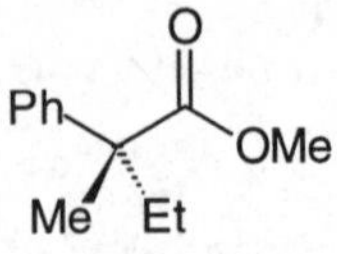

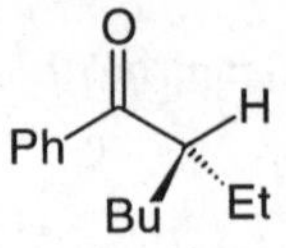

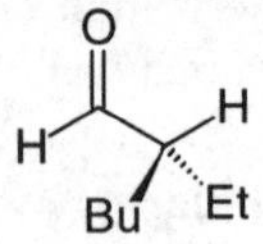

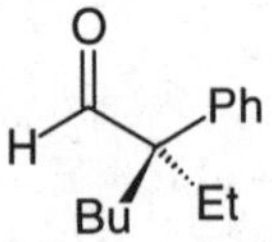

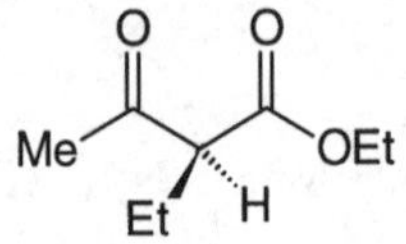

3. The alcohol shown below does not form a conglomerate. Suggest two methods for its resolution, one involving the formation of diastereomers and the other involving a kinetic resolution.

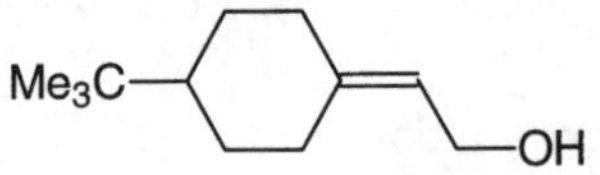

4. The acid shown below can be resolved using (−)-quinine (a naturally occurring, optically active amine). The resolution is carried out in isopropanol and the salt of the (R)-enantiomer of the acid crystallizes from the solution. If the isopropanol solution is then treated with sodium ethoxide, more of the salt of the (R)-enantiomer of the acid is obtained. Draw both a reaction scheme and energy diagrams to account for this chemistry. What would happen if (+)-quinine was used instead of (−)-quinine?

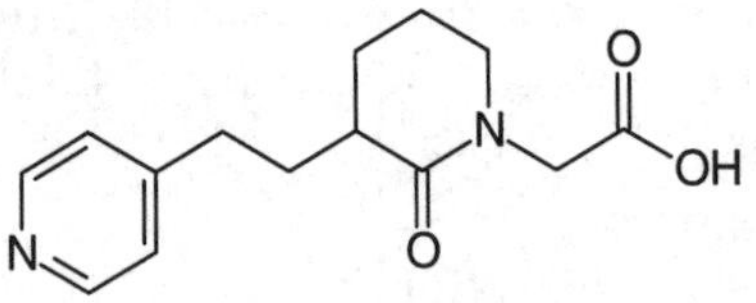

5. In section 5.4.3, it was stated that the reaction of an enantiomerically pure substrate with a racemic chiral derivatizing agent was as effective in proving that no kinetic resolution occurred during the diastereomer formation as the reaction of a racemic substrate with an enantiomerically pure derivatizing agent. Prove that this statement is true.

6. The racemic amino ester shown below can be resolved using (R,R)-dibenzoyl tartaric acid in an ethanol/water solvent system. Under these conditions, the salt of the (S)-enantiomer of the amino ester crystallizes

from the solution. Treatment of the ethanol/water solution with benzaldehyde at 40°C then results in further precipitation of the salt of the
(S)-enantiomer of the amino ester. This sequence can be repeated to
give >70% yield of the salt of the (S)-enantiomer of the amino ester.
Explain what is happening in this process.

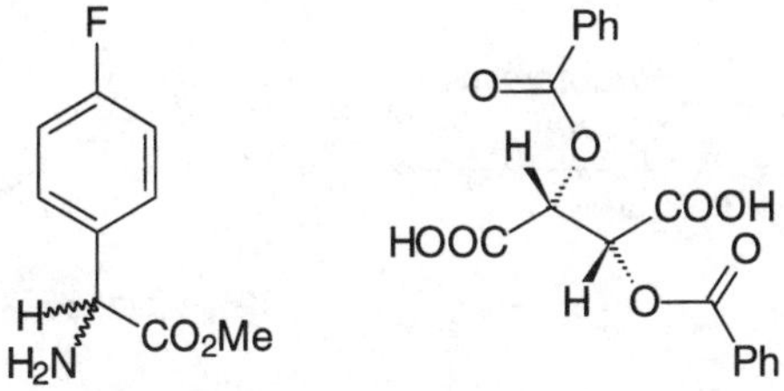

(R,R)-dibenzoyl tartaric acid

7. The hydrogen labelled H_a on compound A shown below is exceptionally
acidic, since enolization forms a heteroaromatic ring system. When
compound A is refluxed in petrol, a single stereoisomer precipitates from
the solution in 73% yield. Explain what is happening during this process.

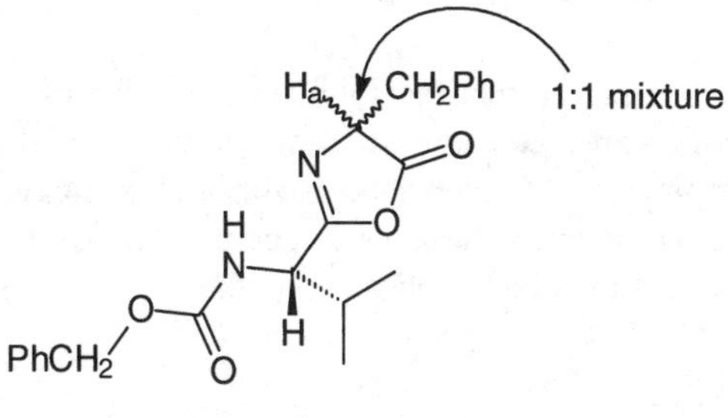

1:1 mixture

A

8. Draw energy/reaction diagrams for: (a) the resolution shown in **Scheme
5.15**; (b) the absorption of a racemic compound onto an achiral chromatography column; (c) the absorption of a racemic compound onto a chiral
chromatography column.

9. Treatment of a mixture of the racemic cyanohydrin and propanal shown
below with the enzyme oxynitrilase gives, at 50% chemical conversion, a
mixture of the (S)-enantiomer of the original cyanohydrin (99% ee) and the
cyanohydrin of propanal. Given that the natural role of the oxynitrilase
enzyme is to catalyse the equilibrium between hydrogen cyanide/benzalde-
hyde and the (R)-enantiomer of the corresponding cyanohydrin, explain this
result.

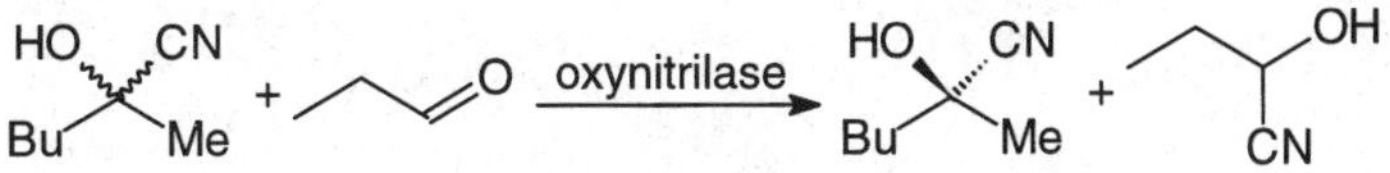

10. This question looks ahead to Chapter 10 but is based upon material covered in this chapter. When ester A or thioester B is treated with an enzyme under the conditions specified, the corresponding carboxylic acid is obtained. Provide an explanation for this transformation, which accounts for both the chemical yield and the enantiomeric excess of the products. **Hint**: for both A and B, H_a is particularly acidic.

A

92% yield; 85% ee

100% yield; 96% ee

B

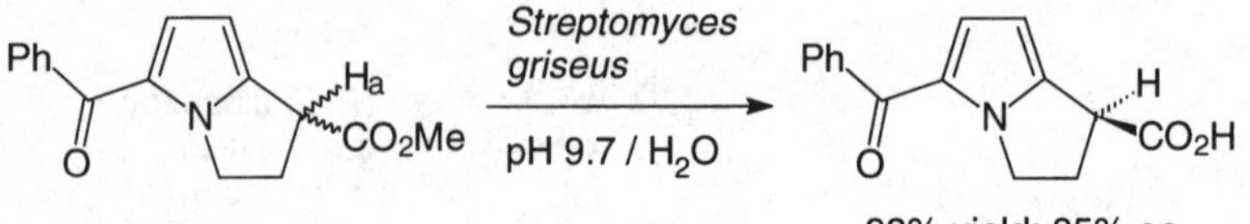

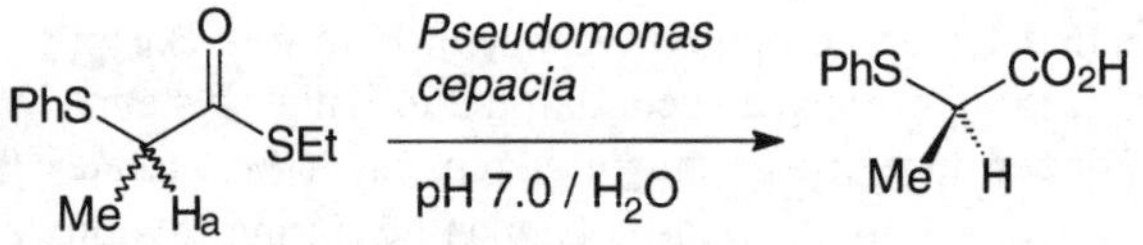

11. It is sometimes observed that the solid state reaction of a racemate with an achiral compound proceeds at a different rate to the solid state reaction of a single enantiomer of the substance with the same achiral compound. Suggest a reason for this difference in chemical reactivity. Could the same effect be observed in solution phase reactions?

6 Molecular symmetry

6.1 Introduction

In the preceding chapters, stereoisomerism has been introduced and the consequences of the presence or absence of stereoisomerism in a molecular system upon the physical and chemical properties of that system have been discussed. This chapter will look more deeply at the arrangement of the groups in chiral and achiral species to answer the question: why are some molecules chiral whilst others are not? It should be remembered (Chapter 4, section 4.1) that the presence of a stereocentre (or any other stereofeature) is *not* sufficient to decide whether or not a molecule is chiral. Compound **6.1**, for example, contains two stereocentres but is an achiral meso compound. One definition of chirality is that a chiral molecule is not superimposable on its mirror image and, in this chapter, the reasons why some molecules are superimposable on their mirror images whilst others are not will be investigated.

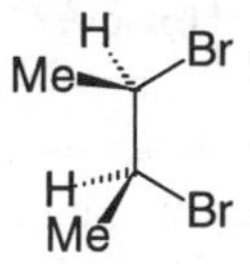

6.1

To address this issue we will investigate the **symmetry** of molecules. Symmetry is a concept which is familiar in everyday life and which can be defined as the regular occurrence of groups within the molecule (or any other object). It will be shown that the regular occurrence of groups within a molecule requires it be possible to carry out operations (rotations and/or reflections) of the molecule which leave the molecule indistinguishable from its original form. These operations are called **symmetry operations**. The geometric features (points, lines or planes) about which symmetry operations are carried out are called **symmetry elements**. It is then possible to group molecules according to the symmetry elements that they contain, to generate a set of **point groups** which classify molecules according to their symmetry.

6.2   Symmetry elements and symmetry operations

6.2.1   Proper axes (C$_n$) and proper rotations (C$_n^k$)

Consider the water molecule **6.2** shown in **Figure 6.1**, in which the two hydrogen atoms have been differently highlighted to allow them to be distinguished. If the molecule is rotated by 180° about the axis indicated in **Figure 6.1**, then the only difference is that the hydrogen atoms exchange location. Since the highlighting of the hydrogen atoms is artificial and they cannot normally be distinguished, the effect of this rotation is to leave the molecule indistinguishable from its original form. The axis shown in **Figure 6.1** is an example of a symmetry element called a **proper axis** and the process of rotation about this axis is a symmetry operation called a **proper rotation**.

The proper axis symmetry element is represented by the symbol C$_n$ where n is 360 divided by the minimum number of degrees through which rotation must occur to produce a structure indistinguishable from the original structure. In the case of water **6.2**, the symmetry element is a C$_2$ axis, since rotation by 180° is required to regenerate the original structure and 360/180 = 2. The corresponding symmetry operation is represented by the symbol C_n^k where k is the number of times the molecule has been rotated through n degrees in a clockwise direction. Thus in the case of the operation shown in **Figure 6.1**, the symmetry operation would be represented by the symbol C_2^1, that is rotate once by 180° clockwise about the C$_2$ axis. A separate symmetry operation would be represented by C_2^2, since this implies rotation about the C$_2$ axis by 2 × 180° (=360°) and so would be equivalent to carrying out no rotation at all. In principle, there are an infinite number of operations associated with a given proper axis, since C_2^3, C_2^4, etc. can all be envisaged. However, since C_2^2 (or more generally C_n^n) returns the molecule to its original orientation, operations corresponding to values of $k > n$ are equivalent to operations corresponding to $k \leq n$. Thus a C$_n$ axis generates n different symmetry operations.

Within the symbol C$_n$, n can take any integer value (and can be infinite), and a molecule may contain more than one proper axis. For example, XeF$_4$ **6.3** is square planar (cf. Chapter 1, section 1.3), and contains a C$_4$ axis as well as four C$_2$ axes as shown in **Figure 6.2** (a molecular model may be helpful in visualizing these). When a molecule contains more than one proper axis, the axis with the highest value of n is referred to as the **principal axis**. Thus in the case of compound **6.3**, the C$_4$ axis would be the principal axis. The C$_4$ axis of compound **6.3** generates four symmetry operations C_4^1, C_4^2, C_4^3 and C_4^4 corresponding to

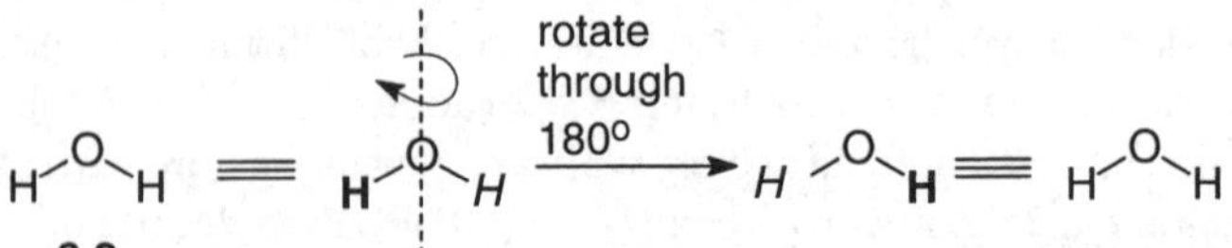

Figure 6.1 The proper rotation axis (C$_2$) in water **6.2**.

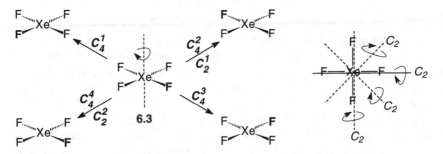

Figure 6.2 The C_4 and four C_2 axes in XeF$_4$ **6.3**.

rotation by 90°, 180°, 270° and 360° respectively around the C_4 axis, as shown by the movement of the emboldened fluorine atom in **Figure 6.2**.

The C_4 axis within compound **6.3** is also a C_2 axis, and the symmetry operation $C_4^2 \equiv C_2^1$; similarly, $C_4^4 \equiv C_2^2$ ($\equiv$ is the mathematical symbol for 'is equivalent to'). Thus both symmetry operations corresponding to the C_2 axis are also operations of the C_4 axis. This will always be the case when two proper axes are coincident, so that for a given axis only the symmetry element of highest n value need be considered. However, when describing symmetry operations, the values of n and k are kept as low as possible so that the symmetry operation C_4^2 would be reported as C_2^1.

Examples of other molecules that contain proper axes are BF$_3$ **6.4**, which is planar (cf. Chapter 1, section 1.3) and contains a C_3 axis as well as three C_2 axes (only one of which is shown in structure **6.4**); benzene **6.5**, which contains a C_6 axis and six C_2 axes (one of which is shown in structure **6.5**); and methane **6.6**, which contains four C_3 axes (one shown in structure **6.6a**) as well as three C_2 axes (one shown in structure **6.6b**). A final example is hydrogen cyanide **6.7**, which is linear and contains a C_∞ axis since rotation by any angle around the axis of the molecule generates a structure indistinguishable from the original. It is strongly recommended that the reader build a model of each of molecules **6.4–6.7** and verify the presence of each of the proper axes.

6.2.2   The identity element (E) and the identical operation (E)

The **identical operation (E)** is the process of doing nothing to a molecule, and the corresponding symmetry element is the **identity element** (E). If no change is made to a molecule (or any other object), then it will always be identical to the original structure, so all molecules (and all objects) possess the identity element and the corresponding identical operation. The words 'identical' and 'indistinguishable' have different meanings when they are used to describe the results of symmetry operations. 'Identical' means that every atom is occupying the same position that it occupied before the operation was carried out, whilst 'indistinguishable' means that some or all of the atoms have moved to positions previously occupied by atoms of the same element (and of the same isotope).

The identity element may seem unnecessary, since all objects possess this symmetry element; however, it is needed in the generation of point groups from symmetry operations as will be discussed in section 6.4. It was shown in section 6.2.1 that a proper rotation operation of C_n^n about a proper axis C_n would return the molecule to its original state, hence $C_n^n \equiv E$.

6.2.3   Plane of symmetry (σ) and reflection operations (σ)

Consider formaldehyde **6.8**, as shown in **Figure 6.3a**. If a mirror is placed along the C=O bond, perpendicular to the plane of the molecule, then the mirror will reflect one half of the molecule onto the other half. In the case of formaldehyde **6.8**, the resulting structure will then be indistinguishable from the original structure other than for the artificial highlighting of the hydrogen atoms in **Figure 6.3a**. Any molecule (or other object) which, when reflected in a plane (the mirror in **Figure 6.3a**), produces a structure which is indistinguishable from the original possesses a symmetry element called a **plane of symmetry** which is represented by the symbol σ. The corresponding symmetry operation is the **reflection operation (σ^n)** where n is the number of times that the molecule has been reflected about the plane. However, if a molecule is reflected twice about

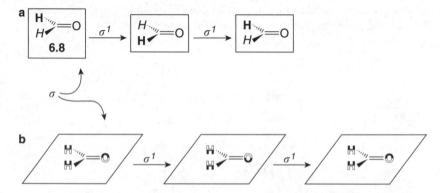

Figure 6.3 The planes of symmetry and reflection operations for formaldehyde **6.8**.

the same plane of symmetry, then all the atoms will be returned to their original positions as shown in **Figure 6.3a**, so $\sigma^2 \equiv E$. Similarly, any value of $n > 2$ will be equivalent to $n = 1$ or $n = 2$, so in practice the reflection operation is restricted to σ^1 which is usually just represented as σ.

A molecule may contain more than one symmetry plane and, indeed, compound **6.8** contains a second symmetry plane, the molecular plane (**Figure 6.3b**). Reflection in this second symmetry plane does not alter the position of any of the atomic nuclei; however, it does alter the position of the electron density which lies above and below the plane of the molecule. This is illustrated in **Figure 6.3b** by the black and white shading of the atoms. The black colouring represents the electron density that is initially above the molecular plane and which is moved by the reflection operation to below the molecular plane. Similarly, the white colouring represents the electron density which is initially below the molecular plane but which the reflection operation moves to above the plane of the molecule. A second reflection in this symmetry plane returns the electron density to its original position.

If a molecule contains both proper axes (cf. section 6.2.1) and symmetry planes, then the planes are classified as σ_v (vertical plane) or σ_d (dihedral plane) if they are parallel to the principal axis or σ_h (horizontal plane) if they are perpendicular to the principal axis. The difference between σ_v and σ_d is that if the molecule contains C_2 axes perpendicular to the principal axis, then a vertical plane also contains one of these axes, whilst a dihedral plane bisects the C_2 axes. One example of each type of symmetry plane is illustrated for benzene in **Figure 6.4**. In total, benzene contains one σ_h, three σ_v and three σ_d planes.

6.2.4 Centre of inversion (i) and the inverse operation (î)

A **centre of inversion** (i) is a point within a molecule (or other object) such that if a line is drawn from any atom within the molecule to the centre of inversion and then continued an equal distance on the other side of the centre of inversion, an atom of the same type is encountered. The centre of inversion is a symmetry element and the corresponding symmetry operation, which moves each atom in a molecule to a point equidistant on the other side of a centre of inversion, is the **inverse operation** ($\hat{i}$).

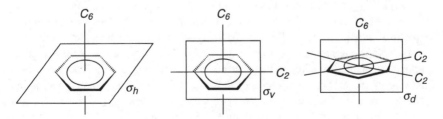

Figure 6.4 Examples of σ_h, σ_v and σ_d planes in benzene.

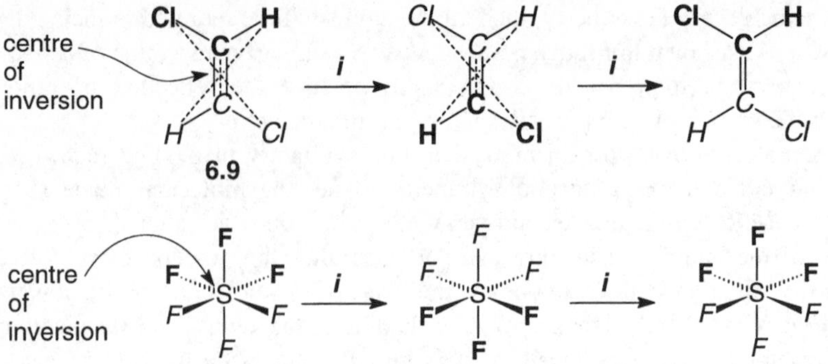

6.9

6.10

Figure 6.5 The centre of inversion and inverse operations for compounds **6.9** and **6.10**.

Two examples of molecules which contain a centre of inversion, (*E*)-1,2-dichloroethene **6.9** and SF_6 **6.10** are shown in **Figure 6.5**. The highlighting of the atoms in structures **6.9** and **6.10** is again artificial to show that particular atoms do move as a result of the inverse operation. It is apparent from **Figure 6.5** that a centre of inversion may or may not coincide with one of the atoms within a molecule, since in the case of compound **6.9** the centre of inversion is the midpoint of the C=C bond whilst in **6.10** the centre of inversion is the sulphur atom. It can also be seen from **Figure 6.5** that $i^2 \equiv E$. That is, that application of the inverse operation twice restores the molecule to its original state. A molecule (or other object) may have only one centre of inversion.

6.2.5 Improper axes (S_n) and improper rotations (S_n^k)

An **improper rotation** (S_n^k) is a symmetry operation in which the molecule (or other object) is rotated k times by $360/n$ degrees about a symmetry element referred to as an **improper axis**, followed by a reflection in a plane perpendicular to that improper axis. Alternatively, the reflection may be carried out before the rotation, i.e. the order in which the two parts of this symmetry operation are carried out is not important in this instance, although this is not generally the case as will be discussed in section 6.4.

An example of a molecule which contains an improper axis is allene **6.11** shown in **Figure 6.6**. In this case, the improper axis is an S_4 axis and is the axis of the C=C=C bonds. Rotation around this axis by 90° does not produce a structure indistinguishable from the starting structure but subsequent reflection in the plane perpendicular to the C=C=C axis does result in a structure indistinguishable from the starting structure (other than for the artificial highlighting of the atoms). An improper axis (S_n) generates n symmetry operations (S_n^k), $k = 1$ to n, if n is even but $2n$ symmetry operations (S_n^k), $k = 1$ to $2n$, if n is odd. The four symmetry operations S_4^1, S_4^2, S_4^3 and S_4^4 of allene **6.11** are all shown in **Figure 6.6**.

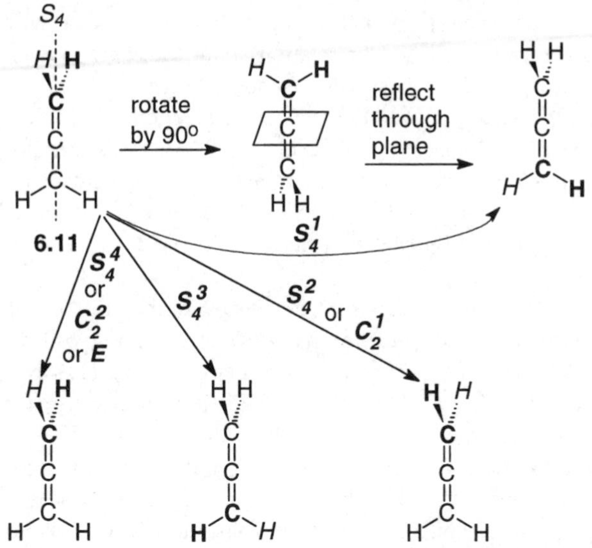

Figure 6.6 The (S_4) improper axis and associated symmetry operations for allene **6.11**.

A number of improper axes are equivalent to other symmetry elements. An S_1 axis, for example, is equivalent to a plane of symmetry $(S_1 \equiv \sigma)$. Since an S_n axis involves rotation by $360/n$ degrees followed by reflection, an S_1 axis involves rotation by $360°$ followed by reflection. However, rotation by $360°$ returns any object to its original orientation, so only the reflection part of the S_1 axis has any effect and $S_1 \equiv \sigma$. Similarly, an S_2 axis is equivalent to a centre of inversion $(S_2 \equiv i$ and $S_2^1 \equiv i)$. This equivalence is best seen by an example, and is illustrated in **Figure 6.7** for alkene **6.9** which was shown in section 6.2.4 to possess a centre of inversion.

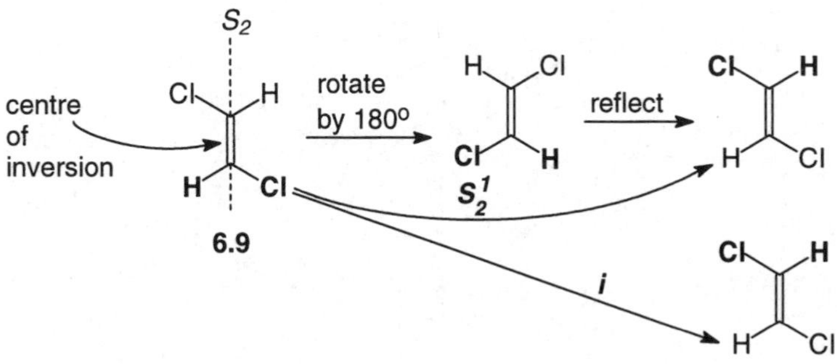

Figure 6.7 The equivalence of the symmetry elements S_2 and i and the corresponding symmetry operations S_2^1 and i illustrated for compound **6.9**.

Carrying out an improper rotation operation twice (S_n^2) will always be equivalent to carrying out a proper rotation about the same axis twice (C_n^2), or more generally, $S_n^k \equiv C_n^k$ if k is even. This equivalence comes about since S_n^2 involves rotation twice by $360/n$ degrees, and reflection twice. However, it was shown in section 6.2.3 that $\sigma^2 \equiv E$, so the reflection parts of an S_n^2 operation cancel out, leaving just the proper rotation C_n^2. This argument can be generalized, since $\sigma^k \equiv E$ provided k is even so $S_n^k \equiv C_n^k$. An example of this equivalence is seen in **Figure 6.6**, where for allene **6.11**, $S_4^2 \equiv C_4^2 \equiv C_2^1$ (cf. section 6.2.1).

If a molecule contains an S_n axis where n is even, then the operation $S_n^n \equiv E$, since both the rotation and reflection parts of the operation will have no effect (rotation by $360°$ and reflection an even number of times). However, if n is odd then $S_n^n \equiv \sigma_h$ and $S_n^{2n} \equiv E$. Thus if n is odd, it is necessary to rotate twice completely around an improper axis to find all of the symmetry operations. This is illustrated in **Figure 6.8**, for PF$_5$ **6.12**, a molecule which has a trigonal bipyramidal geometry (cf. Chapter 1, section 1.3) and possesses an S_3 axis through the phosphorus and axial fluorines. Inspection of the structures shown in **Figure 6.8** reveals the following equivalencies in the symmetry operations: $S_3^2 \equiv C_3^2$; $S_3^3 \equiv \sigma_h$; $S_3^4 \equiv C_3^4 \equiv C_3^1$; and $S_3^6 \equiv C_3^6 \equiv C_3^3 \equiv E$. Thus only S_3^1 and S_3^5 correspond to unique symmetry operations.

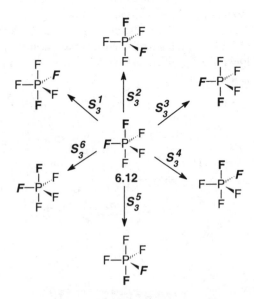

Figure 6.8 The S_3^n operations of PF$_3$ **6.12**.

6.3   Symmetry of molecules with multiple conformations

So far in this chapter, only the symmetry of compounds with a single conformation has been considered. However, most chemicals can exist in more than one
conformation and, in this case, each conformation may have different symmetry
elements. As an example, consider ethane **6.13**, for which there are an infinite
number of conformations (cf. Chapter 1, section 1.5.2) which fall into three
groups each with different symmetry elements as shown in **Figure 6.9**. The
three equivalent staggered conformations **6.13a** each contain the symmetry
elements (E, C_3, S_6, $3C_2$, i, $3\sigma_d$), whilst the three equivalent eclipsed conformations **6.13b** contain the symmetry elements (E, C_3, S_3, $3C_2$, $3\sigma_v$, σ_h). Any of the
infinite number of conformations which are neither staggered nor eclipsed is
represented by structure **6.13c** and each contains just the symmetry elements (E,
C_3, $3C_2$). A molecular model will be very useful in visualizing these symmetry
elements.

Thus, whenever the symmetry of a substance with multiple conformations is
being considered, the symmetry of each conformation must be considered as a
separate static structure. Even this is an approximation since molecules vibrate
and this can lead to conformations interconverting. A more accurate analysis,
however, requires a quantum mechanical approach which is beyond the scope of
this text.

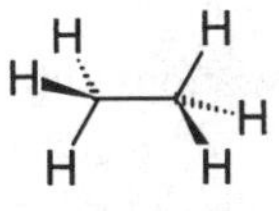

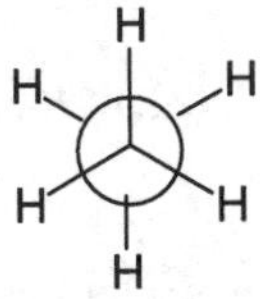

6.13a

S_6 and C_3 along C-C bond
i at midpoint of C-C
σ_d containing each planar H-C-C-H unit
C_2 through midpoint of C-C and
    perpendicular to each H-C-C-H plane

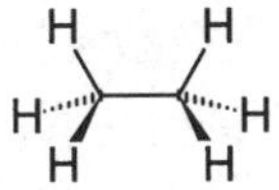

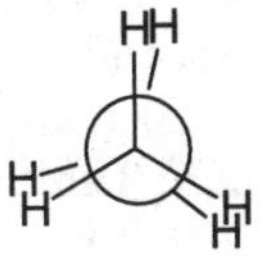

6.13b

S_3 and C_3 along C-C bond
σ_v containing each planar H-C-C-H unit
σ_h perpendicular to C-C through
    midpoint of C-C
C_2 through midpoint of C-C and
    contained within each H-C-C-H plane

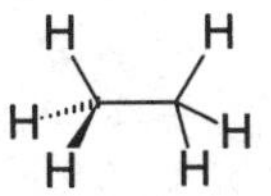

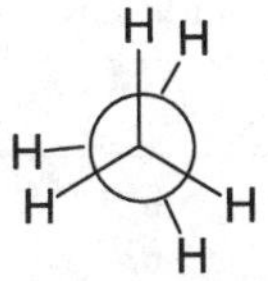

6.13c

C_3 along C-C bond
C_2 through midpoint of and perpendicular
    to C-C and bisecting each H-C-C-H

Figure 6.9 The symmetry elements present in the staggered, eclipsed and neither staggered nor
eclipsed conformations of ethane **6.13**.

6.4 Point groups

Point groups are a way of classifying molecules according to the symmetry operations that can be carried out on them. This is useful since chemically diverse molecules contain the same symmetry elements. For example, water **6.2**, formaldehyde **6.8**, dichloromethane **6.14** and (Z)-1,2-dibromoethene **6.15** all contain the symmetry elements E, C_2 and $2\sigma_v$ and so would be grouped together in the same point group, which in this case is called $\mathbf{C_{2v}}$.

6.2   **6.8**   **6.14**   **6.15**

A point group is the collection of all possible symmetry operations that can be carried out on a system. In order to be valid the group must obey four rules:

1. The product of any two symmetry operations must also be a member of the group. Thus if P and Q represent symmetry operations present in the group, then carrying out operation P followed by operation Q must give a symmetry operation R where R is also a member of the group. Or in the form of an equation, $QP = R$. Note that, in general, the order in which the two operations is carried out is important, i.e. $QP \neq PQ$ in most cases.
2. The group must contain the identical operation E, such that the product of E and any other operation P which is a member of the group must leave P unchanged. i.e. $EP = PE = P$.
3. The associative law must hold. That is, if P, Q and R are all operations within the group, then $P(QR) = (PQ)R$.
4. Each operation R must have an inverse operation R^{-1} which is also a member of the group. The inverse operation is defined such that $R^{-1}R = E$. Note that the symmetry operations E, i and σ are their own inverses.

To see how these four rules are applied, consider the symmetry operations of any of the four molecules **6.2**, **6.8**, **6.14** or **6.15**, each of which contains the symmetry operations E, C_2^1, σ_v and σ_v' where σ_v, and σ_v' represent the reflection operations in the two different planes. A group multiplication table can be constructed for these four operations as shown in **Table 6.1**.

Table 6.1 The group multiplication table for point group C_{2v}. The first operation is given along the horizontal axis, and the second operation along the vertical axis

C_{2v}	E	C_2^1	σ_v	σ_v'
E	E	C_2^1	σ_v	σ_v'
C_2^1	C_2^1	E	σ_v'	σ_v
σ_v	σ_v	σ_v'	E	C_2^1
σ_v'	σ_v'	σ_v	C_2^1	E

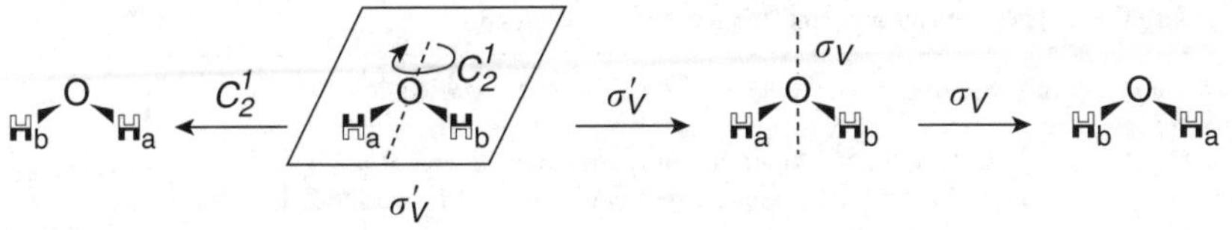

Figure 6.10 An illustration that for water **6.2**, $\sigma_v \times \sigma_v' = C_2^1$.

For this to be a valid group, it must obey the four rules given above. The group appears to obey the first rule, since the product of any two members of the group is always E, C_2^1, σ_v or σ_v', which are the four members of the group. An example showing that $\sigma_v \times \sigma_v' = C_2^1$ is shown for the water molecule **6.2** in **Figure 6.10**. In **Figure 6.10**, the hydrogen atoms have again been coloured partly black and partly white to illustrate the movement of the electron density which is initially above and below the plane of the molecule as occurs during both the σ_v' and C_2^1 operations. Similar diagrams can be constructed to show that the other operation products given in **Table 6.1** are also valid. Note that for this point group the order in which the two operations is carried out is not important, i.e. $\sigma_v \times \sigma_v' = C_2^1$ and $\sigma_v' \times \sigma_v = C_2^1$, etc. However, this is not generally true in other point groups as will be seen in the problems set at the end of this chapter.

The group also obeys the second rule since it contains the identical operation E, and for any of the four symmetry operations (R) that constitute the group, the product of R and E is R. Thus $E \times C_2^1 = C_2^1 \times E = C_2^1$; $E \times \sigma_v = \sigma_v \times E = \sigma_v$; and $E \times \sigma_v' = \sigma_v' \times E = \sigma_v'$. It can also be seen from the group multiplication table that the group obeys the third rule, since, in determining the product of any three operations, the order in which they are carried out is irrelevant. For example, the result of carrying out σ_v followed by σ_v' is C_2^1. The subsequent application of C_2^1 is then E. This series of operations would be represented by $C_2^1(\sigma_v' \times \sigma_v)$. To carry out the sequence $(C_2^1 \times \sigma_v') \times \sigma_v$, the result of carrying out σ_v' followed by C_2^1 is first determined and found to be σ_v. Thus $(C_2^1 \times \sigma_v') \times \sigma_v \equiv \sigma_v \times \sigma_v = E$. Hence, the order in which the three operations are carried out does not matter. The same independence of the order in which they are carried out would have been obtained whichever three operations were chosen.

Finally, the inverse of each operation is also a member of the group, so the group obeys the fourth rule. It was shown in section 6.2.3 that $\sigma^2 \equiv E$, so σ_v and σ_v' are their own inverses. Similarly, it was shown in section 6.2.4 that $i^2 \equiv E$, so i is its own inverse. A C_2^1 operation is also its own inverse since C_2^1 involves a rotation by 180°, so $C_2^1 \times C_2^1 \equiv C_2^2 \equiv E$. Thus, in this case, each member of the group is its own inverse, and the point group C_{2v} obeys all of the rules for a valid group.

The point group C_{2v} is only one example of a point group; a list of all possible point groups is given in **Table 6.2**. Also given in **Table 6.2** is a list of the symmetry elements which a molecule must contain to belong to that point

Table 6.2 The point groups and their symmetry elements

Point group	Symmetry elements
C_n	E, C_n (if $n = 1$, then the only symmetry element is E)
D_n	E, C_n, and n C_2 axes perpendicular to C_n (n must be >1)
C_s	E, σ
C_i	E, i
C_{nv}	E, C_n, and n σ_v (n must be >1 and can be infinite)
C_{nh}	E, C_n, σ_h (n must be >1)
D_{nh}	E, C_n, n C_2 axes perpendicular to C_n, σ_h (n must be >1 and can be infinite)
D_{nd}	E, C_n, n C_2 axes perpendicular to C_n, n σ_d (n must be >1)
S_n	E, S_n (n must be even, otherwise the group is equivalent to C_{nh})
T_d	E, 4 C_3, 3 S_4, 6 σ_d
O_h	E, 4 S_6, 3 S_4, 6 C_2, i, 3 σ_h, 6 σ_d
I_h	E, 6 C_5, 10 C_3, 15 C_2, 15 σ

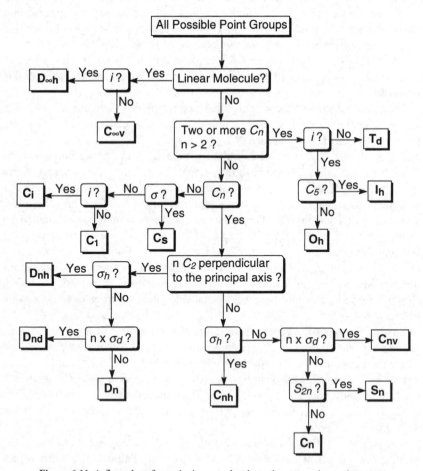

Figure 6.11 A flow chart for assigning a molecule to the appropriate point group.

group. The symmetry elements given in **Table 6.2** may not be the only ones possessed by a particular point group; however, they are the ones that are sufficient to distinguish the point group from all other point groups. The easiest way to assign a molecule to the appropriate point group is to use a flow chart such as that shown in **Figure 6.11**.

6.5   Symmetry and chirality

The preceding sections have introduced some of the important concepts associated with molecular symmetry and point groups. This was done since it is possible to use symmetry considerations to determine whether or not a molecule will be chiral. By definition, a chiral object is not superimposable on its mirror image, where the mirror image of an object is its reflection. However, for any molecule which contains a plane of symmetry (σ), the corresponding symmetry operation (the reflection operation σ) will superimpose the molecule on its reflection (section 6.2.3). Thus, any molecule which contains a plane of symmetry must be achiral.

In fact, any molecule which contains an improper axis of rotation (S_n axis) will be achiral. The reason for this is that an improper axis of rotation is the product of a proper rotation (C_n) followed by a reflection (σ) (section 6.2.5). It was also shown in section 6.2.5 that a plane of symmetry (σ) and centre of inversion (i) are just special cases of an S_n axis (S_1 and S_2 respectively). The reflection part of these operations will always cause the molecule to be superimposable on its mirror image and hence achiral. This provides an alternative definition of chirality: **Any molecule (or other object) which in all accessible conformations does not contain an improper axis of rotation (S_n, $n > 0$) is chiral.**

A consequence of the dependence of chirality on the absence of an S_n axis is that any molecule that belongs to a point group which requires the presence of an S_n axis (or a plane of symmetry or centre of inversion) will be achiral, and a molecule which belongs to any other point group will be chiral. As the only point groups which do not require the presence of an S_n axis are the C_n (including C_1) and D_n groups, all chiral molecules must belong to one of these groups and all species which do belong to one of these point groups will be chiral.

Cis-1,3-dichlorocyclobutane **6.16** is an example of a molecule which contains an S_1 axis (i.e. a plane of symmetry) and so is achiral. In structure **6.16**, the cyclobutane ring is drawn in a planar conformation for clarity. It will be shown in Chapter 8 that this ring is not actually planar but, in this case, the molecule will still contain a plane of symmetry and so be achiral. Similarly, the six-membered ring within compound **6.17** is planar, so this molecule contains an S_2 axis (i.e. a centre of inversion) provided the hydrogen atoms on the methyl groups are correctly orientated or not considered. Whilst S_1 and S_2 axes are easy to identify, other improper axes of rotation are easily overlooked. For example, the ammonium salt **6.18** contains an S_4 axis and so is achiral.

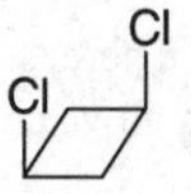

6.16 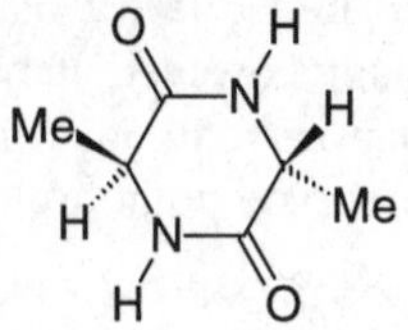   **6.17** 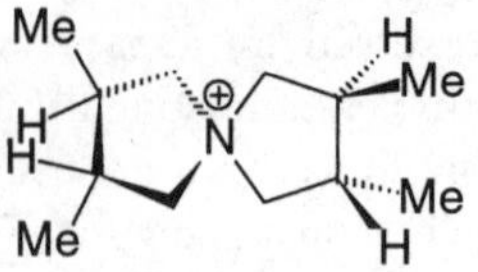   **6.18**

To illustrate the way in which symmetry considerations apply to meso compounds (cf. Chapter 4, section 4.1), consider compound **6.1** which is achiral despite containing two stereocentres. The conformation **6.1a** contains a plane of symmetry and so is achiral; however, this is an eclipsed conformation (cf. Chapter 1, section 1.5.2) which will have only a very small population. The three staggered conformations which are the three minimum energy conformations and hence are the three conformations with appreciable populations of compound **6.1** are shown in structures **6.1b–d**. Conformation **6.1c** contains a centre of inversion and so is achiral. Conformations **6.1b** and **6.1d**, however, do not possess an S_n axis and so are chiral. These two conformations are, however, mirror images of one another, and hence enantiomers. In general, if any accessible conformation of a molecule contains an S_n axis, then this is sufficient to make the molecule achiral, so compound **6.1** like all meso compounds is achiral. A meso compound will always have at least one accessible conformation which possesses an S_n axis and, for all accessible conformations which do not possess an S_n axis, the enantiomeric conformation will also be accessible.

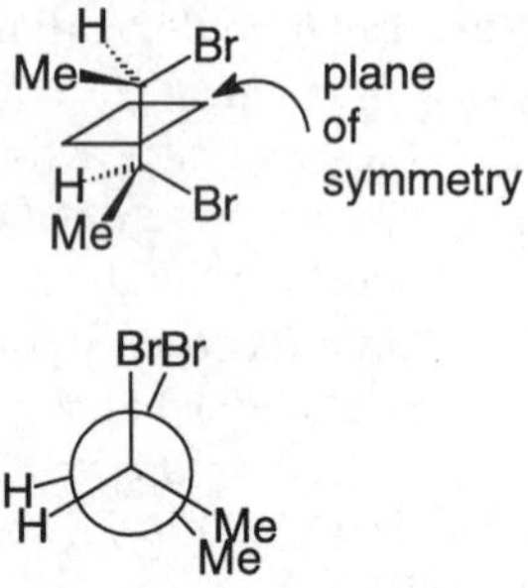

6.1a 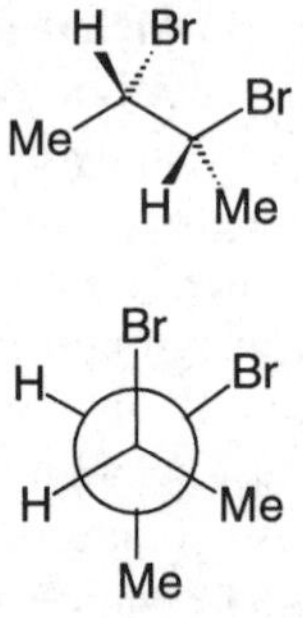   **6.1b** 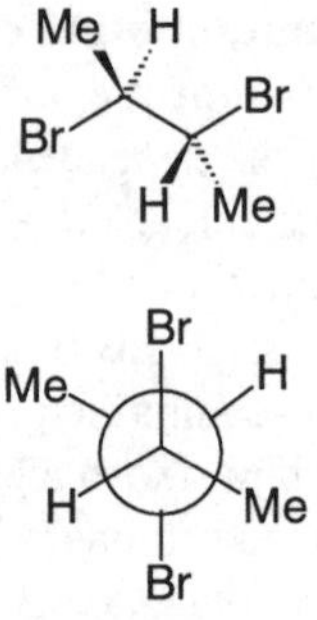   **6.1c** 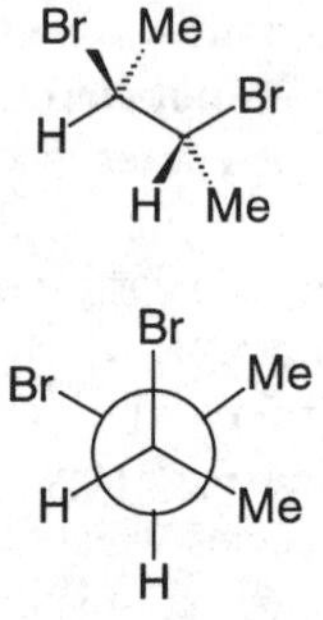   **6.1d**

As a final example of the relationship between symmetry and chirality, consider biphenyl derivative **6.19**, a molecule which is chiral and possesses a stereoaxis (cf. Chapter 3, section 3.8.1). In this case, the two enantiomeric conformations **6.19a** and **6.19b** do not possess an S_n axis, and the planar conformations **6.19c** and **6.19d** which do contain a plane of symmetry are at such a high energy as to be unpopulated at ambient temperatures. Hence, in this case (and in the case of all atropisomers), none of the conformations which are

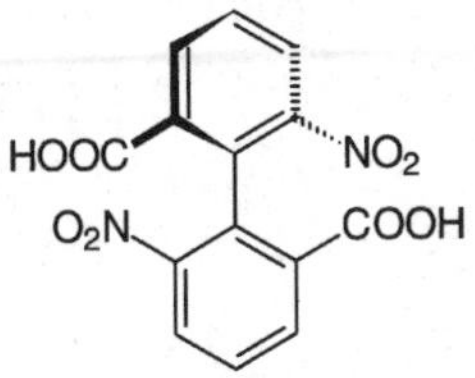

6.19a 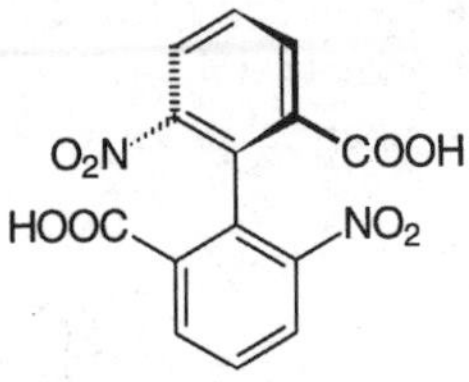   **6.19b**

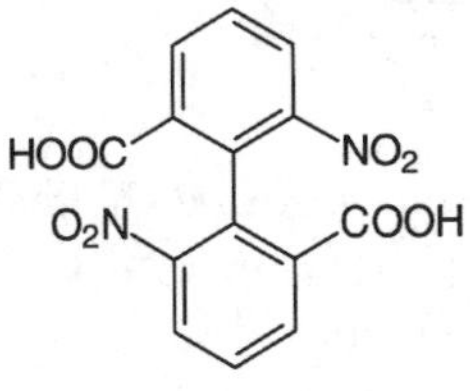

6.19c 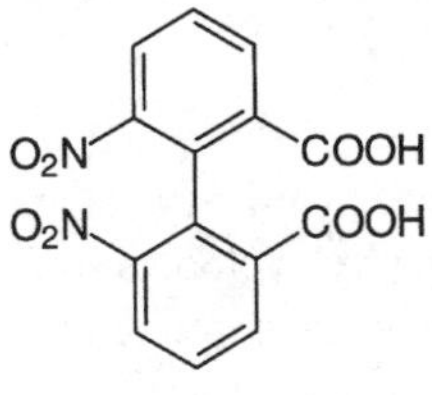   **6.19d**

accessible at ambient temperature contains an S_n axis, so the compounds are chiral. At higher temperatures, however, the planar conformations **6.19c** and **6.19d** will become populated, so two of the conformations available to the molecule will possess an S_n axis, and the compound will thus be achiral.

6.6 Further reading

General

Stereochemistry of Organic Compounds E.L. Eliel and S.H. Wilen. Wiley: London, 1994, chapter 4.

Group Theory and Chemistry D.M. Bishop. Oxford University Press: London, 1973, chapters 1–3.

Chemical Applications of Group Theory 3rd edn, F.A. Cotton. Wiley: Chichester, 1990, chapters 1–3. Chapter 11 of this text also extends the concepts of molecular symmetry to crystallographic symmetry.

Symmetry and Structure S.F.A. Kettle. Wiley: Chichester, 1985.

6.7 Problems

1. For each of the following molecules identify each symmetry element and the corresponding symmetry operations; hence determine the point group to which the structure belongs. (a) Bromobenzene; (b) 1,2-difluorobenzene; (c) 1,3-dichlorobenzene; (d) 1,4-dibromobenzene; (e) 4-fluoro-bromobenzene; (f) 1,3,5-tribromobenzene; (g) 1,2,3-trifluorobenzene; (h) 1,2,4-trichlorobenzene; (i) $POCl_3$; (j) CH_4; (k) CH_3Cl; (l) CH_2Cl_2; (m) $CHCl_3$; (n) CCl_4; (o) PF_5; (p) BHDCl; (q) $Mo(CO)_6$; (r) *cis*-$PtBr_4Cl_2$.

2. Determine the symmetry elements and hence the point groups of ferrocene A
   and *bis*(benzene)chromium B if the two aromatic rings are eclipsed, staggered, and neither eclipsed nor staggered.

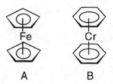

A   B

3. Determine the symmetry elements present in 1,3-dichloroallene A and
   1,3,5,7-tetramethylcyclooctatetraene B, and hence determine whether or not
   these molecules are chiral. In the case of compound B, ignore the hydrogen
   atoms in the methyl groups.

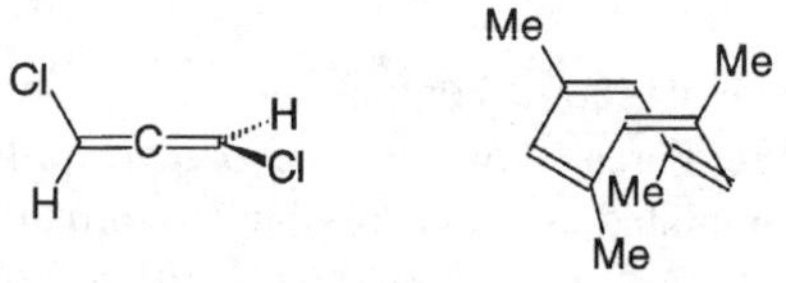

A   B

4. Show that ammonia contains the following symmetry elements: E, C_3, and
   $3 \times \sigma$. Hence determine which point group ammonia belongs to and construct the group multiplication table for this point group. In section 6.4, it was
   shown that for the point group C_{2v} $PQ = QP$ if P and Q are both symmetry
   operations contained in the group. Does this relationship hold true for any or
   all of the symmetry operations in the point group to which ammonia belongs?
   Use your group multiplication table to determine the inverse operation of
   each of the symmetry operations within this point group.
5. Show that each of the chiral molecules discussed in Chapter 3 belongs to a
   C_n or D_n point group.
6. The compound shown below was discussed in Chapter 4. Determine the point
   group to which it belongs and hence whether it is chiral or not. Assume that
   the two triphenylphosphine and the two but-2-yne ligands adopt identical
   conformations.

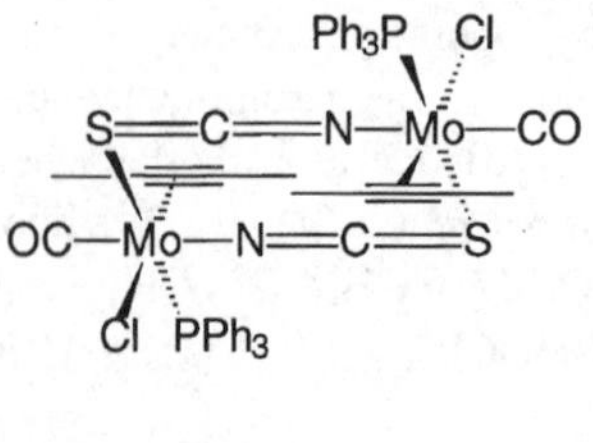

7. Consider a point at cartesian coordinates (x,y,z). What will the coordinates of the point be after the application of each of the following symmetry operations?
   (a) C_2^1 about the z-axis;
   (b) C_2^1 about the y-axis;
   (c) C_2^1 about the x-axis;
   (d) σ in the x,y-plane;
   (e) σ in the x,z-plane;
   (f) σ in the y,z-plane;
   (g) i if the centre of inversion is at the origin $(0,0,0)$.
   Use these mathematical relationships to prove that $S_2^1 \equiv i$; and that $S_4^2 \equiv C_2^1$.

7 Topism and prostereogenicity

So far in this book, the stereochemical relationships between molecules have been discussed. In this chapter, however, we will examine the stereochemical relationships between groups within the same molecule.

7.1 Homotopic, enantiotopic and diastereotopic groups

Two groups of identical constitution (i.e. containing the same atoms connected by the same bonds) within a molecule are said to be **homotopic** if making a change to first one of the groups then, separately, to the other group gives two structures which are indistinguishable from one another. Similarly, two groups within a molecule are said to be **enantiotopic** if making a change to first one of the groups then to the other group gives two structures which are enantiomers of one another. Finally, two groups within a molecule are said to be **diastereotopic** if making a change to first one of the groups then to the other gives two structures which are diastereomers of one another. The term **heterotopic** is used to describe groups of identical constitution which are not homotopic, i.e. which are either enantiotopic or diastereotopic.

These definitions are best understood by a series of examples. Consider first dichloromethane **7.1**: if first one of the hydrogen atoms (H_a) of dichloromethane is changed to a group X, then structure **7.2** is obtained as shown in **Figure 7.1**. If the other hydrogen atom (H_b) is changed to the same group X, then structure **7.2** is again obtained. Thus the two hydrogen atoms of dichloromethane are homotopic. The same argument will show that the two chlorine atoms within

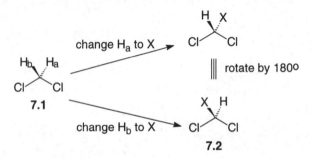

Figure 7.1 Illustration that the hydrogen atoms in dichloromethane are homotopic.

dichloromethane are also homotopic. Note that this procedure of changing the two groups to some other group X is a thought experiment; there is no requirement that the chemistry should actually be possible.

As a second example, consider ethanol **7.3**. The three hydrogens of the methyl group are all homotopic, since changing in turn each of them to a group X will give structures which are interconvertible by rotation around the carbon–carbon bond. The two hydrogens attached to the methylene group (H_a and H_b) are, however, not homotopic since replacement of each of them in turn by a group X (where X ≠ H, OH or CH_3) gives structures **7.4** and **7.5** which are enantiomers of one another as shown in **Figure 7.2**. Thus the two hydrogens attached to the methylene group are enantiotopic.

The carbon atom of the methylene group of ethanol is called a **prochiral centre**. In general, a prochiral centre is an atom to which two heterotopic groups are attached (or which possesses two heterotopic faces, cf. section 7.2) and which becomes a stereocentre when one of the two heterotopic groups is changed to a different group not already present in the molecule (or when a new group is added to one of the heterotopic faces). In a similar way, it is possible to define a **prochiral axis** and a **prochiral plane**. Thus the two hydrogen atoms of allene **7.6** are enantiotopic, and the C=C=C bond is a prochiral axis. Similarly, the hydrogens shown in paracyclophane **7.7** are enantiotopic and the plane of the aromatic rings is a prochiral plane. In the examples shown in structures **7.3**, **7.6** and **7.7**, the enantiotopic groups have been single atoms – hydrogen atoms. This is not always the case, any two groups if appropriately located within a molecule may be enantiotopic. Thus the two methyl groups of 2-bromopropane **7.8** are also enantiotopic.

Consider next the relationship between the two hydrogen atoms (H_a and H_b) which are part of the methylene group in (S)-aspartic acid **7.9**. Replacement of each of these two hydrogen atoms in turn by a group X (where X is different to any of the other three ligands attached to the carbon atom of the methylene

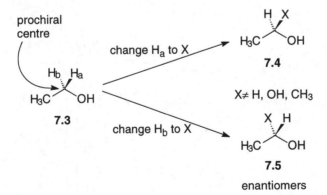

Figure 7.2 Illustration that the hydrogen atoms attached to the methylene group in ethanol are enantiotopic.

7.6   **7.7**   **7.8**

group) generates structures **7.10** and **7.11** which are diastereomers of one another as shown in **Figure 7.3**. Thus the two hydrogens within the methylene group are diastereotopic and the carbon atom of the methylene group is a prochiral centre. In this case, the two hydrogen atoms (H_a and H_b) are diastereotopic because of the presence of a stereocentre elsewhere in the molecule. However, the configuration of the stereocentre is not important in determining whether or not the hydrogen atoms are diastereotopic: the corresponding hydrogens in (R)-aspartic acid and in the racemate of aspartic acid would also be diastereotopic.

It is also possible to find diastereotopic groups in achiral molecules, as the hydrogen atoms labelled H_a and H_b in cyclobutyl bromide **7.12** and propene **7.13** illustrate. In these cases, however, the carbon atom bearing the diastereotopic groups is not a prochiral centre since it is not converted into a stereocentre when one of the diastereotopic groups is changed to a new group X. The phrase **prostereogenic centre** (and the analogous **prostereogenic axis** and **prostereogenic plane**) is used to describe an atom which is attached to two heterotopic ligands whether or not the atom becomes a stereocentre when one of the heterotopic ligands is changed to a different group. Thus the carbon atoms attached to H_a and H_b in structures **7.12** and **7.13** are prostereogenic centres even though they are not prochiral centres.

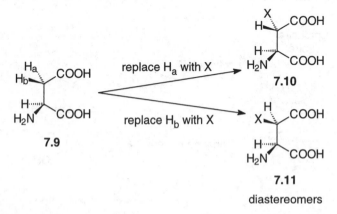

diastereomers

Figure 7.3 Illustration that the hydrogen atoms attached to the methylene group in (S)-aspartic acid are diastereotopic.

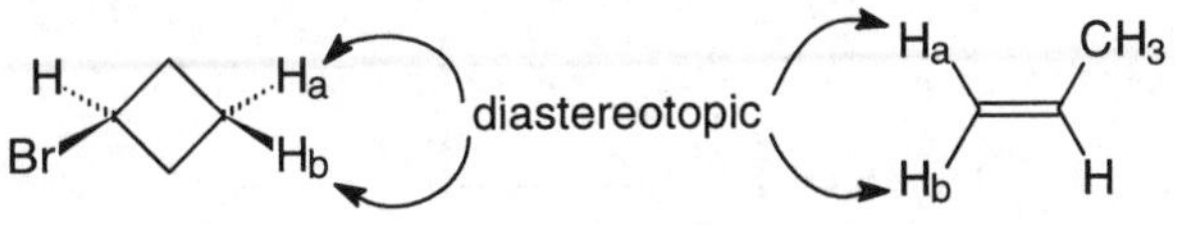

7.12   **7.13**

In the above discussion, it has been shown that the topicity of two or more groups can be determined by a substitution process in which the groups are separately changed to a new group X. An alternative method for determining the topicity of a set of groups is to consider the symmetry elements present within the molecule, and the effect of the corresponding symmetry operations (cf. Chapter 6) on the groups whose topicity is to be determined. Any set of groups of identical constitution will be homotopic if their locations can be interchanged by a rotation around a proper axis (C_n). A set of groups will be enantiotopic if their locations cannot be interchanged by a rotation around a proper axis, but can be interchanged by a rotation around an improper axis (S_n) including the special cases of improper axes, a plane ($\sigma \equiv S_1$) or a centre of inversion ($i \equiv S_2$). Finally, a set of groups will be diastereotopic if their locations cannot be interchanged by any symmetry operations.

To illustrate the application of molecular symmetry to the determination of topicity, consider again dichloromethane **7.1**. It has already been shown, using the substitution method, that both hydrogens in dichloromethane are homotopic, as are both chlorines. Dichloromethane belongs to the point group $\mathbf{C_{2v}}$ and so contains a C_2 axis (as well as two planes of symmetry and the identity element). The C_2 axis is illustrated in **Figure 7.4**, and it can be seen that rotation by 180° around this axis (i.e. carrying out the symmetry operation C_2^1) interchanges the position of the two hydrogens and the position of the two chlorines, thus making the two hydrogen atoms and the two chlorine atoms homotopic.

Just because a molecule contains a proper axis of rotation, it does not follow that all groups of identical constitution within the molecule will be homotopic; rotation around the proper axis must interchange the positions of the groups. A good example of this is seen in compound **7.14**. This molecule again contains a C_2 axis, rotation around which interchanges the positions of the two hydrogens shown in normal type (**Figure 7.5**), so these two hydrogens are homotopic. However, H_a and H_b are not homotopic since rotation around the C_2 axis does not interchange their positions. A 180° rotation around the C_2 axis does,

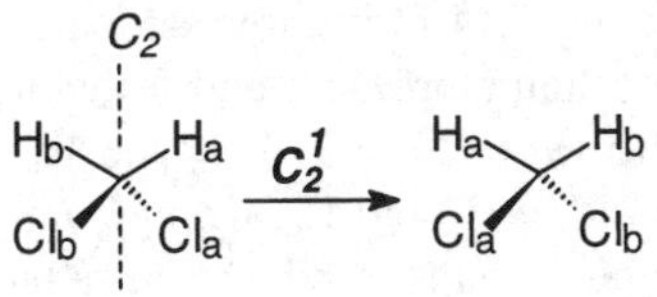

7.1

Figure 7.4 The effect of the symmetry operation C_2^1 on the positions of the atoms in dichloromethane.

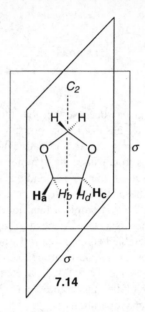

Figure 7.5 The relationships between the hydrogen atoms in compound **7.14**.

however, interconvert the positions of H_a and H_c so these two hydrogen atoms (shown in bold) are homotopic, and similarly H_b and H_d (shown in italics) are homotopic as their positions are also interchanged. Compound **7.14** also contains two planes of symmetry, the first of which is in the plane of the paper. Reflection in this plane of symmetry interchanges the positions of H_a and H_b so these two hydrogens are enantiotopic. Reflection in the same plane also interchanges the positions of H_c and H_d so these two hydrogens are also enantiotopic. The positions of the two hydrogen atoms shown unemphasized are also interchanged by reflection in this plane, but these two hydrogens have already been shown to be homotopic. This illustrates that for two groups to be enantiotopic they must be interconverted by rotation around an improper axis but not by rotation around a proper axis.

The second plane of symmetry in compound **7.14** is located along the C_2 axis but is orthogonal to the first plane. Reflection in this plane of symmetry interchanges the positions of H_a and H_d and also interchanges the positions of H_b and H_c making each of these pairs of hydrogens also enantiotopic. Hence, the hydrogen atoms in compound **7.14** form three sets: the two hydrogens shown in normal typeface which are homotopic; H_a and H_c which are homotopic to one another but enantiotopic to H_b and H_d; and H_b and H_d which are homotopic to one another but enantiotopic to H_a and H_c. Compound **7.14** also illustrates that homotopic and enantiotopic groups need not be attached to the same atom within a molecule, and the same is true of diastereotopic groups.

The two methylene hydrogens (H_a and H_b) in (S)-aspartic acid **7.9** cannot be interconverted by any symmetry operations and so are diastereotopic. In this case, the molecule belongs to point group C_1 and so does not contain any symmetry elements other than the identity element. Molecules **7.12** and **7.13** both possess a single plane of symmetry (and belong to point group C_s); however, reflection in this plane does not interconvert the positions of the hydrogen atoms labelled H_a and H_b, so these atoms are again diastereotopic.

7.2   Homotopic, enantiotopic and diastereotopic faces

The previous section focused upon the topicity of groups of atoms usually attached to tetrahedral, sp^3 hybridized atoms. However, it is also possible to describe the two faces of a molecule (which will almost always contain a double bond) as being homotopic, enantiotopic or diastereotopic. Whereas the topicity of groups was investigated by substitution of the groups by another group X, the topicity of faces is investigated by the addition of a group X to first one face of the molecule, then separately to the other face. Alternatively, exactly the same symmetry considerations used to determine the topicity of groups can be used to determine the topicity of faces. Three examples will illustrate the topicity of molecular faces.

Consider first propanone **7.15**: the addition of a group X to both ends of the C=O bond from the top face of the molecule generates structure **7.16**. The addition of group X to both ends of the C=O bond from the bottom face of the molecule generates a structure which is indistinguishable from **7.16**, after a 180° rotation as shown in **Figure 7.6**. Hence, the two faces of propanone are homotopic. The same conclusion would have been reached by realizing that propanone possesses a C_2 axis (along the C=O bond), and that a 180° rotation about this C_2 axis interconverts the top and bottom faces of the molecule.

As a second example, consider ethanal **7.17**. The addition of a group X to both ends of the C=O bond from the top face generates structure **7.18**, whilst the corresponding addition to the bottom face of **7.17** forms structure **7.19**. However, structures **7.18** and **7.19** do not represent the same structure (provided X ≠ H, CH_3, or OX), rather they represent a pair of enantiomers as shown in **Figure 7.7**. Thus the two faces of ethanal are enantiotopic and the carbon atom

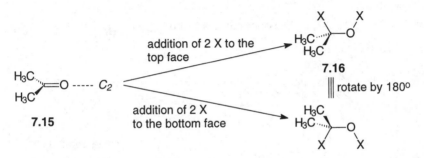

Figure 7.6 Illustration that the top and bottom faces of propanone **7.16** are homotopic.

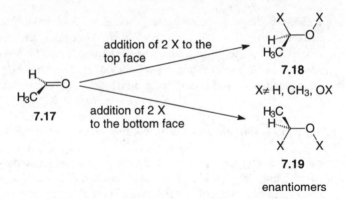

Figure 7.7 Demonstration that the two faces of ethanal 7.17 are enantiotopic.

of the carbonyl bond is a prochiral centre. Note, however, that, whilst the oxygen atom is also a prostereogenic centre, it is not a prochiral centre since, although it possesses two enantiotopic faces, it is not converted into a stereocentre by the addition of group X. That the two faces of ethanal are enantiotopic could also have been determined by considering the symmetry of the molecule. Compound 7.17 does not possess a C_n axis, so the two faces cannot be homotopic. The molecule does, however, contain a plane of symmetry (the molecular plane), and reflection in this plane exchanges the positions of the top and bottom faces, hence the two faces are enantiotopic.

The final example to be considered here is 3-bromo-cyclobutanone 7.20. The addition of a group X to the top or bottom face of the C=O of compound 7.20 generates structures 7.21 and 7.22 respectively, which are diastereomers of one another as shown in Figure 7.8. Hence the two faces of compound 7.20 are diastereotopic. Neither the carbon nor the oxygen atom of the C=O group is a prochiral centre in this case, since neither of them is converted into a stereocentre in compounds 7.21 and 7.22. Both atoms in the C=O bond are, however, prostereogenic centres. The same conclusion would have been reached by considering the symmetry of compound 7.20, since there is no symmetry

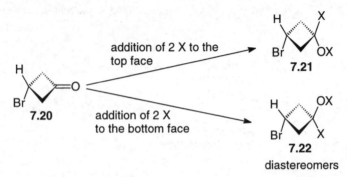

Figure 7.8 Demonstration that the two faces of compound 7.20 are diastereotopic.

operation that can be carried out on this compound which would interconvert the two faces of the carbonyl bond.

7.3 Physical and chemical properties of heterotopic groups and faces

The relationship between homotopic, enantiotopic and diastereotopic groups or faces is closely analogous to the relationship between achiral, enantiomeric and diastereomeric compounds. Thus enantiomeric groups or faces just like enantiomers cannot be distinguished by any physical or chemical technique in an achiral environment. Diastereotopic groups, however, just like diastereomers can normally be distinguished by spectroscopic techniques, most notably by NMR as will be discussed below. Diastereotopic groups and faces may also undergo chemical reactions with achiral reagents at different rates, a point which will be discussed in detail in Chapters 9 and 10. In particular, in Chapter 10 the various methodologies which can be used for asymmetric synthesis (the synthesis of non-racemic, chiral products from achiral starting materials) will be discussed. These methods all work by converting enantiotopic groups or faces into diastereotopic groups or faces before a chemical reaction is carried out.

7.3.1 The NMR spectra of compounds which contain diastereotopic groups

Consider the ^{1}H NMR spectrum of (S)-1-naphthylalanine hydrochloride **7.23** recorded in D_2O so as to eliminate the signals due to the NH_3^+ and COOH groups and any couplings to these hydrogens. A simplistic prediction of the appearance of the region of the spectrum corresponding to H_a, H_b and H_c might then suggest that H_c should be a triplet since it is adjacent to two other hydrogens (H_a and H_b), and that H_a and H_b should appear as a doublet (due to coupling to H_c) integrating to two hydrogens. However, this analysis would overlook the fact that H_a and H_b are not homotopic but are diastereotopic and so distinguishable by spectroscopic techniques including ^{1}H NMR. The appropriate region of the 250 MHz ^{1}H NMR spectrum of compound **7.23** recorded in D_2O is shown in **Figure 7.9**. All three hydrogens are observed as separate signals each of which is a doublet of doublets due to coupling to the other two hydrogens.

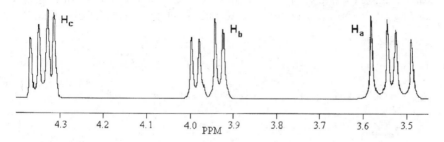

Figure 7.9 An extract of the ^{1}H NMR spectrum of compound **7.23**.

Thus the signal for H_a is seen at 3.52 ppm with coupling constants of magnitude 14 and 9 Hz, the signal for H_b occurs at 3.95 ppm with coupling constants of magnitude 14 and 5 Hz, and the signal for H_c is found at 4.27 ppm with coupling constants of magnitude 9 and 5 Hz. From these data, it can be concluded that $J_{ab} = 14$ Hz, which is typical for geminal protons, that $J_{ac} = 9$ Hz, and that $J_{bc} = 5$ Hz.

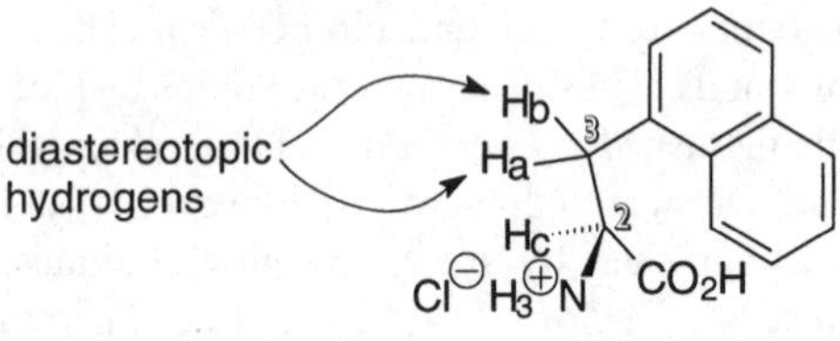

7.23

The non-equivalence of hydrogens H_a and H_b in compound **7.23** is best seen in the Newman projections about the C2–C3 bond for the three staggered conformations as shown in **Figure 7.10**. In conformation **7.23a**, H_a is orientated between the amino and acid groups attached to the stereocentre, whilst H_b is orientated between the amino group and H_c. In conformation **7.23b**, H_b occupies the location that H_a occupied in conformation **7.23a**. However, in conformation **7.23b** H_a does *not* occupy the position that H_b occupied in conformation **7.23a**. Hence, the distribution of the electron density in conformation **7.23a** will be different to the electron density distribution in conformation **7.23b**. Since it is the distribution of the electron density in a molecule which determines the degree to which a nucleus is shielded from the applied magnetic field, and hence the chemical shift of a nucleus, H_a and H_b will have different chemical shifts even when they occupy the same location relative to the substituents on C2. Examination of all three Newman projections shows this argument to be true for any of the staggered conformations. Since the observed chemical shift for H_a and H_b will be the average of the chemical shift they would exhibit in each of conformations **7.23a–c** weighted with respect to the relative populations of each conformation, H_a and H_b will, in general, have different chemical shifts.

The analysis discussed above for compound **7.23** can be used to show that a pair of diastereotopic groups in any molecule will, in principle, be distinguishable by NMR spectroscopy. It is not necessary that the diastereotopic groups be

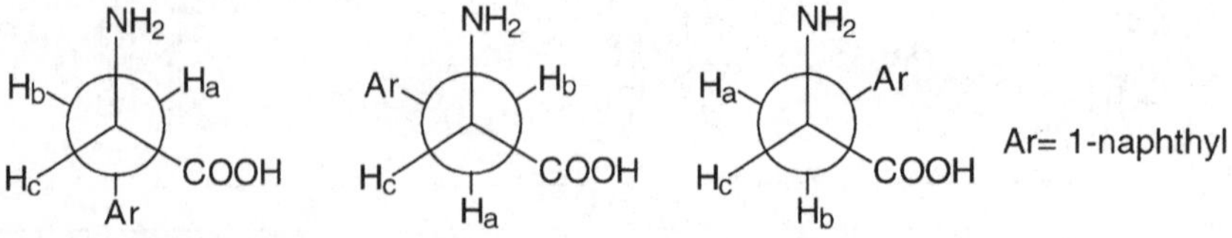

Figure 7.10 Newman projections about C2–C3 for compound **7.23**.

located directly adjacent to a stereocentre or that a stereocentre is even present in the molecule. The analysis is also not limited to ^{1}H NMR; if the diastereotopic groups contain other suitable nuclei (^{13}C, ^{31}P, ^{19}F, etc.) then the non-equivalence of the groups will be seen in these NMR spectra. For example, the ^{13}C NMR spectrum of (S)-leucine **7.24** shows separate signals for the carbon atoms in the two methyl groups, since the methyl groups are diastereotopic. In this case, the ^{1}H NMR spectrum also shows the diastereotopicity of the methyl groups, as two doublets each integrating to three hydrogens are observed in the spectrum.

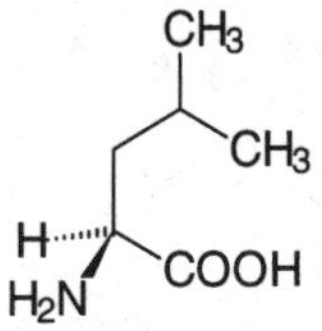

7.24

7.4 Nomenclature for heterotopic groups and faces

It is possible to use the Cahn, Ingold, Prelog priority rules to develop nomenclature systems to allow heterotopic groups and faces to be distinguished. When a prochiral centre is attached to two heterotopic groups, the two heterotopic groups are designated as *pro-R* and *pro-S*. The *pro-R* group is the group which when replaced by a different group of just slightly higher priority (it is often convenient to use an isotope for this) converts the prochiral centre into a stereocentre of *R*-configuration. Similarly, the *pro-S* group is the group which when replaced by a different group of just slightly higher priority generates a stereocentre of *S*-configuration.

The application of the *pro-R* and *pro-S* labels to the enantiotopic hydrogens of ethanol **7.3** is illustrated in **Figure 7.11**. Replacement of the emboldened hydrogen atom in **Figure 7.11** by a group (deuterium) which is of higher priority than hydrogen but which is still lower in priority than the OH or CH$_3$ groups generates (*R*)-1-deuteroethanol **7.25**, so this hydrogen is the *pro-R* group. Similarly, replacement of the italicized hydrogen atom generates (*S*)-1-deuteroethanol **7.26**, so this hydrogen is the *pro-S* group.

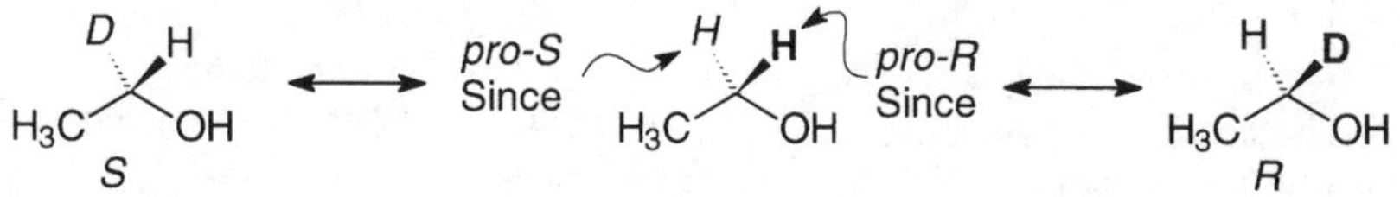

Figure 7.11 Application of the *pro-R/pro-S* nomenclature system to the enantiotopic hydrogens in ethanol.

Sometimes the heterotopic groups are attached to a prostereogenic centre which is not a prochiral centre. Examples of such cases are seen in compounds **7.12**, **7.13** and **7.27**. To provide a nomenclature system for these cases, the *pro-* prefix can be used in conjunction with any stereochemical descriptor. Thus the hydrogen atoms labelled H_a and H_b in compound **7.12** could be classified as *pro-cis* and *pro-trans*, whilst those in compound **7.13** could be referred to as *pro-E* and *pro-Z*. In compound **7.27**, the prostereogenic centre becomes a pseudoasymmetric centre (cf. Chapter 4, section 4.2) when one of the heterotopic hydrogen atoms is replaced by a different substituent, so the correct descriptors are *pro-r* and *pro-s*.

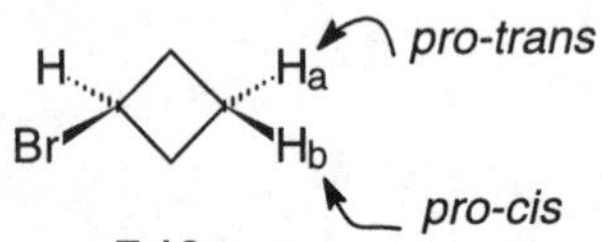

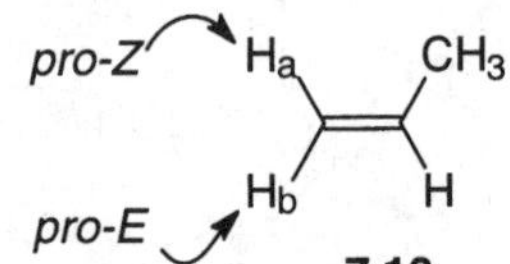

The nomenclature system for molecules with heterotopic faces uses the descriptors *Re* and *Si*. The molecule is drawn in the plane of the paper and the priorities of the three groups attached to the prochiral centre are assigned. If these decrease in order of priority in a clockwise direction, then the front face of the molecule as viewed is called the *Re*-face and the back face is referred to as the *Si*-face. Conversely, if the three groups attached to the prochiral centre decrease anti-clockwise, then the front face is the *Si*-face and the back face is the *Re*-face. Occasionally, the two heterotopic faces will be associated with a prostereogenic centre which is not a prochiral centre. In these cases, the same procedure can be used to classify the two faces, but the lower case descriptors *re-* and *si-* are used to indicate that an achiral system is being considered.

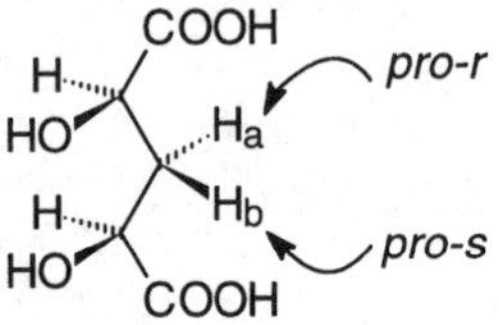

An example of the use of the *Re-* and *Si-* descriptors is shown in **Scheme 7.1**. In this example, cyanide is shown reacting from the *Re*-face of an aldehyde to

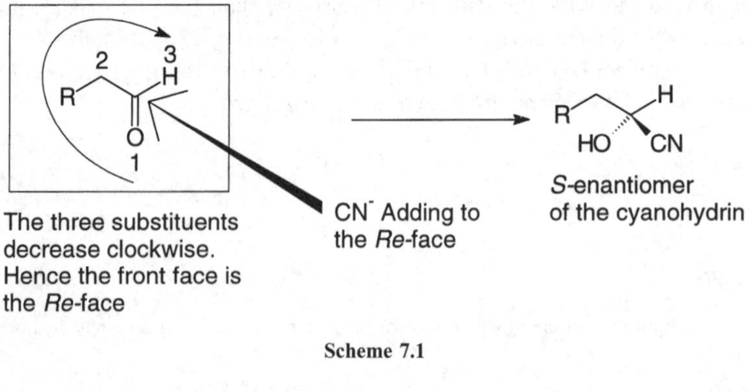

Scheme 7.1

produce the (S)-enantiomer of a cyanohydrin. Note, however, that this is only an illustration; in reality, unless there is some chiral influence present in the reaction mixture, it is equally likely that achiral cyanide will attack the enantiotopic *Re-* or *Si*-face of the aldehyde, giving equal amounts of the (S)- and (R)-enantiomers of the cyanohydrin respectively (cf. Chapter 3, section 3.6.1). The example shown in **Scheme 7.1** also illustrates that there is no correlation between the descriptors *Re-* and *Si-* for the heterotopic faces of a molecule and the absolute configuration R or S of the product resulting from addition to that face; addition of cyanide to the *Re*-face of the aldehyde produces the (S)-enantiomer of the cyanohydrin.

When the *Re-* and *Si-* descriptors are applied to systems such as alkenes where two prostereogenic centres may be present, it is necessary to specify which prostereogenic centre the descriptors are being applied to. Thus, in the case of (Z)-2-pentene **7.28**, the front face of the molecule around C2 is the *Si*-face, but around C3 the front face is the *Re*-face.

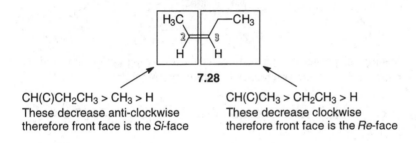

7.28

CH(C)CH$_2$CH$_3$ > CH$_3$ > H
These decrease anti-clockwise
therefore front face is the *Si*-face

CH(C)CH$_3$ > CH$_2$CH$_3$ > H
These decrease clockwise
therefore front face is the *Re*-face

7.5 Further reading

General
Stereochemistry of Organic Compounds E.L. Eliel and S.H. Wilen. Wiley: London, 1994, chapter 8.

Enantiotopic and diastereotopic groups
Topics in Stereochemistry Vol. 1, K. Mislow and M. Raban (N.L. Allinger and E.L. Eliel eds). Wiley: London, 1967, chapter 1.

7.6 Problems

1. For each of the following compounds, determine whether the highlighted groups are homotopic, enantiotopic or diastereotopic.

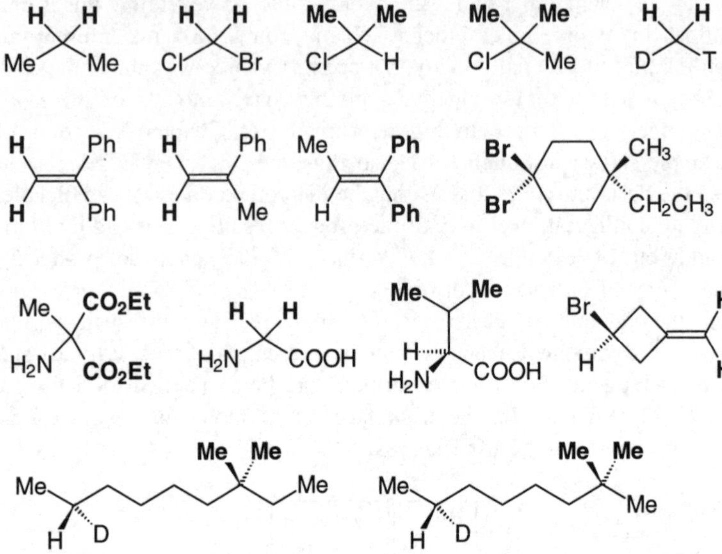

2. Identify all prostereogenic features in the following compounds and determine which of the prostereogenic features are prochiral.

$$CH_2(OMe)_2 \qquad CH_3CH_2CH_2CH_2CH_3 \qquad CH_3CH_2CH(CH_3)CH_2CH_3$$

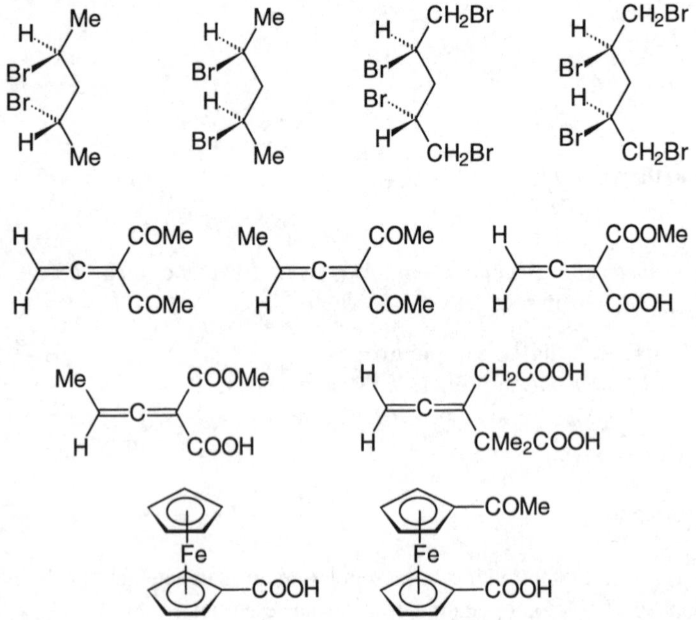

3. Assign the topism of the highlighted group in each of the following molecules.

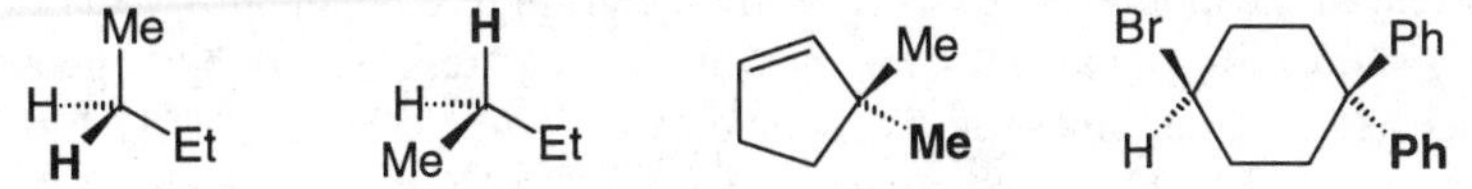

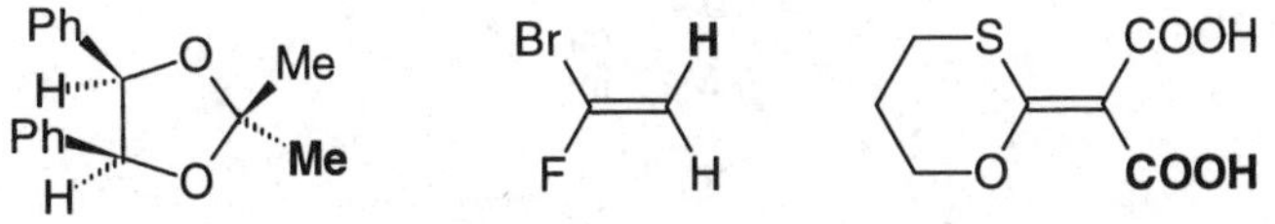

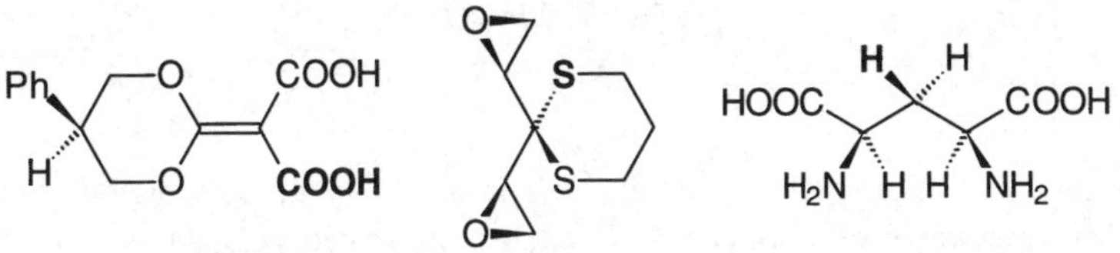

4. Use the CIP rules to assign the topism of the front face of the double bond in each of the following molecules. Where necessary, assign the topism at both ends of the double bond.

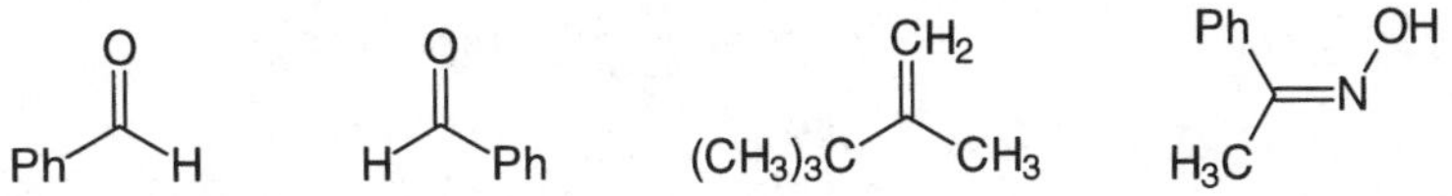

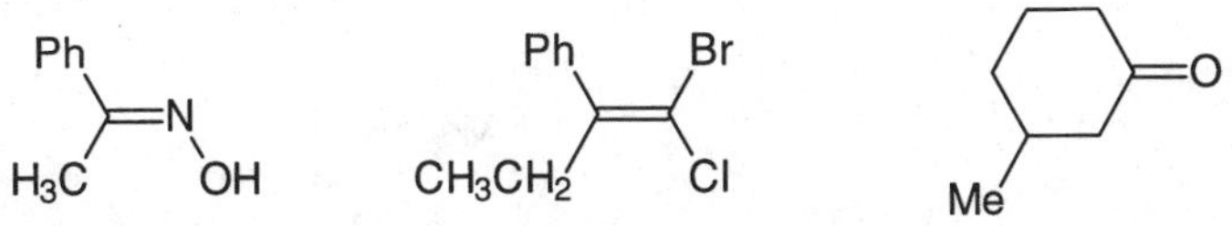

5. Compound A shows 12 lines in its proton decoupled ^{13}C NMR spectrum. Account for the presence of this number of lines and predict the number of lines in the proton decoupled ^{13}C NMR spectrum of compounds B–D.

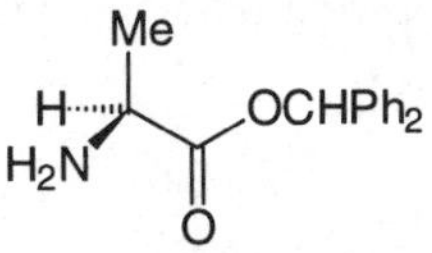

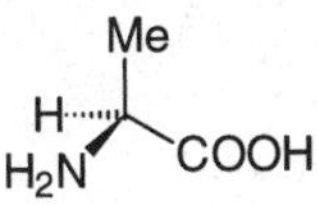

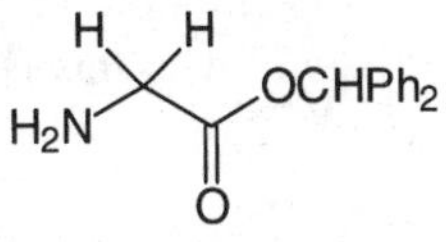

6. The ^1H NMR signal of the OCH_2 groups of compound A occurs as a quartet, but the corresponding signal for compound B consists of 16 lines. Account for this difference.

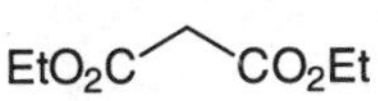

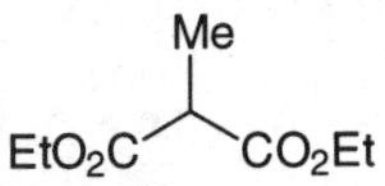

7. How many pairs of heterotopic groups and how many prochiral centres are present in compound A? The ^{1}H NMR spectrum of compound A (recorded in D_2O) shows a signal at 5.3 ppm which integrates to two hydrogens and consists of three lines at frequencies of 1341, 1323 and 1312 Hz with intensities of 1 : 2 : 1 respectively. Suggest which hydrogens give rise to this signal and account for its appearance.

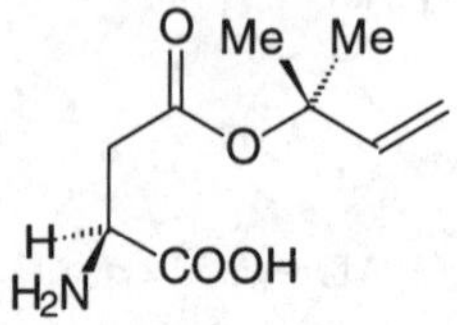

A

8. Predict the number of signals and the multiplicity of each signal in the ^{1}H NMR spectra of compounds A–D each recorded in D_2O so as to eliminate signals due to protons attached to oxygen and nitrogen atoms.

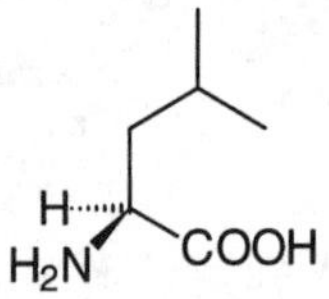

A 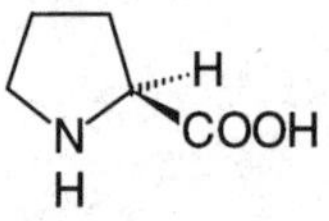   B 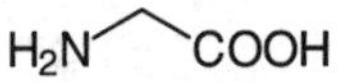   C

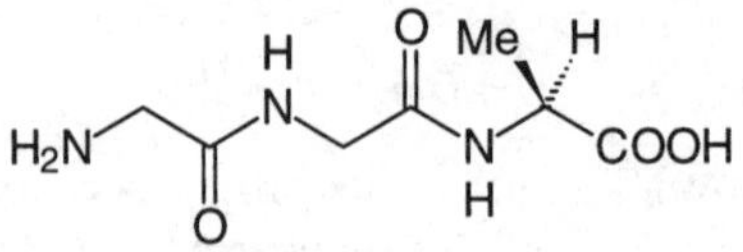

D

9. All heterotopic groups or faces in chiral molecules must be diastereotopic. Explain why this is so.

8 Conformations of acyclic and cyclic molecules

8.1 Introduction

The previous seven chapters of this book have mostly been concerned with configurational isomers, that is stereoisomers which cannot be interconverted without breaking at least one chemical bond. This chapter, however, will focus upon conformational isomers, these being defined as stereoisomers which can be interconverted without breaking a chemical bond. The topic of molecular conformation was briefly introduced in Chapter 1 (section 1.5.2), where it was emphasized that there is some overlap between the classifications of configurational and conformational isomers. The stereoisomers of some compounds (such as amides) could be classified as either conformational or configurational isomers. In other cases, stereoisomers which cannot interconvert without breaking a chemical bond at low temperature may do so at a higher temperature. A number of examples of these limitations in the classification system were illustrated in Chapters 2 and 3.

This chapter will be concerned only with stereoisomers which are unambiguously classified as conformational isomers. Hence, only stereoisomers which differ due to rotation around a bond with no π-character and which lack any large substituents or other factors to hinder this interconversion substantially will be considered. In Chapter 1, section 1.5.2, the conformations of ethane, the simplest molecule to exhibit conformational isomerism, were discussed. We will start by extending this discussion to other acyclic molecules and will then focus on the conformations of cyclic compounds containing three to six-membered rings.

8.2 Conformations of acyclic molecules

It was shown in Chapter 1, section 1.5.2 that there are an infinite number of conformations for ethane, which lie along the curve in the energy versus torsional angle diagram shown in **Figure 8.1**. The two extreme conformations, however, are the staggered **8.1** and eclipsed **8.2** conformations. The staggered conformation is the energy minimum and the eclipsed conformation is an energy maximum. Conformations which occur at minima on an energy versus torsional angle curve are called conformers or conformational isomers.

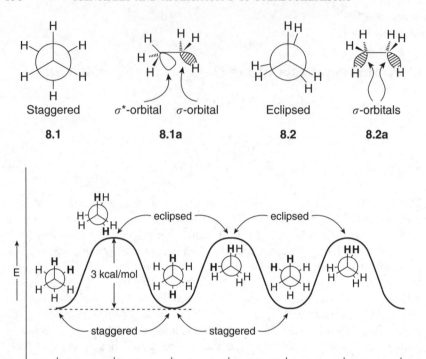

Figure 8.1 Energy profile for rotation about the carbon–carbon bond of ethane.

The obvious reason for the eclipsed conformation of ethane (and eclipsed conformations in general) being an energy maximum whilst the staggered conformation is an energy minimum is that the hydrogen atoms on adjacent carbon atoms are closer together in the eclipsed than in the staggered conformation, causing a steric (van der Waals) repulsion between them. However, the eclipsing hydrogens in structure **8.2** are only slightly closer together than the sum of their van der Waals radii. Calculations suggest that this steric repulsion can account for only about 10% of the observed energy difference between the staggered and eclipsed conformations. Steric repulsion is, however, important in determining the energy difference between staggered and eclipsed conformations when groups larger than hydrogen atoms are involved as will be discussed below. The main factors in determining the energy difference between the staggered and eclipsed conformations of ethane are thought to be favourable interactions between the bonding and antibonding orbitals in the staggered conformation **8.1a**, and unfavourable interactions between bonding orbitals in the eclipsed conformation **8.2a**.

Propane contains two carbon–carbon bonds about which rotation can occur;

however, as these two bonds are identical, the conformational situation is the same for both bonds. Once again, there are an infinite number of conformations and two extreme conformations: a minimum energy, staggered conformation **8.3** and maximum energy, eclipsed conformation **8.4** respectively. The energy versus torsional angle diagram for rotation around either of the carbon–carbon bonds in propane is almost identical to that of ethane shown in **Figure 8.1**. The only difference is that the energy difference between the staggered and eclipsed conformations is higher in propane (3.4 kcal mol^{-1}; 14.2 kJ mol^{-1}). This increase in the energy difference between the two extreme conformations is due to the eclipsing of a hydrogen atom and a methyl group in the eclipsed conformation **8.4** of propane as opposed to the eclipsing of just hydrogen atoms in the eclipsed conformation of ethane **8.2**. Since a methyl group has a larger van der Waals radius than a hydrogen atom, it generates a larger repulsive interaction with the eclipsing hydrogen than is observed between the two hydrogen atoms of ethane.

8.3   8.4

For butane, the situation is more complex as there are two distinct carbon–carbon bonds around which rotation can occur. The situation as regards rotation around either of the CH_3–CH_2 bonds is again analogous to the situation discussed above for ethane and propane with triply degenerate staggered and eclipsed conformations. For rotation around the central CH_2–CH_2 bond, however, there are four extreme conformations, two staggered conformations and two eclipsed conformations as shown in structures **8.5–8.8**. Rotation around the central carbon–carbon bond then produces a graph of the form shown in **Figure 8.2** in which the energy minima and maxima are no longer all of equal energy.

The relative energies of structures **8.5–8.8** can be easily understood by considering just steric effects, starting from conformation **8.5** which is called the **antiperiplanar conformation** since the two methyl groups are at 180° to one another. Conformation **8.5** has the lowest energy of any of the conformations of butane, since all of the atoms on adjacent atoms are staggered to one another, and the two largest groups (the methyl groups) are as far away from one another as possible, thus minimizing the steric repulsion between them. Rotation around the central carbon–carbon bond of butane then starts to bring the groups on C2 and C3 which were all staggered closer together and so raises the energy of the system until, after a 60° rotation, an eclipsed conformation called the **anticlinal conformation** (the two largest groups at 120° to one another) **8.6** is obtained. For butane, conformation **8.6** is calculated to be 3.8 kcal mol^{-1} (16 kJ mol^{-1}) higher in energy than conformation **8.5**.

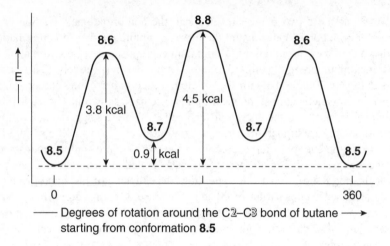

Figure 8.2 Energy profile for rotation about the central carbon–carbon bond of butane.

Continued rotation around the central carbon–carbon bond causes the groups that were eclipsing in conformation **8.6** to move apart, so the energy falls until, after a further 60° rotation, a second staggered conformation called the **synclinal conformation** (the two largest groups at 60° to one another) **8.7** is obtained. Although conformation **8.7** is a staggered conformation and an energy minimum, it still has a higher energy than conformation **8.5** since the two methyl groups are now only 60° from one another so there is some repulsion between them. For butane, the energy difference between conformations **8.5** and **8.7** is calculated to be 0.9 kcal mol^{-1} (3.8 kJ mol^{-1}). Conformations such as **8.7** which are energy minima, but not the overall minimum energy conformation, are called **local energy minima**. Conformation **8.5** is correspondingly called the **global energy minimum**.

Further rotation once again causes the groups on C2 and C3 to come closer together until, after a further 60° rotation (a total rotation of 180° from conformation **8.5**), a second eclipsed conformation, the **synperiplanar con-formation** (the two largest groups eclipsing one another) **8.8** is obtained. This conformation has a higher energy even than the eclipsed conformation **8.6**, since in conformation **8.8** the two largest groups (the methyl groups) are forced to eclipse one another whereas in conformation **8.6** the methyl groups were only eclipsing hydrogen atoms. For butane, the energy difference between conformations **8.5** and **8.8** is calculated to be 4.5 kcal mol^{-1} (18.9 kJ mol^{-1}). Continued rotation around the central carbon–carbon bond in butane does not generate further new conformations, but rather produces structures which are degenerate with conformations **8.7** and **8.6**, until, after a 360° rotation, conformation **8.5** is again obtained.

The same methodology detailed above for butane can be applied to any acyclic compound, although as the number of rotatable bonds increases, the number of possible conformations increases very rapidly. For an unbranched hydrocarbon chain (of less than 18 carbon atoms), the global minimum energy conformation will be the extended structure in which the antiperiplanar conformation is adopted about each rotatable bond as shown in structure **8.9** for pentane. This does not mean, however, that this conformation will necessarily have the highest population. If a higher energy conformer has a greater degeneracy than the global minimum energy conformation, then the higher energy conformation may have the greater population. For example, for pentane there are four degenerate conformations having a synclinal arrangement about either the C2–C3 or C3–C4 bonds **8.10**. Although these conformers have a higher energy (0.55 kcal mol^{-1}, 2.3 kJ mol^{-1}) than the global minimum energy structure, at room temperature approximately 55% of the molecules of pentane will be present in conformations represented by structure **8.10**, compared to just 35% in conformation **8.9**.

8.9   8.10

The energy barriers between the many conformers of an acyclic compound are usually relatively low, so that interconversion between the conformers is rapid. Most spectroscopic techniques cannot then detect the individual conformers and so produce only a single spectrum, corresponding to the average spectrum of each of the conformers weighted according to its population. The remainder of this chapter will be concerned with cyclic molecules where the requirements for the formation of a cyclic structure may significantly reduce the number of available conformations and affect the ease with which they can interconvert.

8.3   Conformations of cyclic alkanes and their derivatives

The relative energies of the conformations which may be adopted by cyclic compounds are determined by the same factors that affect the relative energies of the conformations of acyclic compounds. For acyclic compounds there will usually be at least one conformation in which all these factors are negligible; such a compound is said to be unstrained. For cyclic compounds, however, the requirement that the atoms adopt the correct geometry to form a ring limits the number of available conformations, and all of the possible conformations may possess some strain energy. The total strain (E_{TOT}) present in a cyclic system can be broken down into four components as shown in **Equation 8.1**:

$$E_{TOT} = E_r + E_\theta + E_\phi + E_s \tag{8.1}$$

The four components of the total strain energy in **Equation 8.1** are defined as follows:

E_r = **Strain due to bond stretching**. This is rarely important since even a small distortion in bond lengths from the preferred values results in a large increase in energy.

E_θ = **Strain due to bond bending**. This can be very large since the strain increases approximately in proportion to the square of the distortion in the bond angle.

E_ϕ = **Torsional strain**. This occurs whenever two groups are forced to eclipse one another.

E_s = **Steric strain**. This is caused whenever two atoms are forced closer together than the sum of their van der Waals radii.

The total strain present in a ring system can be determined experimentally by measuring its heat of combustion, the results for cycloalkanes of formula $(CH_2)_n$ ($n = 3$–16) being shown in **Figure 8.3**. The graph shown in **Figure 8.3** is

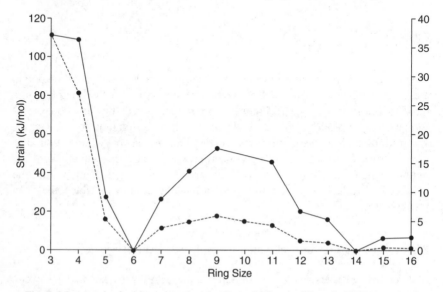

Figure 8.3 A graph of strain against ring size for cycloalkanes of formula $(CH_2)_n$, $n = 3$–16.

obtained by measuring the heat of combustion of each cycloalkane and then subtracting from this value n times the heat of combustion of an unstrained CH_2 group, which has been calculated to be $157 \, \text{kcal mol}^{-1}$ ($659 \, \text{kJ mol}^{-1}$). The remaining heat of combustion must have come from the strain energy present in the ring system. In the graph shown in **Figure 8.3**, the cycloalkane which possesses the lowest strain, cyclohexane, is arbitrarily assigned a strain of 0, and the strain present in all other cycloalkanes is measured relative to this. Two graphs are shown: the solid curve read against the left y-axis indicates the total strain in the ring system, whilst the broken curve read against the right y-axis gives the strain per CH_2 unit; both graphs follow the same basic shape. From **Figure 8.3**, the strain associated with ring systems can be seen to vary with ring size in the following ways:

3–4 membered rings are referred to as **small rings**, and are highly strained since the bond angles are less than 109° causing E_θ to be large.

5–7 membered rings are referred to as **normal rings**, and are relatively unstrained since the bond angles are approximately equal to 109° so E_θ is small.

8–11 membered rings are referred to as **medium rings**, and are strained (although not as strained as small rings) since the bond angles are greater than 109° causing E_θ to be large.

12 membered rings and larger are referred to as **large rings**, and are unstrained since a ring of this size is flexible enough to adopt one or more conformations in which all of the bond angles equal 109°.

Having highlighted the factors that are important in determining the conformations adopted by cyclic systems, the conformations of cycloalkanes containing three to six-membered rings will be examined in detail. Although the remainder of this chapter will focus on cycloalkanes, the conformations of cyclic compounds containing heteroatoms in the ring are similar to those of the cycloalkanes discussed here. For each ring system, in addition to discussion of the conformation of the unsubstituted ring, the influence that substituents on the ring have on the formation of both conformational and configurational stereoisomers will be surveyed.

8.4 Cyclopropane

Since cyclopropane **8.11** is known to be a cyclic molecule, all of the C–C–C internuclear angles must equal 60° since the molecule has no choice but to adopt a conformation in which the three carbon atoms are coplanar and form an equilateral triangle. Hence, there is a large E_θ since each of the C–C–C internuclear angles is distorted by almost 50°. In addition, all of the CH_2 groups eclipse one another (cf. the Newman projection **8.11a**) so E_ϕ is also large. This results in cyclopropane being a highly strained structure, although the experimental value ($27.5 \, \text{kcal mol}^{-1}$, $115 \, \text{kJ mol}^{-1}$) for the strain energy is not as high as would have been predicted by **Equation 8.1**. Quantum mechanical calculations

Table 8.1 The C–H stretching frequencies of cyclopropane, ethene and ethane

Structure	ν_{max} (C–H stretch)
Cyclopropane	3000–3050 cm^{-1}
Ethene sp^2 hybridized (33% s-character)	3010–3100 cm^{-1}
Ethane sp^3 hybridized (25% s-character)	2850–2960 cm^{-1}

show that in cyclopropane, the carbon atoms do not hybridize their 2s and 2p orbitals to form sp^3 hybrid orbitals. Rather, the carbon–carbon bonds in cyclopropane actually have more p-character than if they were sp^3 hybridized, and since p-orbitals are orientated at 90° to one another, unlike the 109° 28′ orientation of sp^3 orbitals, this helps to relieve the strain present in the ring system. The nature of the bonding in the carbon–carbon bonds of cyclopropane is illustrated in structure **8.11b**; these bonds are sometimes referred to as banana bonds.

Since the carbon–carbon bonds have more p-character than expected, then the carbon–hydrogen bonds must have less p-character and hence more s-character than they would have had if they were sp^3 hybridized. **Table 8.1** tabulates the C–H stretching frequencies observed in the infrared spectra of cyclopropane **8.11**, ethene and ethane. It can be seen from the data in **Table 8.1** that the carbon–hydrogen bonds of cyclopropane resemble those of ethene which are sp^2 hybridized and so composed of 33% s-character rather than the carbon–hydrogen bonds of ethane which are sp^3 hybridized and so possess only 25% s-character. Hence, there is more s-character (and so less p-character) in the carbon–hydrogen bonds of cyclopropane than would have been expected if they were sp^3 hybridized. Another illustration of the large amount of p-character in the carbon–carbon bonds of cyclopropane is seen in the chemistry of the cyclopropane ring. Cyclopropane undergoes addition reactions (e.g. addition of bromine as shown in **Scheme 8.1**) just like an alkene, rather than the substitution reaction that would have been typical of an alkane. The chemistry of an alkene is dominated by the π-bond, which has 100% p-character.

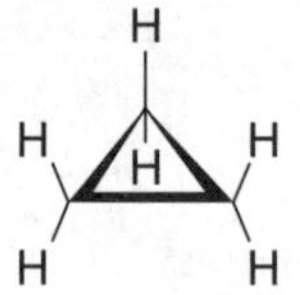

8.11 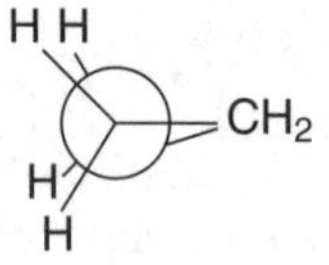   8.11a 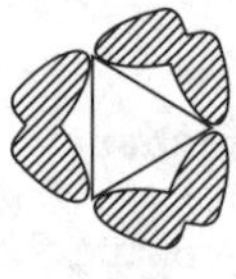   8.11b

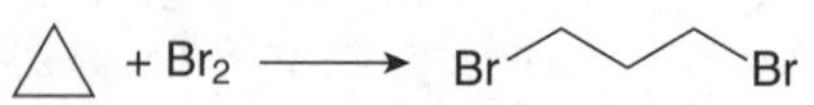

8.11

Scheme 8.1

8.4.1 Isomerism in cyclopropane derivatives

1,2-Disubstituted cyclopropanes exhibit cis–trans isomerism (cf. Chapter 2) as shown in structures **8.12** and **8.13**. However, both the *cis*- and *trans*-isomers possess two stereocentres, so there are a maximum of four possible stereo-isomers of a 1,2-disubstituted cyclopropane. Two of the stereoisomers (a pair of enantiomers) will be the *cis*-isomer (structure **8.12** and its enantiomer), whilst the other two stereoisomers will be the *trans*-isomer (structure **8.13** and its enantiomer). The two *cis*-isomers are diastereomers of the two *trans*-isomers and vice versa. This is a good illustration of the fact that cis–trans isomers are also diastereomers.

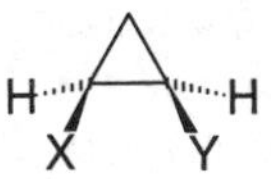

Cis Trans
8.12 **8.13**

As with any other diastereomers, when the two stereocentres contain the same substituents, then the symmetry of the molecule may cause it to possess fewer than the maximum possible number of stereoisomers (cf. Chapter 4). This is illustrated in structures **8.14** and **8.15** for cyclopropane-1,2-dicarboxylic acid. The *cis*-isomer contains a plane of symmetry (cf. Chapter 6) shown by the broken line, hence it is superimposable on its mirror image and there is only one stereoisomer **8.14** with the *cis*-geometry. Compound **8.14** is achiral and is a meso compound. The *trans*-isomer, however, does not contain a plane of symmetry and hence it is not superimposable on its mirror image, thus there are two stereoisomers (**8.15** and its enantiomer) which have the *trans*-geometry. Overall, there are three stereoisomers of cyclopropane-1,2-dicarboxylic acid.

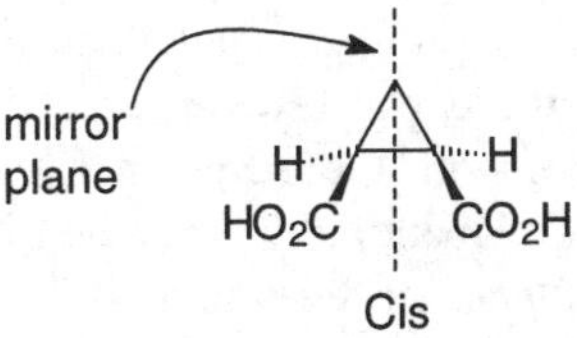

Cis Trans
8.14 **8.15**

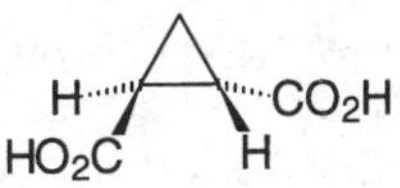

8.5 Cyclobutane

If cyclobutane were to adopt a flat, square conformation then the structure would be highly strained as, in addition to the high E_θ due to all of the bond angles being 90°, there would be a very large contribution from E_ϕ as all of the adjacent CH_2 groups would eclipse one another. In order to avoid this, cyclobutane adopts the non-planar, puckered conformation shown in structure **8.16** to reduce the overall strain. This non-planar conformation actually increases E_θ, as the bond angles are now less than 90°. However, the non-planar conformation allows the CH_2 groups to adopt a more staggered conformation with respect to one another, and so considerably reduces E_ϕ. In the non-planar conformation, the opposite carbon atoms are just 3 Å apart whilst the sum of their Van der Waals radii is 4 Å, thus some E_s is also introduced in the non-planar conformation. Overall, however, the increases in E_θ and E_s are more than compensated for by the decrease in E_ϕ, so the non-planar conformation of cyclobutane **8.16** is more stable than the planar conformation would be. As **Figure 8.3** shows, the actual strain energy in cyclobutane is, however, still only slightly lower than that of cyclopropane. There are actually two non-planar conformations of cyclobutane, since C1 and C3 can be below or above C2 and C4. For cyclobutane itself, these two conformations are of equal energy and the energy barrier between them is quite low, hence (at room temperature) each molecule rapidly inter-converts between the two conformations, giving a **fluxional structure**.

8.16

When considering cis–trans isomerism in four-membered rings, the ring can be considered as being flat to emphasize the *cis*- or *trans*-relationship between the substituents. For 1,2-disubstituted cyclobutanes, the situation is exactly as discussed in section 8.4.1 for 1,2-disubstituted cyclopropanes, i.e. if the two substituents are different, then there will be a total of four possible stereo-isomers: a pair of enantiomers with the *cis*-geometry, and a pair of enantiomers with the *trans*-geometry. If, however, the two substituents are the same, then there will only be a single stereoisomer with the *cis*-geometry, a meso com-pound, although there will still be a pair of enantiomers with the *trans*-geometry. However, for 1,3-disubstituted cyclobutanes no enantiomerism is possible, as both the *cis*- and *trans*-isomers have a plane of symmetry as shown in structures **8.17** and **8.18**.

Trans (achiral)   plane of symmetry   Cis (achiral)

8.17   **8.18**

8.6   Cyclopentane

If cyclopentane were planar, there would be almost no angle strain E_θ, since all the bond angles would be 108°. However, all the adjacent CH_2 groups would eclipse one another resulting in a large E_ϕ. To reduce this, cyclopentane adopts one of two non-planar structures called the **envelope 8.19** and **half chair** (also referred to as the **twist conformation**) **8.20** structures respectively. Structures **8.19** and **8.20** differ in that the envelope conformation has four atoms coplanar with the fifth offset above or below the plane of the other four, whilst the half chair structure has only three atoms coplanar, the remaining two atoms being offset one above and one below the plane. Both structures result in a large reduction in the torsional strain E_ϕ, without introducing any significant angle strain E_θ.

Envelope   Half Chair

8.19   **8.20**

Both the envelope and half chair conformations of cyclopentane are fluxional, i.e. within the envelope and half chair structures the atoms are constantly interchanging between the planar and offset locations. In addition, a particular molecule will be constantly exchanging between an envelope and half chair conformation. These conformational changes occur very rapidly, so that on the time scale of most spectroscopic techniques (and during chemical reactions) all five carbon atoms and all 10 hydrogen atoms appear to be equivalent. Thus, for example, both the ^{13}C and ^1H NMR spectra of cyclopentane show just a single line.

When considering cis–trans isomerism in disubstituted cyclopentanes, the ring can be considered to be flat and the situation is the same as discussed for cyclopropane in section 8.4.1. Thus, whether the two substituents have a 1,2- or 1,3-relationship to one another, *cis*- and *trans*-isomers may be formed both of which will exist as a pair of enantiomers unless the two substituents are identical, in which case the *cis*-isomer will be an achiral meso compound.

8.7   Cyclohexane

A planar six-membered ring has bond angles of 120° and so, if composed of sp^3 hybridized carbon atoms, would have appreciable angle strain E_θ. In addition, all of the adjacent CH_2 groups would eclipse one another giving a large E_ϕ.

To avoid these unfavourable interactions, cyclohexane and its derivatives adopt non-planar structures. Benzene, however, is composed of sp^2 hybridized carbon atoms for which the minimum energy bond angle is 120°. This is one reason why benzene does adopt a planar conformation. The various conformations which cyclohexane and its derivatives may adopt will be examined in some detail, both because these conformations are well defined and because a very large number of organic compounds contain six-membered rings, so the analysis of the conformations of these rings is of some importance.

8.7.1 The chair conformation

The **chair conformation** of cyclohexane is shown in structure **8.21**. In this conformation, all the C–C–C bond angles are approximately 109°, so there is no angle strain E_θ. All the adjacent CH_2 groups are staggered with respect to one another as highlighted in the Newman projection **8.21a**, so there is no torsional strain E_ϕ. In addition, there are no close interactions, so there is no strain energy at all in the chair form of cyclohexane.

8.21   **8.21a**

The name 'chair conformation' arises because (with a little imagination) the structure resembles a chair. Four of the carbon atoms are coplanar and form the seat of the chair whilst the other two are located above and below this plane respectively. The carbon atom which is displaced above the plane of the seat of the chair can be imagined as forming a head rest, whilst the carbon atom which is displaced below the plane forms a foot stool. One important point that must be realized, however, is that all six carbon atoms in the chair conformation of cyclohexane are equivalent. Depending upon the direction from which the molecule is viewed, any of the six carbon atoms may be considered as being part of the seat, head rest or foot stool of the chair. A molecular model will be of great help in appreciating this.

Whilst all the carbon atoms in the chair conformation of cyclohexane are equivalent, the same is not true of the hydrogen atoms. The chair conformation contains two types of hydrogen: **axial hydrogens** (shown emboldened in structures **8.21** and **8.21a**) which are perpendicular to the plane formed by the seat of the chair, and **equatorial hydrogens** (shown italicized in structures **8.21** and **8.21a**) which are located at angles of ±30° to the plane of the seat of the chair. Despite the presence of two types of hydrogen atoms, the ^1H NMR spectrum of cyclohexane at room temperature shows only one peak. This is because a rapid (at room temperature) dynamic equilibrium exists between two energetically degenerate chair conformations as shown in **Scheme 8.2**. This equilibrium has the effect of converting all the axial hydrogens in one chair structure into equatorial hydrogens in the other and vice versa, thus only a single average peak is seen in the NMR spectrum.

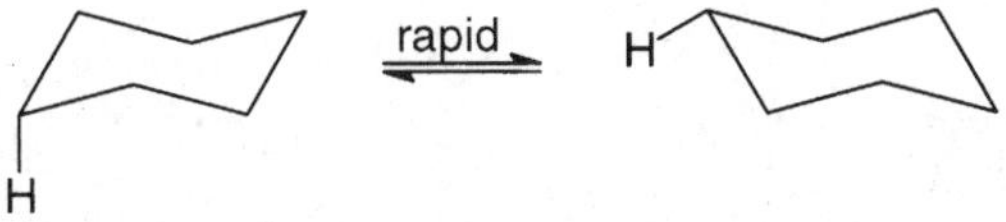

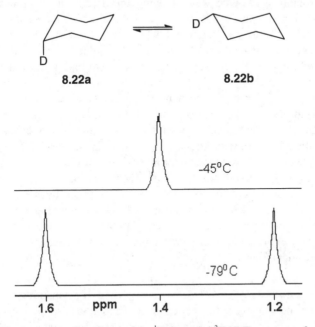

Scheme 8.2

At low temperature, the rate at which the two chair conformations inter-convert is reduced, and it is possible to see two peaks in the NMR spectra corresponding to the two types of hydrogens in the chair conformation. This is illustrated in the NMR spectra shown in **Figure 8.4**, where one of the hydrogen atoms on cyclohexane has been labelled by replacing it with a deuterium atom giving conformations **8.22a** and **8.22b**. At −45°C and above, the proton decoupled ^{2}H NMR spectrum (deuterium has a spin of 1 and so can give an NMR spectrum) shows a single peak corresponding to an average signal of the rapidly interconverting chair conformations containing an axially and equatorially located deuterium atom respectively. However, at −79°C the proton decoupled ^{2}H NMR spectrum shows two lines of equal intensity. At such a low

8.22a   **8.22b**

-45°C

-79°C

1.6   ppm   1.4   1.2

Figure 8.4 The temperature dependence of the ^{1}H decoupled ^{2}H NMR spectrum of monodeutero-cyclohexane **8.22a,b**.

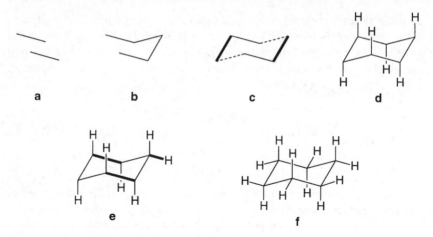

Figure 8.5 How to draw the chair conformation of cyclohexane.

temperature, the rate of interconversion between the two chair forms is slowed sufficiently that the axially and equatorially located deuterium atoms can be detected.

The chair conformation of cyclohexane is of such importance in organic chemistry that it is important to be able to sketch rapidly an accurate representation of structure **8.21**. One way of doing this is shown in **Figure 8.5**. The starting point is a pair of slightly offset parallel lines drawn at an angle of approximately 30° to the horizontal which will form the seat of the chair (**Figure 8.5a**). This pair of parallel lines are then joined by two additional lines to form the head rest (**Figure 8.5b**) and a further two lines to form the foot stool (**Figure 8.5c**). The completed carbon framework should consist of three pairs of parallel lines as shown by the normal, emboldened and hashed lines in **Figure 8.5c**. Next, the axial hydrogen atoms are added as shown in **Figure 8.5d**. These are located by vertical bonds drawn alternately above and below the ring, starting with a line drawn vertically upwards from the head rest. Finally, the equatorial hydrogens are added starting again with the head rest position (**Figure 8.5e**). The bonds to each of the equatorial hydrogens should be parallel to two of the carbon–carbon bonds as illustrated by the bold bonds in **Figure 8.5e**. Once all six equatorial hydrogen atoms have been located (**Figure 8.5f**) the structure is complete. With a little practice, it becomes relatively easy to draw the chair conformation of a cyclohexane ring in a variety of different orientations.

8.7.1a Monosubstituted cyclohexane derivatives. If a substituent is placed on the cyclohexane ring then the two chair conformations are no longer equal in energy. The lower energy conformation is usually the one in which the substituent occupies an equatorial position since this is less hindered than an axial position

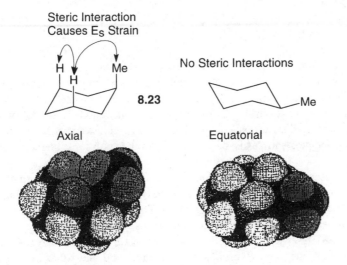

Steric Interaction
Causes E$_S$ Strain

No Steric Interactions

8.23

Axial

Equatorial

Figure 8.6 The two chair conformations of methylcyclohexane **8.23**.

as shown in **Figure 8.6** for methylcyclohexane **8.23**. The steric interaction between axial groups (E_s) indicated in **Figure 8.6** is called a 1,3-diaxial interaction, and the magnitude of this interaction increases as the size of the groups in the axial positions increases. The space filling diagrams also shown in **Figure 8.6** illustrate this 1,3-diaxial interaction more clearly. The energy difference is such that at 20°C methyl cyclohexane exists 95% in the conformer with an equatorial methyl group and 5% in the conformer with an axial methyl group.

If a very large substituent (such as a *tert*-butyl group) is introduced onto a cyclohexane ring, then the 1,3-diaxial interactions in the conformation with the substituent in an axial position become very large. This results in the conformation with the axially located substituent having a much higher energy and, hence, a much smaller population than the conformation with the substituent in an equatorial position. In the case of *tert*-butylcyclohexane **8.24**, the effect is so large that at room temperature the population of the conformation with an axial *tert*-butyl group is just 0.01%. For this reason, a *tert*-butyl substituent is often referred to as a **conformational anchor** which locks the cyclohexane ring into a single conformation. This term is, however, misleading since the equilibrium shown in **Scheme 8.3** does still exist; it is just the

8.24

Scheme 8.3

position of the equilibrium which is moved far in favour of the conformation with an equatorial substituent.

8.7.1b   Cyclohexane derivatives with two or more substituents.   When two or more substituents are located on a cyclohexane ring, the preferred conformation will depend upon the position of the two substituents on the ring and on their stereochemical relationship (*cis* or *trans*) to one another. In the following discussion, only the various isomers of dimethylcyclohexane will be considered, but the principles can easily be extended to other polysubstituted cyclohexane derivatives.

In the case of 1,1-dimethylcyclohexane **8.25**, the two chair conformations are of equal energy, as in both conformations one methyl group must be axial and the other one equatorial as shown in **Figure 8.7**. There are two possible diastereomers of 1,2-dimethylcyclohexane, the *cis*- and *trans*-isomers. In the case of the *cis*-isomer **8.26**, the two chair conformations are again equal in energy, because in each, one of the methyl groups is equatorial and the other is axial. *Cis*-1,2-dimethylcyclohexane is an achiral meso compound as is apparent in the planar representation of compound **8.26** in **Figure 8.7** which contains a plane of symmetry. However, the planar structure is not a realistic conformation of a cyclohexane ring and neither of the two chair conformations of compound **8.26** contains a plane of symmetry. It will be shown in section 8.7.2, however, that there is another conformation of cyclohexane derivatives, the boat conformation and in this conformation compound **8.26** will possess a plane of symmetry. The stereochemical analysis of compound **8.26** is thus exactly analogous to that of meso-2,3-dibromobutane discussed in Chapter 6 (section 6.5).

For the *trans*-isomer of 1,2-dimethylcyclohexane **8.27**, the two possible conformations are no longer equal in energy, as in one conformation both methyl groups adopt equatorial positions, whilst in the other the two methyl groups occupy axial positions. As substituents prefer to adopt equatorial positions, the conformer with both methyl groups equatorial is the lower energy conformation. Compound **8.27** is chiral and exists as a pair of enantiomers.

For 1,3-dimethylcyclohexane, there are again two possible diastereomers, the *cis*- and *trans*-isomers. In this case, however, the two possible conformations of the *cis*-isomer **8.28** have different energies since both methyl groups can occupy equatorial positions or both can occupy axial positions. As would be expected, the conformation with both methyl groups in equatorial positions will have the lower energy. *Cis*-1,3-dimethylcyclohexane is another achiral, meso compound, although in this case a plane of symmetry is apparent in the two chair conformations. For the *trans*-diastereomer **8.29**, however, the two chair conformations are of equal energy since in both conformations, one methyl group must adopt an axial position and the other an equatorial location. *Trans*-1,3-dimethylcyclohexane is a chiral molecule which exists as a pair of enantiomers.

Finally, there are two diastereomeric 1,4-dimethylcyclohexanes, the *cis*- and *trans*-isomers, both of which are achiral. For the *cis*-isomer **8.30** the two chair

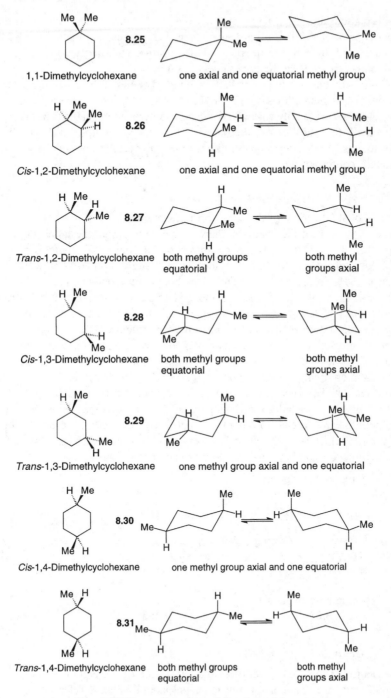

Figure 8.7 The conformations of dimethylcyclohexanes.

conformations are of equal energy since one methyl group must be axial and the other equatorial, whilst for the *trans*-isomer **8.31** either both methyl groups can be equatorial, or both can be axial. The conformer with the two methyl groups both equatorial will be the lower energy conformer.

Cis-1,3-dimethylcyclohexane **8.28** is the most stable of the *cis*-isomers as it is the only one which allows both methyl groups to adopt equatorial positions, and the 1,3-series is the only case where the *cis*-isomer is more stable than the *trans*-isomer. If the substituents are larger than methyl groups, then, if all the substituents cannot adopt equatorial positions, the cyclohexane ring may adopt a different conformation.

8.7.2 The boat conformation

If the foot stool carbon of the chair form of cyclohexane is moved up above the plane of the seat of the chair, an alternative conformation, the **boat conformation 8.32** is obtained. In this conformation all the bond angles are again approximately equal to 109°, so there is no angle strain E_θ. However, the adjacent CH_2 groups at the bottom of the boat eclipse one another as shown in the Newman projection **8.32a**, so there is torsional strain E_ϕ. The hydrogen atoms attached to the two non-coplanar carbon atoms are referred to as the flagpole and bowsprit hydrogens as shown in structure **8.32**. There is a non-bonded interaction between the two flagpole hydrogens giving rise to steric strain E_s. The sum of these various strain components results in the boat conformation being 7.1 kcal mol^{-1} (30 kJ mol^{-1}) less stable than the chair conformation. This means that for cyclohexane at room temperature, only about 0.01% of the molecules will adopt a boat conformation. However, some compounds are forced to adopt a boat conformation, an example being bicyclo[2.2.2]octane **8.33** which is a very rigid molecule in which each of the six-membered rings is forced to adopt a boat conformation.

flagpole hydrogens

bowsprit hydrogens

8.32   **8.32a**   **8.33**

8.7.3 The twist boat conformation

The boat conformation is very flexible and if one of the carbon–carbon bonds forming the base of the boat is twisted about its mid point, a different conformation called a **twist boat** (or just **twist**) **8.34** is obtained. Compared to the boat conformation, the adjacent CH_2 groups in the twist boat eclipse one another to a lesser degree and, the flagpole hydrogens are further apart and so repel one another less. Thus for cyclohexane the twist boat conformation is only 5.5 kcal mol^{-1} (23 kJ mol^{-1}) higher in energy than the chair conformation. Examples of molecules that exist in a twist boat conformation are *cis*-1,4-di-*tert*-butylcyclohexane **8.35** and cyclohexene **8.36**. In the case of *cis*-1,4-di-*tert*-butylcyclohexane, a chair conformation would force one of the *tert*-butyl groups to adopt an axial position which is very unfavourable as discussed in section 8.7.1a. For cyclohexene, it is the requirement that the four carbon atoms in the C–C=C–C subunit all be coplanar that results in a twist boat conformation being adopted.

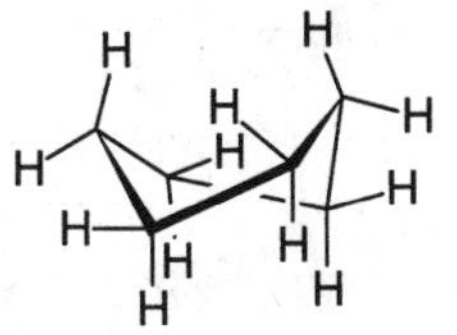

8.34

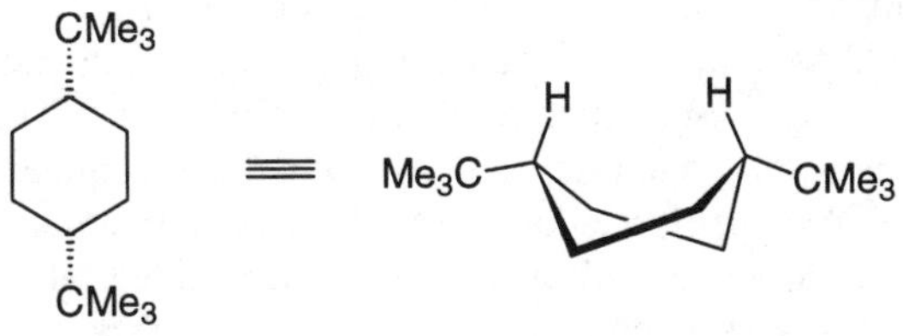

8.35

8.36

8.7.4   *Interconversion of cyclohexane conformations*

For simple cyclohexanes, each individual molecule will be constantly inter-converting between the chair, boat and twist boat conformations, but to do so it must pass through an additional high energy conformation. This is the **half chair conformation 8.37** in which four adjacent carbon atoms are planar. For cyclohexane, this conformation is 10.8 kcal mol^{-1} (45 kJ mol^{-1}) higher in energy than the chair conformation. The high energy of the half chair conforma-tion is due to the distortion of the bond angles in the planar part of the structure away from the 109° preferred for a tetrahedral system, and to the eclipsing of the CH_2 groups in the planar part of the molecule.

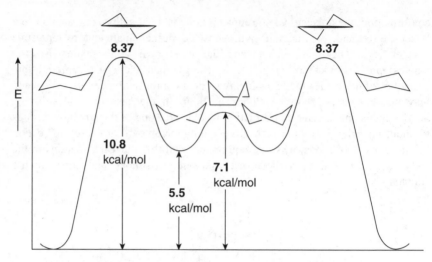

Figure 8.8 A conformation/energy diagram for cyclohexane.

A conformation versus energy diagram for cyclohexane showing how the various conformations interconvert is shown in **Figure 8.8**. Note that for cyclohexane itself, the boat conformation is not a minimum energy conformation but a maximum energy structure, i.e. a transition state. Thus for the two chair conformations of cyclohexane to interconvert, the molecule must pass through two half chair conformations and a boat conformation, each of which are energy maxima, as well as two twist boat conformations, which are at energy minima on the potential energy diagram.

8.7.5   Electronic effects in the conformational analysis of cyclohexane derivatives

Occasionally, substituted cyclohexanes adopt conformations which would not have been predicted purely on the basis of steric effects. An example of this is *cis*-1,3-dihydroxycyclohexane **8.38**, which adopts a chair conformation with both hydroxyl groups in axial positions since this structure is stabilized by an intramolecular hydrogen bond. The intramolecular hydrogen bond could not have been formed in the chair conformation in which both hydroxyl groups are equatorial. A second example is *trans*-1,2-dibromocyclohexane **8.39**, which adopts the expected chair conformation with both bromine atoms in equatorial positions **8.39a** in polar solvents, but in non-polar solvents the other chair conformer with both bromine atoms in axial positions **8.39b** is preferred. The explanation for this is that, with the two bromine atoms both in axial positions, the dipoles that they induce oppose one another and cancel each other out. In non-polar solvents, this lack of charge build up is more important than the steric cost of having the two substituents in axial positions. In polar solvents, however, charge build up is not so important so the molecule adopts the conformation with the lowest steric hindrance.

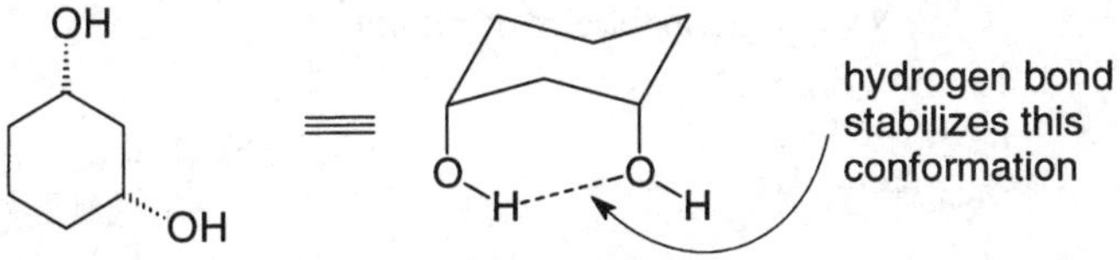

8.38

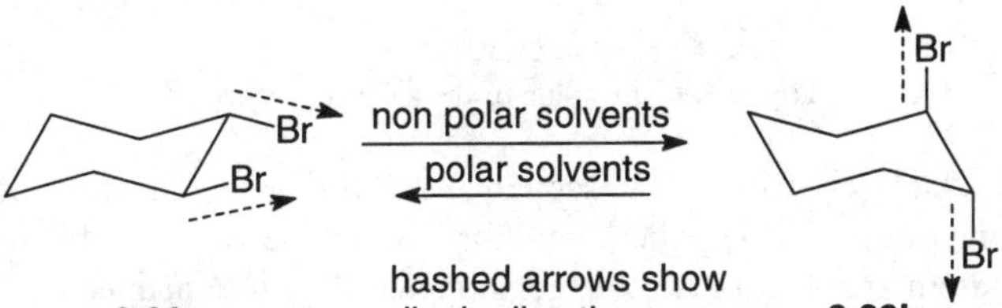

8.39a   hashed arrows show dipole directions   **8.39b**

The most common electronic effect in six-membered ring conformations is the increased preference of heteroatoms in the 2-position of a six-membered ring heterocycle to adopt an axial position. This effect is referred to as the **anomeric effect,** and is illustrated in structure **8.40**. The origin of the anomeric effect is that when the substituent (Y) is in an axial position, one of the lone pairs (which are held in sp^3-orbitals) on the heteroatom (X) within the six-membered ring will be synperiplanar to the σ^* orbital of the C–Y bond. This allows the lone pair of electrons on heteroatom (X) to be delocalized into the σ^* orbital leading to a stabilizing interaction which lowers the energy of the structure. This is illustrated for X=O in **Figure 8.9**. In valence bond terms, this interaction leads to the formation of the two resonance structures **8.41a** and **8.41b** shown in **Figure 8.9**. If, however, the substituent (Y) is in an equatorial position, then the lone pairs on heteroatom (X) are not synperiplanar to the σ^* orbital of the C–Y bond, so delocalization of the lone pair of electrons cannot occur.

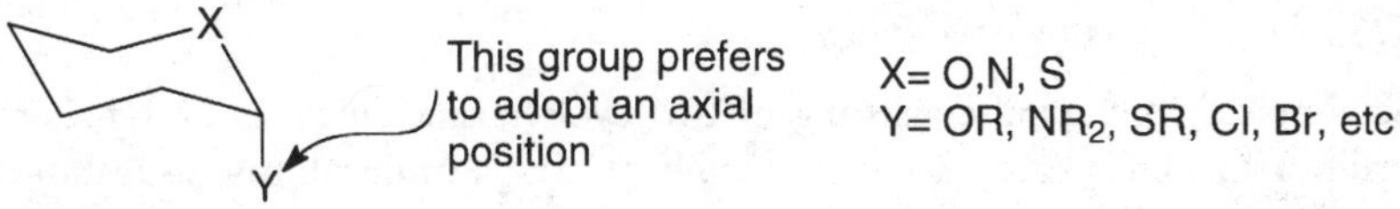

This group prefers to adopt an axial position

X= O,N, S
Y= OR, NR$_2$, SR, Cl, Br, etc

8.40

The most common occurrence of the anomeric effect is in carbohydrate chemistry, where the predominant structure of many sugars consists of an oxygen containing six-membered ring with a hydroxyl group (or a derivative of a hydroxyl group) on all but one of the carbon atoms. An example is methyl

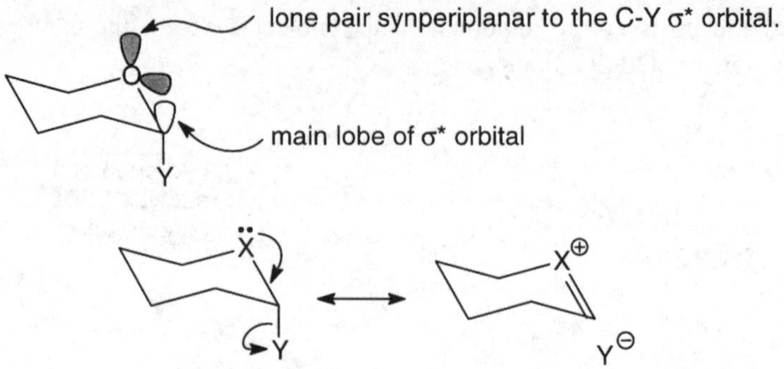

Figure 8.9 The origin of the anomeric effect.

D-glucoside **8.42** (cf. Chapter 3, section 3.2.1 for an explanation of the D/L stereochemical descriptors), where equilibration of the two possible epimers at the carbon atom attached to two oxygens (called the **anomeric centre**) leads to a 2 : 1 preponderance of the structure with an axially orientated OMe group **8.42a** over the corresponding structure with an equatorial OMe group **8.42b** as shown in **Scheme 8.4**. The anomeric effect is the best known example of what are referred to as **stereoelectronic effects**, a term which is used when the structure or reactivity of a system is determined by the orientation of the electron density of the system. The role of stereoelectronic effects in determining the stereochemical consequences of chemical reactions will be discussed in Chapters 9 and 10.

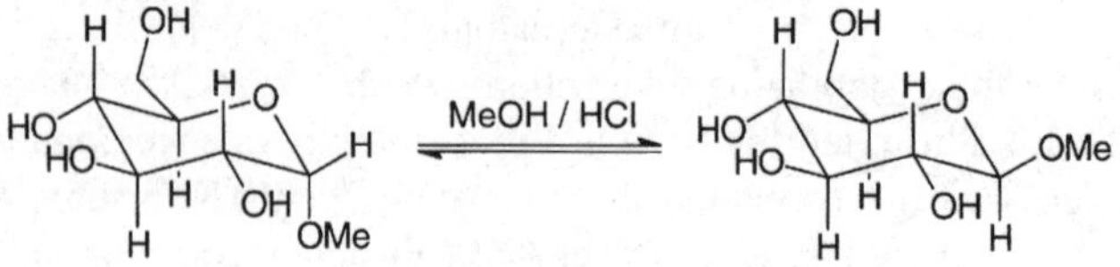

Scheme 8.4

8.7.6 Fused six-membered rings

The simplest compound consisting of two cyclohexane rings fused together is decalin **8.43**. Since each of the six-membered rings in decalin is disubstituted, there are two different stereoisomers (diastereomers or cis–trans isomers) of this compound, which differ in the configuration at the ring fusion. In both diastereomers, both rings can adopt a chair conformation. *Trans*-decalin **8.44** is an achiral meso compound and is a very rigid molecule since the carbon–carbon bonds at the ring junctions have to occupy the equatorial positions for steric reasons. This prevents the cyclohexane rings from undergoing chair–chair interconversion, and places any substituents into well defined axial or equatorial positions.

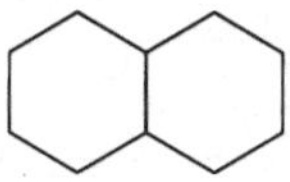

Decalin **8.43** 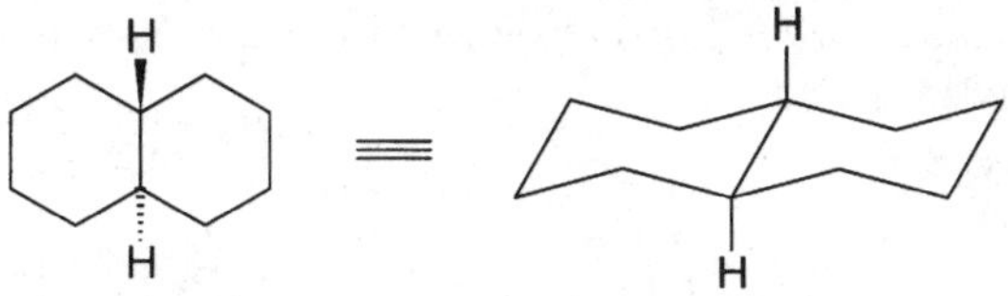   *Trans*-Decalin **8.44**, both hydrogens axial

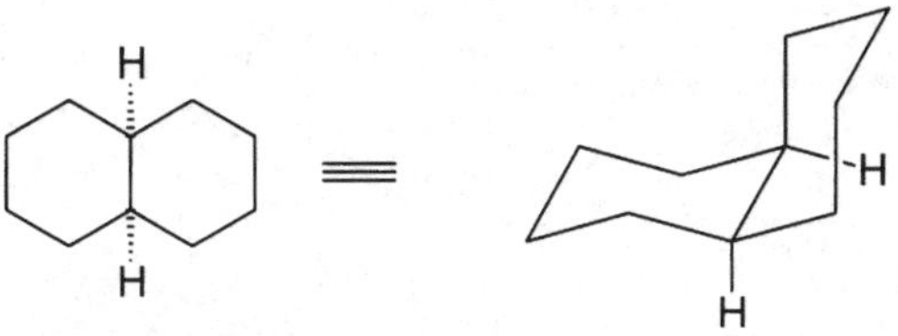

Cis-Decalin **8.45**, one hydrogen axial and one equatorial

Cis-decalin **8.45**, however, is a curved molecule with an accessible outer face and a hindered inner face. It is more flexible than the *trans*-isomer as only one of the carbon–carbon bonds at the ring junction is in an equatorial position, the other being axially orientated. Thus the six-membered rings can now flip from one chair form to the other as shown in **Scheme 8.5**. The two conformers of *cis*-decalin **8.45** shown in **Scheme 8.5** are non-superimposable mirror images of one another, so *cis*-decalin is a chiral molecule. However, *cis*-decalin **8.45** cannot be obtained in optically active form as the chair–chair interconversion allows the two enantiomers to interconvert rapidly. The stereochemical situation with *cis*-decalin is thus closely analogous to that of simple amines of formula R(R′)(R″)N discussed in Chapter 3, section 3.7.3. The chair–chair interconversion of *cis*-decalin also allows substituents to adopt preferentially an equatorial position just as for substituents on cyclohexane.

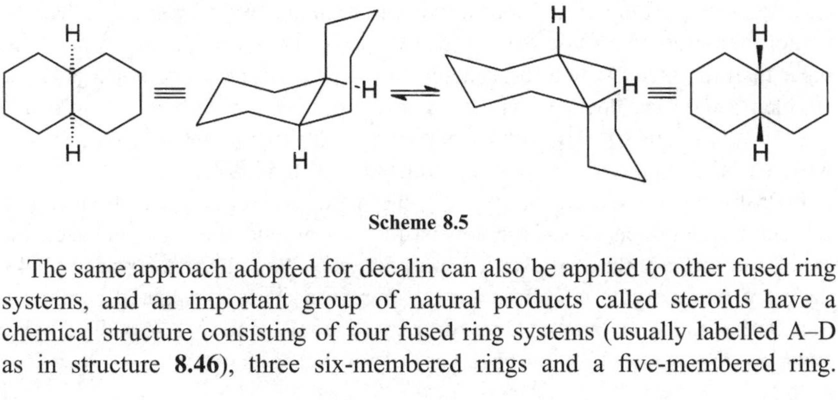

Scheme 8.5

The same approach adopted for decalin can also be applied to other fused ring systems, and an important group of natural products called steroids have a chemical structure consisting of four fused ring systems (usually labelled A–D as in structure **8.46**), three six-membered rings and a five-membered ring.

Substituents can be present anywhere on the steroid ring system (including replacing the indicated hydrogens), the rings can be unsaturated, and the A-ring can be aromatic. In most cases, the five- and six-membered rings found in steroids are *trans*-fused as shown in structure **8.46** though occasionally *cis*-fused rings are found.

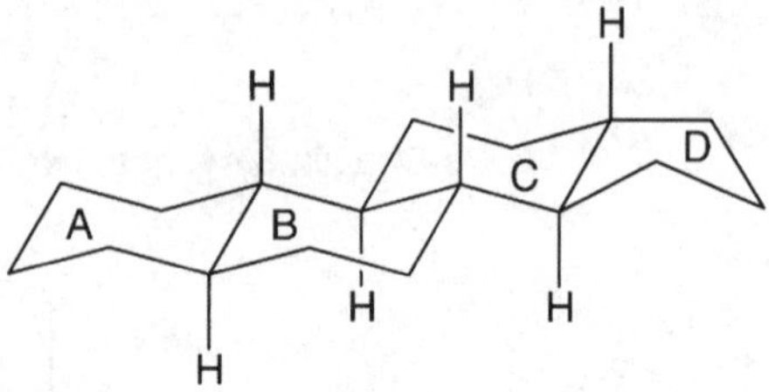

8.46

8.8   Medium and large rings

Rings that contain more than six atoms can adopt a large number of different conformations, and which of the many possibilities is adopted usually depends upon the exact structure of the compound in a way that is not easily determined. In many cases, the structure will consist of an equilibrium between many different conformations of similar energy.

8.9   Conformations of inorganic and organometallic compounds

The conformations of inorganic and organometallic compounds are determined by the same factors that are responsible for the conformations of organic compounds. However, the nature of the bonding in organometallic compounds can lead to situations which are not encountered with organic compounds. If a transition-metal complex contains a five- or six-membered ring chelate, then the conformation of the chelate will be determined by the hybridization of the atoms other than the metal which form the chelate. For example, in complex **8.47** the six-membered ring chelates will be planar since all the carbon, nitrogen and oxygen atoms are sp^2 hybridized. In contrast, the five-membered ring chelate will be non-planar since the carbon atoms are sp^3 hybridized.

In many organometallic compounds, the metal is bound to a carbon–carbon π-bond and, in such cases, rotation is allowed around the bond between the metal and the π-bond as illustrated in structure **8.48**. Thus, for ferrocene **8.49**, there are two extreme conformations: a minimum energy staggered conformation **8.49a** and a maximum energy eclipsed conformation **8.49b**. The same situation occurs in chromium *bis*(arene) complexes **8.50**.

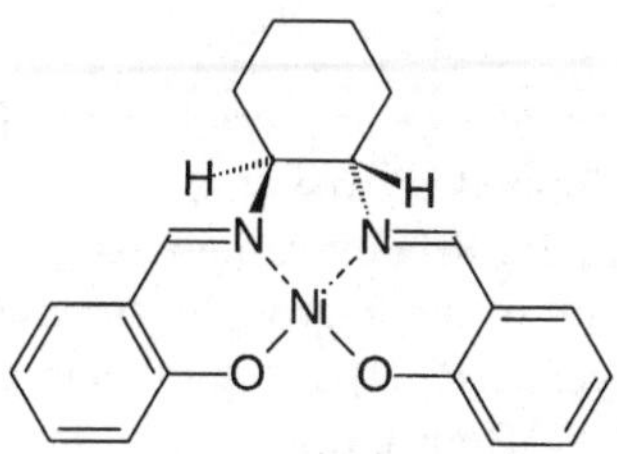

8.47

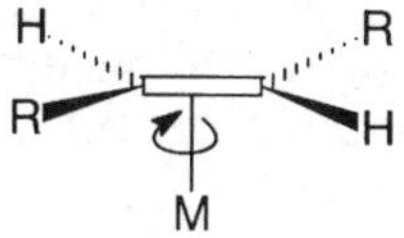

8.48

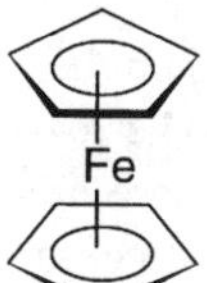

8.49a 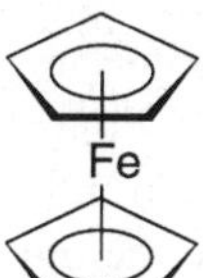   **8.49b** 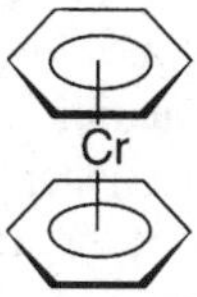   **8.50**    **8.51**

Finally, many inorganic compounds are fluxional which can mask the observation of conformational effects. Thus, PF_5 adopts a trigonal bipyramidal structure **8.51** (cf. Chapter 1, section 1.3) in which three of the fluorine atoms are in equatorial positions, whilst the other two are in axial positions. Hence, there are two different types of fluorine atom in PF_5, yet at room temperature the phosphorus decoupled ^{19}F NMR spectrum of PF_5 exhibits a single resonance. The explanation of this is that PF_5 is a fluxional molecule in which the axial and equatorial fluorine atoms are exchanging position more rapidly than NMR can detect them, so only an average signal is seen. At low temperatures, however, the spectrum does show the two expected signals (a 2F quartet and a 3F triplet) since the rate at which the fluorines exchange position is reduced at low temperatures so that NMR can detect the fluorines in axial and equatorial positions. This situation is very similar to that described in section 8.7.1 for the axial and equatorial hydrogens in cyclohexane, though the mechanisms of the two processes are different.

8.10   Conformations of biopolymers

There are three main classes of biopolymers: proteins, polysaccharides and nucleic acids. Many biopolymers have a preferred conformation which is directly responsible for the biological properties of the chemical. Thus for a biopolymer, the shape of the molecule is as important as its configuration or chemical composition. The conformations of biopolymers are largely determined by hydrogen bonding as will be apparent in the following discussion of each of the three classes of these compounds.

8.10.1   *Proteins*

Proteins have a chemical structure consisting of one or more chains of amino acids **8.52** linked together by amide bonds formed between the amino and acid groups as shown in structure **8.53**. There are 20 different amino acids that may be found in proteins which differ in the nature of the R-group in structure **8.52**. All amino acids found in proteins except glycine (R=H) are chiral, and have the L-configuration (cf. Chapter 3, section 3.2.1). Both the hydrogen and oxygen atoms of the amide bonds can be involved in hydrogen bond formation, and the way in which the hydrogen bonds are formed directly determines the conformation of the protein. If each NH of the protein forms a hydrogen bond to the carbonyl oxygen, four residues towards the *N*-terminal end, then the protein will adopt an α-helical conformation as shown in **Figure 8.10**. The α-helix always has the *P*-configuration (cf. Chapter 3, section 3.8.3) and there are 3.6 amino acid residues in each turn of the helix. There are other helical conformations of proteins but the α-helix is by far the most common.

N-terminus ⟶ *C*-terminus

8.52   **8.53**

The other common conformation found in proteins is the β-sheet in which hydrogen bonds are formed between two chains of amino acids. A β-sheet may be parallel or anti-parallel depending upon the relative orientation of the two amino acid chains as shown in **Figure 8.11**. The two amino acid chains may be covalently connected or held together only by the hydrogen bonds between

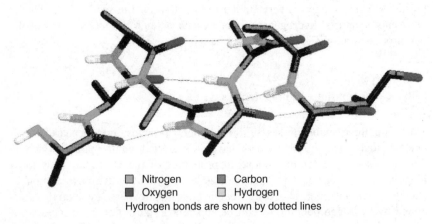

■ Nitrogen   ■ Carbon
■ Oxygen   □ Hydrogen
Hydrogen bonds are shown by dotted lines

Figure 8.10 The α-helix conformation of a protein.

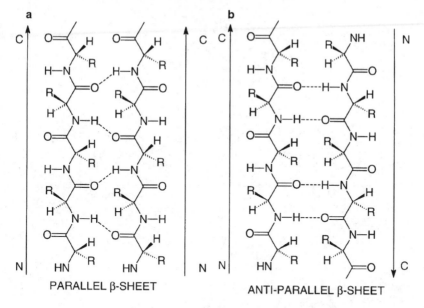

Figure 8.11 The **a**, parallel and **b** anti-parallel β-sheet conformations of a protein.

them. Some proteins are entirely α-helical or entirely a β-sheet but, more often, some regions of the protein will be α-helical, whilst others will adopt a β-sheet conformation, and other parts of the protein will adopt neither of these two main conformations.

8.10.2   Polysaccharides

Polysaccharides are a large and varied group of biopolymers. Two of the most common polysaccharides are cellulose and starch, both of which are composed of a single monomer unit, glucose **8.54**. In both cases, the polymer is built up by ether bond formation between the hydroxyl groups on Cl and C4. The main difference between these two polymers is in the stereochemistry of the Cl–O bond; in cellulose **8.55** this bond is equatorial, whilst in starch **8.56** it is axial. Cl is the anomeric centre of glucose, so axial substituents here are stabilized by stereoelectronic effects as discussed in section **8.7.5**. The change in configuration at this one stereocentre has a major effect on the conformation and physical properties of cellulose and starch. Since all of the C–O bonds in cellulose are equatorial, they all lie in the plane of the six-membered ring and cellulose adopts an essentially flat conformation as shown in structure **8.55**. The flat structure of cellulose allows the molecules to pack well together and, as a result of this, a large number of hydrogen bonds can be formed between individual polymer chains which would have to be broken if cellulose were to dissolve in a solvent such as water. Thus cellulose is a highly insoluble material.

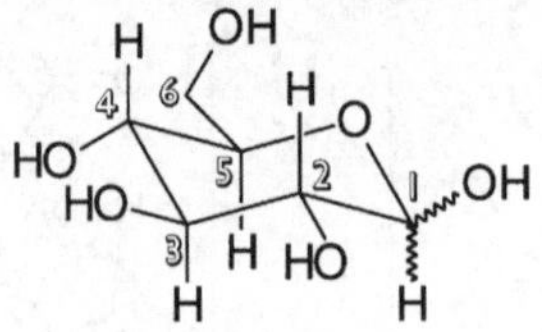

8.54

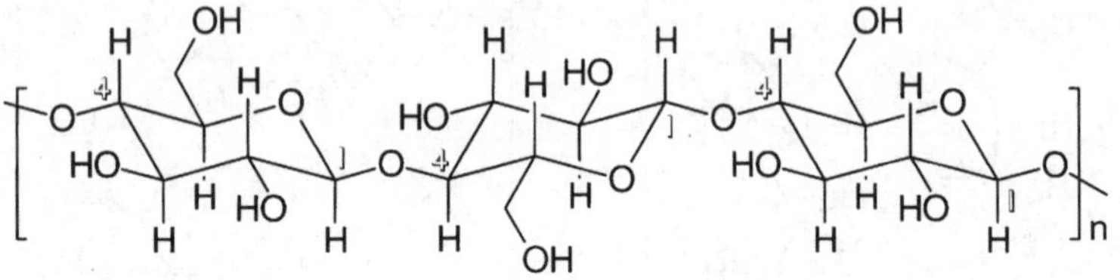

8.55

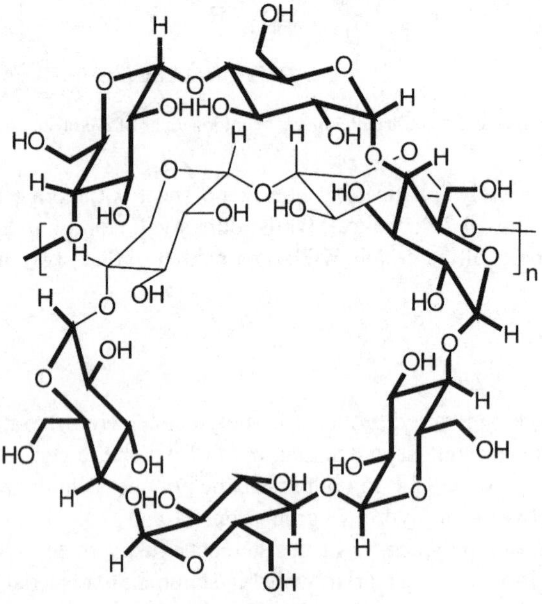

For clarity, the
hydrogen atoms
on C2 and C3
are not shown

8.56

Starch, however, adopts a helical conformation as shown in structure **8.56**, since the C1–O bond is not in the plane of the glucose ring. The helical structure of starch allows more intramolecular hydrogen bonds to be formed, and the helical molecules do not pack together as well as the flat cellulose molecules so fewer intermolecular hydrogen bonds are formed. This results in starch being more water soluble than cellulose. The helical conformation of starch also accounts for its other well known chemical property, the blue colour that is formed when iodine is added to a solution of starch. The iodine molecules can fit inside the hollow helix of starch, resulting in the formation of a starch/iodine complex which absorbs red light, resulting in the blue colour that is characteristic of the complex.

8.10.3   Nucleic acids

Nucleic acids are biopolymers which are constructed from three types of monomer unit: phosphate, a sugar and four bases, joined together as shown in structure **8.57**. The two naturally occurring nucleic acids DNA and RNA differ both in the structure of the sugar, which is ribose (X = OH in structure **8.57**) in RNA but 2-deoxyribose (X = H in structure **8.57**) in DNA, and in the structure of one of the bases, which is thymine (T, **8.58a**) in DNA but uracil (U, **8.58b**) in RNA. The other three bases, adenine (A, **8.59**), cytosine (C, **8.60**) and guanine (G, **8.61**) are common to both DNA and RNA, as is the phosphate. The base is attached to the C1 position of the sugar by the formation of a carbon–nitrogen bond using the nitrogen atom indicated in structures **8.58–8.61**.

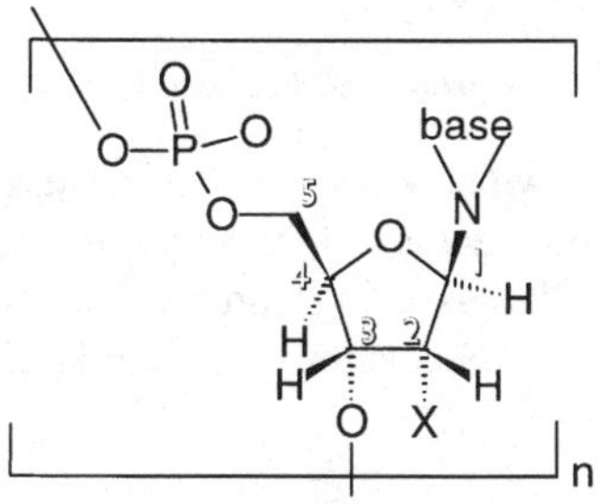

8.57

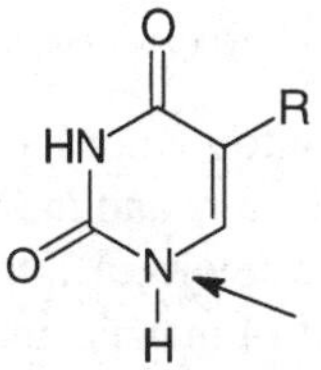

8.58a, R = CH$_3$
b, R = H

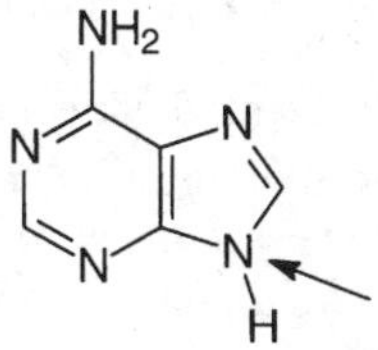

8.59

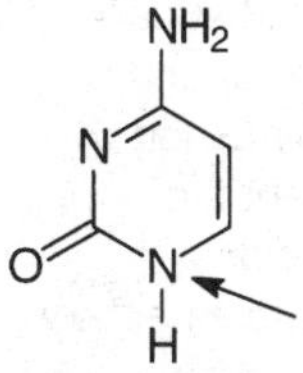

8.60

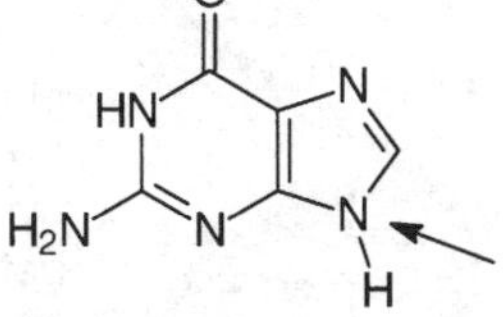

8.61

The conformations of DNA and RNA are determined by hydrogen bonding between pairs of the bases. The structures of the bases **8.58–8.61** are such that the base pair A and T (or U) are capable of forming two hydrogen bonds with one another, whilst the bases C and G can form three hydrogen bonds with one another as shown in **Figure 8.12**. In DNA, the hydrogen bonding between the matched (or complementary) bases occurs intermolecularly between two strands of DNA aligned anti-parallel to one another, and causes the characteristic double helix conformation of DNA shown in **Figure 8.13**. There are actually many

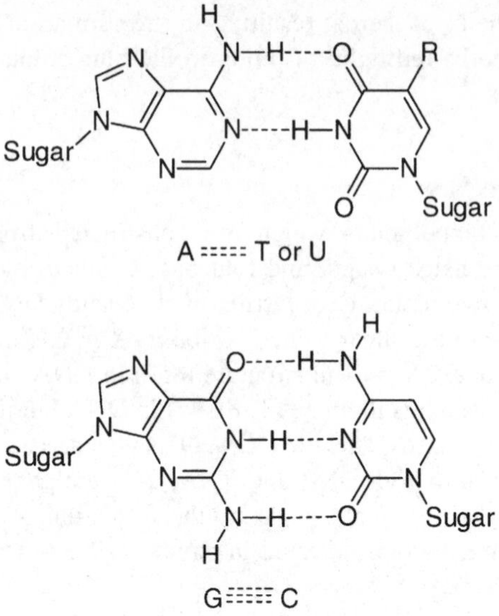

A ===== T or U
G ≡≡≡≡≡ C

Figure 8.12 Formation of hydrogen bonds between base pairs in nucleic acids.

different double helical conformations of DNA, which differ in the relative positioning of adjacent base pairs and in the torsional angles in the sugar. RNA, however, tends to form intramolecular hydrogen bonds using complementary bases along the same strand of nucleic acid. Thus RNA does not form a double helix conformation.

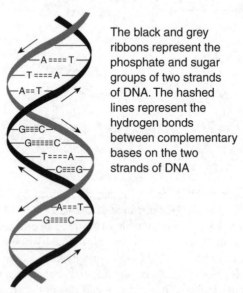

A====T
T====A
A==T
G≡≡≡C
G≡≡≡≡≡C
T====A
C≡≡≡G
A===T
G≡≡≡≡C
The black and grey ribbons represent the phosphate and sugar groups of two strands of DNA. The hashed lines represent the hydrogen bonds between complementary bases on the two strands of DNA

Figure 8.13 The double helix conformation of DNA.

8.11 Methods for determining molecular conformation

There are two widely used experimental methods for determining molecular conformation: X-ray crystallography and NMR spectroscopy. It is also possible to predict the relative energies of the various conformations of a molecular system using computational techniques. The merits and limitations of each of these techniques will be discussed in turn. Circular dichroism and optical rotary dispersion (cf. Chapter 3, section 3.4.2) have also been used to determine molecular conformation, particularly for cyclic ketones and biopolymers such as proteins, polysaccharides and nucleic acids. These more specialized methods will not be discussed here, though references to them are included in the further reading.

8.11.1 X-ray crystallography

X-ray crystallography was introduced in Chapter 2 (section 2.6.2), where its use to investigate molecular configuration was discussed. Since an X-ray analysis determines the location of each of the atoms in a molecule, it automatically determines the conformation of the molecule. An example of the use of X-ray crystallography to determine both the configuration and conformation of a molecule is shown in **Figure 8.14** for compound **8.62**. When the analysis was carried out, the stereocentre in the five-membered ring was known to have the (*S*)-configuration, but the relative and absolute configurations of the two stereocentres on the cyclopropane ring were unknown. The X-ray structure not only

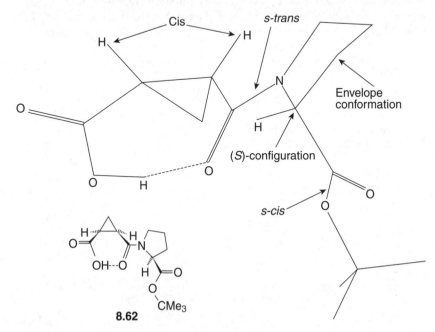

Figure 8.14 X-ray analysis of compound **8.62**.

showed that the two substituents on the cyclopropane ring were *cis* to one another but also showed that the five-membered ring adopted an envelope conformation and that the amide bond was in the *s-trans* conformation whilst the ester adopted an *s-cis* conformation (cf. Chapter 2, section 2.5).

X-ray crystallography is, however, not an ideal method for determining molecular conformation. Firstly, it is necessary to obtain flawless crystals of the compound to be investigated and many compounds do not crystallize. Additionally, the conformation determined by X-ray analysis is the solid state conformation, which may be different to the conformation the molecule would adopt in solution. This is because in the solid state each molecule is surrounded by and interacts with other identical molecules whilst in solution the intermolecular interactions will be with solvent molecules. These differing intermolecular interactions may change the relative energies of the possible conformations.

8.11.2   NMR spectroscopy

NMR spectroscopy (usually though not necessarily ^{1}H NMR) can be used to investigate molecular conformation in two ways; the vicinal coupling constants (^3J) give information on the torsional angle (θ) between two hydrogens, and the nuclear Overhauser effect (nOe) (cf. Chapter 2, section 2.6.1) provides information on which hydrogens are close to one another. The coupling constant between two hydrogen atoms on adjacent atoms depends on the torsional angle between them as described by the **Karplus equation (Equation 8.2)**, which is plotted graphically in **Figure 8.15**. In **Equation 8.2**, A, B and C are constants which depend upon the other substituents near to the two hydrogen atoms and,

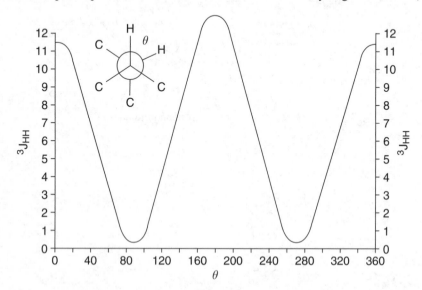

Figure 8.15 A graphical representation of the Karplus equation.

in particular, on the electronegativity of these substituents. Values for the constants A, B and C can only be found experimentally and will differ from system to system, although tables of suitable constants are now available for many common situations.

$$^{3}J = A \cos^2\theta + B \cos\theta + C \qquad (8.2)$$

It is apparent from **Figure 8.15** that most ^3J coupling constants do not correspond to a single torsional angle. However, it is possible to rule out many torsional angles and hence many conformations by measuring the coupling constant. If more than one ^3J coupling constant can be measured in a cyclic molecule then the number of conformations which are compatible with all the coupling constants may be reduced until only a single conformation remains. It should be remembered, however, that acyclic molecules (and many cyclic species) interconvert between different conformations rapidly on the NMR timescale. In this case, all the ^3J coupling constants will tend to be around 7 Hz (the average value for ^3J in **Figure 8.15**), and no information about the individual conformations can be obtained from the coupling constants.

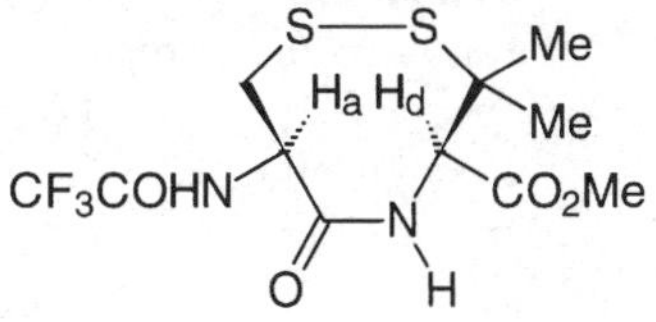

8.63a

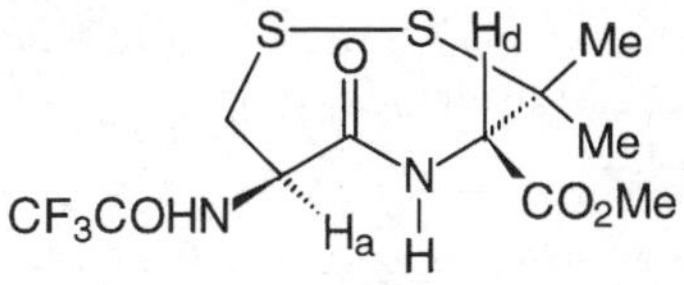

8.63b

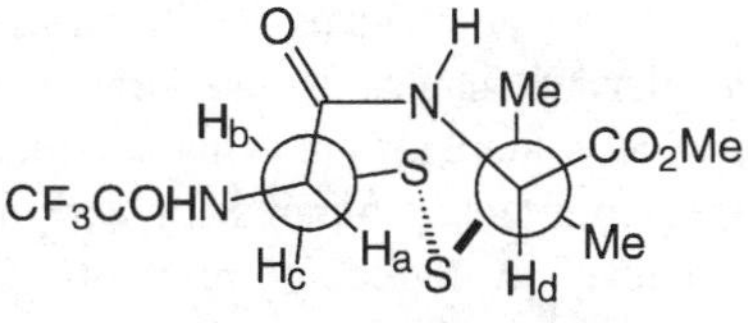

M-helical disulphide

8.63c

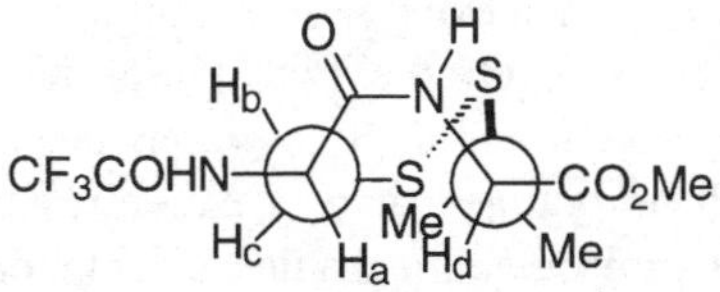

P-helical disulphide

8.63d

The magnitude of the nOe between two nuclei (usually ^{1}H) decreases as the distance between the nuclei (r) increases (it is proportional to r^{-6}). Hence, by determining the magnitude of the nOe between the various hydrogen atoms in a molecule, information on their relative orientations and hence the conformation of the molecule can be obtained. As an example of the application of both coupling constants and the nOe to conformational analysis, consider compound **8.63** which contains an eight-membered ring containing two sulphur atoms and an amide bond. In determining the conformation of this ring system, the first step was to determine the conformation of the amide bond which could be s-cis **8.63a** or s-trans **8.63b** (cf. Chapter 2, section 2.5). That the amide bond in compound **8.63** possessed the *s-cis*-conformation was readily determined by nOe experiments, since irradiation of either H_a or H_d resulted in a large enhancement in the signal for the other hydrogen, indicating that they are close together. This result is consistent only with the presence of an *s-cis*-amide bond **8.63a**.

The other factor that determines the conformation of compound **8.63** is the helicity (cf. Chapter 3, section 3.8.3) of the disulphide bond. Newman projections of the *M* and *P* helical disulphide conformations of compound **8.63** are shown in structures **8.63c** and **8.63d** respectively. These two conformations can be distinguished by the ^3J coupling constants between H_a and H_b and H_c. In both cases, the torsional angle between H_a and H_b is approximately 180° and the coupling constant between H_a and H_b is about 12 Hz as predicted by **Figure 8.15**, so no useful information can be obtained from this coupling constant. In structure **8.63c**, the torsional angle between H_a and H_c, however, is approximately 50°, which corresponds to a predicted coupling constant of around 5 Hz, whilst in structure **8.63d** the torsional angle is about 75°, which would be predicted to give a coupling constant of around 2 Hz. The observed coupling constant between H_a and H_c is 2.5 Hz, which indicates that the conformation of compound **8.63** is as shown in structure **8.63d** with an *s-cis*-amide bond and a *P*-helical disulphide.

8.11.3   Computer assisted conformational analysis

It is possible to use computational techniques to predict the conformation(s) that will be adopted by a molecular system. Two methodologies have been developed to achieve this, quantum mechanical based approaches and molecular mechanics. Of the two approaches, molecular mechanics is the simpler to understand and apply, and has been more widely adopted; only this technique will be outlined here. Relatively inexpensive computer programs for carrying out molecular mechanics calculations are widely available for both PC and Macintosh computers as well as for more powerful workstations.

Molecular mechanics methodology treats atoms as spheres, and bonds as springs; classical mechanics based on Newton's laws of motion is then used to determine the energy of the molecular system. Mathematical routines are employed to adjust the positions of the spheres (representing atoms) so as to minimize the overall energy of the system. At the heart of any molecular mechanics package is a force field which consists of a collection of parameters relating to each type of atom that may be present in the molecule. It was shown in section 8.3 that the total energy (E_{TOT}) of a molecular system could be broken down into a number of components (**Equation 8.1**); the force field contains parameters to allow each of these components of the total energy to be calculated.

For each possible type of bond (C–C; C=C; N–O; S–S; etc.) the preferred bond length will be included in the force field, along with a force constant specifying by how much the energy of the system should be raised when the actual bond length is shorter or longer than the preferred value. Similarly, for each bond angle and torsional angle the preferred values along with the force constant for deviation from these values will be listed. Non-bonded interactions can be calculated from the van der Waals radii of the atoms. Most force fields also include an additional energy component (E_{es}) to take electrostatic effects into account, thus giving **Equation 8.3**, and some force fields include a number of other parameters, for example, to calculate hydrogen bonds.

$$E_{TOT} = E_r + E_\theta + E_\phi + E_s + E_{es} \qquad (8.3)$$

To see how molecular mechanics works in practice, consider methyl cyclo-hexane **8.23**. The computer program must first be told the structure of the molecule and this is usually achieved by drawing the molecule on the computer screen, giving a structure such as **8.64**. Such a drawing will usually be two dimensional and have severely distorted bond lengths and angles. The molecular mechanics package will then calculate all of the bond lengths in structure **8.64** and compare them with the preferred bond lengths stored in the force field to calculate E_r for structure **8.64**. Each of the other components of E_{TOT} can similarly be calculated and the sum of all of these components gives E_{TOT} for the structure **8.64**. A mathematical algorithm is then used to change the positions of the atoms so that the bond lengths, bond angles, etc. are brought nearer to their preferred values; this may produce a structure such as **8.65**. The energy of this modified structure is then calculated and should be lower than that of the starting structure **8.64**. The whole process is then repeated in an iterative process until after a few thousand cycles no further reduction in E_{TOT} is possible. At this stage, a minimum energy conformation has been found and the final structure **8.66** is produced. The whole process of producing structure **8.66** from structure **8.64** may take from a few seconds to about 5 minutes depending on the power of the computer used.

8.64   **8.65**   **8.66**

There are, however, a number of potential problems with molecular mechanics calculations. The first of these is that a very large number of parameters are needed to describe all possible bonding arrangements, but most of these parameters are unknown. Thus whilst preferred bond lengths and angles are well known for most possible bonds, the energetic cost of deviations from these preferred values have been determined in only a few cases. In particular, very few parameters have been determined for metals, so most applications of molecular mechanics tend to be concerned with the modelling of organic compounds. Force fields are now available that will accommodate any element in the Periodic Table. However, most of the parameters in these force fields have been guessed so there are likely to be large errors in the energies of structures minimized using these parameters.

The other problem with the molecular mechanics approach outlined above is that whilst the final structure **8.66** produced for methylcyclohexane is a minimum energy conformation it is not the global minimum energy conformation since the methyl group is in an axial rather than equatorial position. This is a general problem; energy minimization of a starting structure will find the minimum energy conformation which happens to be nearest in structure to the starting structure, and this may or may not happen to be the global minimum energy conformation. Usually, it is desired to either find the global minimum energy conformation of a structure, or to find all minimum energy conformations and their relative energies so that the populations of each conformation can be determined. Unfortunately, there is no methodology which other than in trivially simple cases can guarantee to find all possible minimum energy conformations, or to find the global minimum energy conformation. There are, however, various methods which can be used to generate multiple minimum energy conformations of a molecule.

The simplest approach (and the only one to be discussed here) to finding minimum energy conformations is to carry out a grid search. Each rotatable bond in a molecule is systematically rotated through a specified angle (often 60°), to generate a set of conformations, each of which is subsequently energy minimized to produce a set of minimum energy conformations. As an example, consider butane for which there are two minimum energy conformations **8.5** and **8.7** (section 8.2). In this case, if the initially produced structure from an energy minimization happened to be **8.7**, then rotation around the central carbon–carbon bond in 60° increments would generate six structures evenly spread across the conformational space available to the molecule (**Figure 8.2**). Energy minimization of each of these six structures would in most cases just reproduce structure **8.7**, but at least one of the six structures will lead to structure **8.5**, thus achieving the desired task of finding all possible minimum energy conformations.

8.7   rotate   8.5
                   in 60°
                   increments

The main problem with this approach is that the number of structures that will be generated (N_s) and need to be energy minimized during a grid search is given by **Equation 8.4** (θ = angle through which each bond is to be rotated), and this value increases exponentially as the number of rotatable bonds (n) increases. Thus for a moderately sized molecule containing 10 rotatable bonds, rotation in 60° increments around each bond would generate 60,466,176 structures each of which may require up to 5 minutes to minimize the energy, and the whole calculation could then take as long as 9 years to complete! As computers become more and more powerful this will become less of a problem, but it is always likely to remain a major limitation in the application of conformational analysis to large and medium sized molecules.

$$N_s = (360/\theta)^n \qquad\qquad (8.4)$$

Despite the difficulties mentioned in this section, computational conformational analysis is a valuable tool for chemists wishing to predict the conformation of a molecular system. It can be used in conjunction with experimental data to predict chemical reactivity, to determine the conformations adopted by cyclic compounds, including those which contain medium and large rings, and to design molecules of a specified shape to fit into a biological receptor to elicit or block a biological response. In this section, it has been possible to give only a very brief overview of the methodology; more detail will be found in the recommended further reading.

8.12   Further reading

General
Stereochemistry of Organic Compounds E.L. Eliel and S.H. Wilen. Wiley: London, 1994, chapters 10 and 11.

Conformations of five-membered rings
Topics in Stereochemistry Vol. 10, B. Fuchs (E.L. Eliel and N.L. Allinger eds). Wiley: Chichester, 1978, chapter 1.

The anomeric effect
The Anomeric Effect and Related Stereoelectronic Effects at Oxygen A.J. Kirby. Springer-Verlag: Berlin, 1983.

Stereoelectronic effects
Stereoelectronic Effects in Organic Chemistry P. Deslongchamps. Pergamon: Oxford, 1983.
Stereoelectronic Effects Oxford Chemistry Primer Number 36, A.J. Kirby. Oxford University Press: Oxford, 1996.

Circular dichroism and optical rotary dispersion

Optical Rotary Dispersion C. Djerassi. McGraw-Hill: London, 1960.
Optical Rotary Dispersion and Circular Dichroism in Organic Chemistry P. Crabbe. Holden-Day: London, 1965.
Optical Circular Dichroism L. Velluz, M. Legrand and M. Grosjean. Verlag Chemie, 1965.

Conformations of biopolymers

Chemistry of Biomolecules: An Introduction R.J. Simmonds. Royal Society of Chemistry: Cambridge, 1992.
Natural Products their Chemistry and Biological Significance J. Mann, R.S. Davidson, J.B. Hobbs, D.V. Banthorpe and J.B. Harborne. Longman: Harlow, 1994.

NMR techniques in conformational analysis

Biomolecular NMR Spectroscopy J.N.S. Evans. Oxford University Press: Oxford, 1995.
Introduction to Organic Spectroscopy Oxford Chemistry Primer Number 43, L.M. Harwood and T.D.W. Claridge. Oxford University Press: Oxford, 1997.

Analysis of compound 8.63

S. Cumberbatch, M. North and G. Zagotto. *Tetrahedron*, 1993, **49**, 9049.

Computational conformational analysis

Computational Chemistry Oxford Chemistry Primer No 29, G.H. Grant and W.G. Richards. Oxford University Press: Oxford, 1995.
Practical Strategies for Electronic Structure Calculations W.J. Hehre. Wavefunction Inc.: Irvine, 1995.

8.13   Problems

1. If you did not or could not attempt question 5 in Chapter 1, you should be able to do so now. Draw similar diagrams for rotation around: (a) C2–C3 in butadiene; (b) the S–S bond in Me–S–S–Me; and (c) C2–C3 in 3-bromo-prop-1-ene.

2. It has been shown that $E_\theta = 0.042(\Delta\theta)^2$ where $\Delta\theta$ is the difference between the actual and preferred bond angles (in degrees) for sp^3 hybridized carbon atoms and E_θ is in kJ mol^{-1}. Use this equation to calculate E_θ for cyclopropane and cyclobutane. It has also been shown that $E_r = 1470(\Delta r)^2$ where Δr is the difference between the preferred and actual carbon–carbon single bond lengths in nanometres and E_r is in kJ mol^{-1}. What distortion of the carbon–carbon bonds in cyclopropane and cyclobutane would result in $E_r = E_\theta$.

3. Predict whether the *cis* or the *trans* isomer of a 1,3-disubstituted cyclobutane will be thermodynamically more stable. Repeat the analysis for the hypothetical case of a cyclobutane ring with a planar conformation. Explain why the following reaction occurs:

4. How many stereoisomers are there of the cyclobutane derivative shown below? Which of the stereoisomers are chiral? How, if at all, would the situation change if the CH_2 was changed to $C(Me)_2$?

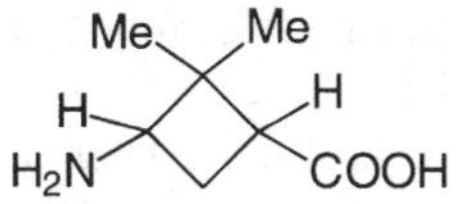

5. There is one conformation of a monosubstituted cyclopentane such as methylcyclopentane which is significantly lower in energy than any other conformation of the molecule. Draw clear diagrams (or build models) of methylcyclopentane with the ring in both envelope and half chair conformations and with the methyl group in all possible locations. For each structure, consider the interactions between the methyl group and the hydrogens and hence predict the structure of the lowest energy conformation.

6. Draw a clear diagram (or build a model) of *cis*-1,2-dimethylcyclohexane in a boat conformation and show that in this conformation the molecule possesses a plane of symmetry. What is the importance of this observation?

7. Suggest a reason why the C–C–C bond angles in cyclohexane are 111° rather than 109°. Your explanation should be consistent with the fact that the C–C–C bond angle in propane is also 111°.

8. Explain the following observations associated with the proton decoupled ^{13}C NMR spectrum of chlorocyclohexane. At room temperature, the spectrum shows four lines, although at −150°C eight lines are observed all at different positions to the four lines seen at room temperature. If the NMR solution is allowed to crystallize at −150°C, then the residual solution (species A) shows just four of the eight lines. Redissolving the crystals (species B), still at −150°C, gives a spectrum consisting of the other four of the eight lines. If either of the NMR solutions is allowed to warm to room temperature then just the original four lines are again observed. What is the relationship between species A and B?

9. Explain why *trans*-4-*tert*-butyl-cyclohexyl acetate (A) is hydrolysed by sodium hydroxide to the corresponding alcohol more rapidly than *cis*-4-*tert*-butyl-cyclohexyl acetate (B) under the same conditions.

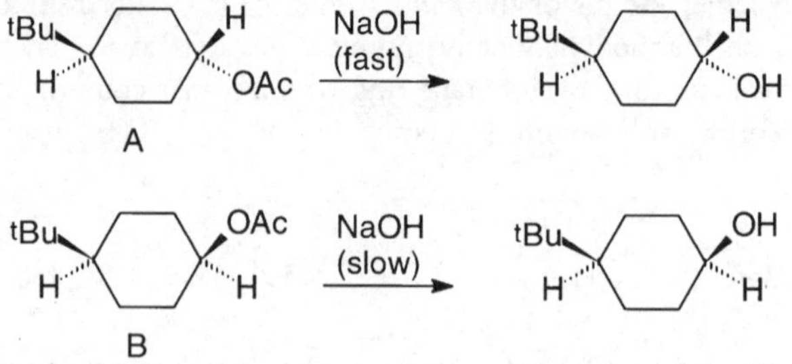

10. Carry out a conformational analysis of all possible isomers of: (a) a cyclohexane with two different substituents; and (b) a cyclohexane with three identical substituents. For each isomer, draw the structure of the minimum energy conformation (assuming that the cyclohexane ring adopts a chair conformation) and decide whether the compound is chiral or not.

11. Why is 60° often used as the torsional angle increment in grid searches when exploring the conformational space available to a molecule?

12. Explain why the ^{1}H NMR spectrum of *cis*-decalin measured at room temperature shows just three signals, whilst the corresponding spectrum of *trans*-decalin consists of five peaks. Would you expect either of these spectra to be temperature dependant?

13. Use the graph of the Karplus equation shown in **Figure 8.15** to predict the magnitude of the coupling constants between H_a and H_c; H_a and H_d; H_b and H_c; and H_b and H_d in the cyclic acetal shown below.

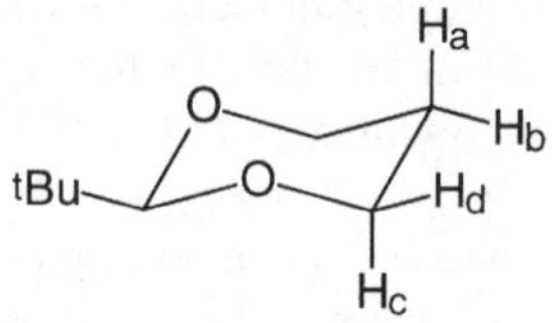

14. In Chapter 3, section 3.7.3, aziridines were stated as undergoing 'umbrella inversion' more slowly than acyclic amines, thus allowing optically active aziridines to be isolated. Suggest a reason why the 'umbrella inversion' of aziridines should be slower than that of acyclic amines. Hint: azetidines also have a slow rate of 'umbrella inversion' whilst five- or six-membered rings containing nitrogen atoms undergo 'umbrella inversion' at a rate similar to that of acyclic amines.

9 Stereochemistry of chemical reactions

9.1 Introduction

So far in this book we have mostly discussed the stereochemistry of individual compounds. Chemistry is, however, concerned with the reactions of chemicals and the stereochemistry of the starting materials or intermediates is often important in determining the outcome of a chemical reaction. Hence, the last two chapters of this book will deal with the stereochemistry of chemical reactions. In this chapter, the stereochemical consequences of reactions which take place at or near an existing stereocentre will be discussed, as will reactions which generate racemic products by addition reactions to alkenes. Chapter 10 will then show how non-racemic chiral products can be obtained from achiral starting materials.

It should be stressed, however, that reaction stereochemistry is an enormous area and these two chapters can only highlight the most important reactions. It is not possible to survey the stereochemical consequences of every possible reaction, nor is it desirable to do so; rather the reader should appreciate the underlying principles behind the stereochemical consequences of the reactions that are illustrated and be able to apply these to other reactions. The emphasis in these two chapters is on the reactions of organic compounds since these have been the most extensively studied and are the best understood.

Before starting a survey of stereochemistry of chemical reactions, two new terms need to be defined. A **stereoselective reaction** is one in which only one of a set of stereoisomers is formed exclusively or predominantly. In a **stereospecific reaction**, one stereoisomer of the starting material produces one stereoisomer of the product whilst another stereoisomer of the starting material gives a different stereoisomer of the product. It follows from these definitions that all stereospecific reactions are also stereoselective, but not all stereoselective reactions are stereospecific. A number of examples of both situations will be seen in the following sections.

9.2 Substitution reactions

When a tetravalent carbon atom is attached to a leaving group (X), a nucleophile (Y$^-$) may displace the leaving group as shown in **Scheme 9.1**. The effect of

Scheme 9.1

such a reaction is that the group X is substituted by the group Y. Such a reaction is called a **substitution reaction**. Substitution reactions of this type can only occur if the carbon atom bearing the leaving group is sp^3 hybridized. If the other three groups attached to the carbon atom are all different (and different to both X and Y), then both the starting material and product will contain a stereocentre, and the substitution reaction will affect the configuration of this stereocentre.

There are two extreme mechanisms by which such a substitution reaction can occur, the mechanisms being referred to as S_N2 and S_N1 respectively. In these labels, the S and N refer to substitution and nucleophilic respectively. The descriptors 1 and 2 describe the **molecularity** of the reaction, that is the number of molecules which are involved in the rate limiting step of the reaction. The two mechanisms have different stereochemical consequences and each will be discussed in turn.

9.2.1 S_N2 reactions

In an S_N2 reaction, the nucleophile and the electrophile (the species containing the leaving group) are both involved in the rate limiting (and only) step of the mechanism as shown in **Scheme 9.2**, giving a bimolecular reaction. Thus the nucleophile (Y^-) 'attacks' the carbon bearing the leaving group (X) at an angle of 180° to the C–X bond, to give in a single step the product via a trigonal bipyramidal transition state. A number of reasons have been proposed to account for the fact that the nucleophile approaches at 180° to the C–X bond. This

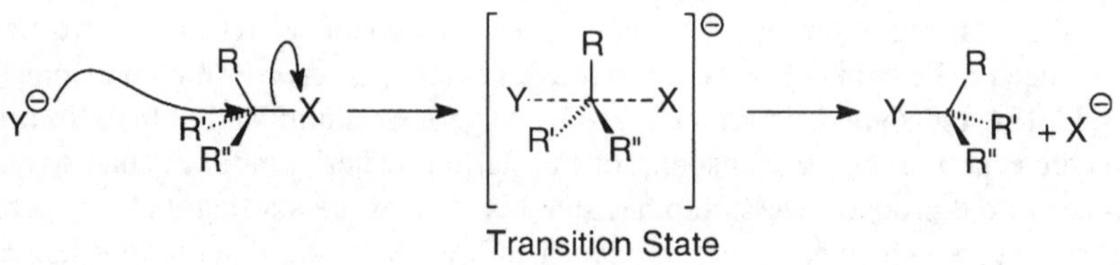

Transition State

Scheme 9.2

σ* orbital

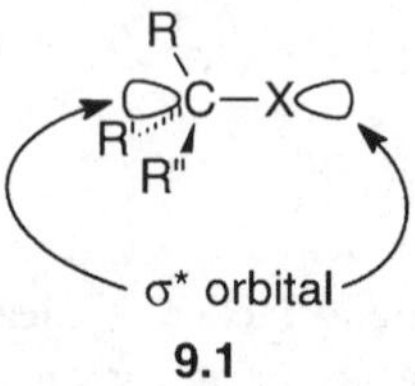

9.1

trajectory minimizes both steric and electronic repulsions between the nucleo-phile and the four substituents already attached to the carbon atom. Probably more importantly, however, the unoccupied C–X σ^* orbital is orientated in this direction as shown in structure **9.1**.

During the substitution reaction, an electron pair is donated by the nucleophile to the electrophile as represented by the curly arrow in **Scheme 9.2**. These electrons can only be donated into an unoccupied molecular orbital and for energetic reasons this will be the **Lowest energy Unoccupied Molecular Orbital (LUMO)** which is the σ^* orbital of the C–X bond. Similarly for energetic reasons, the electrons will be donated from the **Highest energy Occupied Molecular Orbital (HOMO)** of the nucleophile. Thus for reaction to occur there must be an interaction between the HOMO of the nucleophile and the LUMO of the electrophile, and it is this that determines the trajectory along which the nucleophile approaches the C–X bond. The orientations and relative energies of the HOMO and LUMOs of chemical species determine the nature of the reactions they undergo. These two orbitals are collectively referred to as **frontier orbitals**. In both **Scheme 9.1** and **Scheme 9.2**, the nucleophile has been shown as being negatively charged; this is not necessarily the case, since many neutral species which contain lone pairs of electrons or electrons held in π-orbitals are also nucleophiles.

An S_N2 reaction inverts the configuration at a stereocentre, as shown in **Scheme 9.2**. This is a direct consequence of the nucleophile approaching at 180° to the leaving group and is an example of the course of a chemical reaction being determined by stereoelectronic effects (cf. Chapter 8, section 8.7.5). The inversion of configuration during an S_N2 reaction is often called **Walden inversion**. This inversion may or may not result in a change in the stereo-chemical descriptor (R or S) of the stereocentre as the descriptor depends upon the relative priorities of the four substituents attached to the stereocentre which may be different in the starting material and the product. An S_N2 reaction is an example of a stereospecific reaction since each enantiomer of the starting material gives a different enantiomer of the product.

Walden inversion was first observed in 1923 during the sequence of reactions shown in **Scheme 9.3**, although it was not until the work of Hughes and Ingold in 1937 that the relationship between Walden inversion and the S_N2 reaction mechanism was established. Since in **Scheme 9.3** the same starting material **9.2** is converted into both enantiomers of the product **9.3** and **9.4**, at least one step in the sequence must involve inversion of configuration at the stereocentre. The only reaction which involves breaking a bond at the stereocentre is the one shown with a hashed arrow, so this is the only step which could invert the configuration of the stereocentre. Since this is known to be an S_N2 reaction, it follows that S_N2 reactions must result in inversion of configuration. It is worth pointing out that when this chemistry was carried out the absolute configuration of each of the compounds shown in **Scheme 9.3** was unknown; all that was known was that compound **9.2** was enantiomerically pure and the products **9.3**

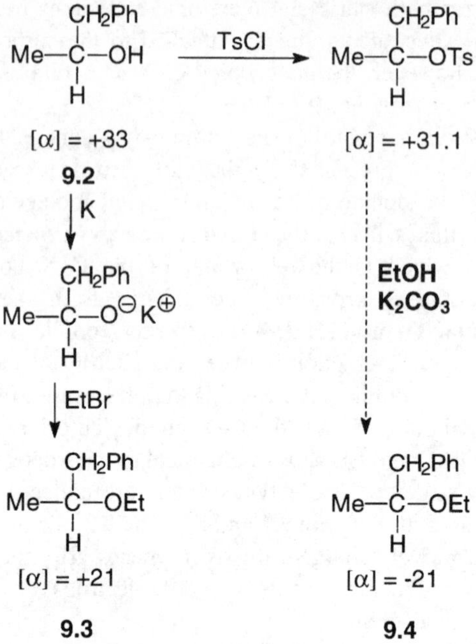

$$\underset{\overset{|}{\underset{H}{|}}}{Me-C-OH} \quad \xrightarrow{\text{TsCl}} \quad \underset{\overset{|}{\underset{H}{|}}}{Me-C-OTs}$$

CH₂Ph (top left)   CH₂Ph (top right)

$[\alpha] = +33$ $[\alpha] = +31.1$

9.2

↓ K

EtOH
K₂CO₃

CH₂Ph
$Me-C-O^{\ominus}K^{\oplus}$
H

↓ EtBr

CH₂Ph   CH₂Ph
$Me-C-OEt$ $Me-C-OEt$
H   H

$[\alpha] = +21$ $[\alpha] = -21$

9.3   **9.4**

Scheme 9.3

and **9.4** were enantiomers of one another. Since 1923, many other examples of S$_N$2 reactions which have been proven to proceed with inversion of configuration have been reported.

It is a requirement of an S$_N$2 reaction that the nucleophile attacks at 180° to the leaving group. However, in some cyclic systems this is not possible since the rest of the molecule is in the way. This is particularly common in bicyclic systems when the leaving group is at the bridgehead, an example being compound **9.5** where the carbon and hydrogen atoms shown emboldened are located directly behind the C–Br bond. Compound **9.5** and related species do not therefore undergo S$_N$2 substitution reactions.

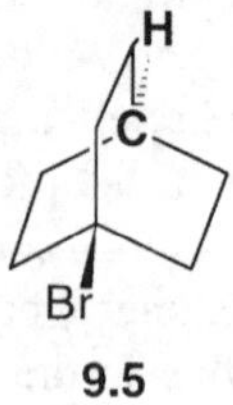

9.5

9.2.2 S$_N$1 reactions

An S$_N$1 reaction proceeds in two discrete steps and only the electrophile (the species bearing the leaving group) is involved in the first, rate limiting step. Thus in an S$_N$1 reaction, the C–X bond first breaks heterolytically producing a

carbenium ion **9.6** (also referred to as a carbocation or a carbonium ion) and X⁻.
The nucleophile (Y⁻) is only involved in the second, faster step where it
combines with the carbenium ion to produce the product as shown in **Scheme
9.4**. It can be shown using VSEPR theory (cf. Chapter 1, section 1.3) that since
the carbenium ion **9.6** consists of a carbon atom surrounded by just three bond
pairs of electrons it will be a planar species. The two faces of this planar
carbenium ion are enantiotopic (assuming no stereocentres are present within R,
R' or R''), so it is equally likely that the nucleophile will react from the *re*- or the
si-face and a racemic mixture of the two enantiomeric products **9.7** and **9.8** will
be formed.

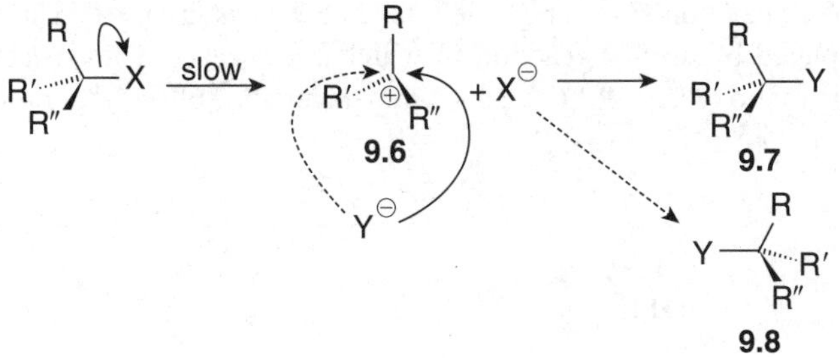

Scheme 9.4

Since an S_N1 reaction proceeds via a planar carbenium ion intermediate, if a
compound cannot form a planar carbenium ion it will not undergo S_N1 reactions.
This is common in bicyclic systems with bridgehead leaving groups such as
compound **9.9**, since the ring systems are so rigid that they cannot deform to form
a planar carbenium ion. A molecular model will make this very clear. These are
the same systems that were shown in section 9.2.1 not to undergo S_N2 reactions
(cf. structure **9.5**). Thus compound **9.9** is completely inert towards substitution
reactions (by an S_N1 or an S_N2 mechanism) as shown in **Scheme 9.5**.

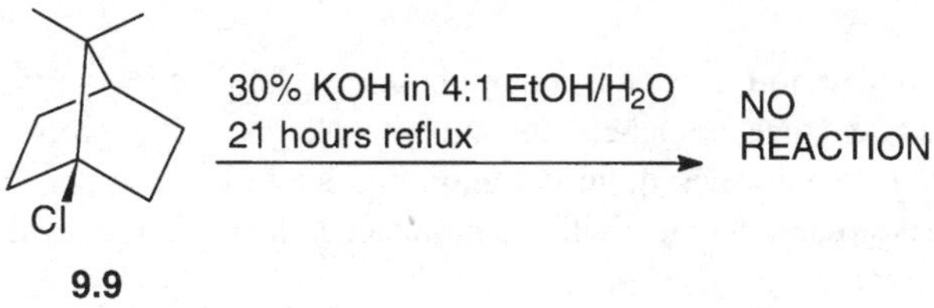

30% KOH in 4:1 EtOH/H_2O
21 hours reflux

NO
REACTION

9.9

Scheme 9.5

The S_N2 and S_N1 mechanisms represent two extreme reaction pathways for
substitution reactions. In many cases, a particular substitution reaction will
proceed in a way which is adequately represented by either an S_N2 or an S_N1
mechanism. In some cases, however, the actual mechanism is intermediate
between these two extremes and in such cases it is common to obtain a product
which is not racemic but which has a lower enantiomeric excess than the starting
material.

9.2.3 Substitution with neighbouring group participation

Occasionally, a substitution reaction is seen to occur with retention of configuration at a stereocentre, an example being the conversion of (*S*)-2-aminopropanoic acid **9.10** into (*S*)-2-hydroxypropanoic acid **9.11** upon treatment with nitrous acid as shown in **Scheme 9.6**. The explanation of this stereochemical outcome is that not one but two substitution reactions are occurring, each of which inverts the configuration of the stereocentre. Thus, after formation of the diazonium salt **9.12**, the carboxylate ion participates in an intramolecular substitution reaction producing α-lactone **9.13** with inversion of configuration at the stereocentre. This step of the mechanism occurs in the same way as an S_N2 reaction, but should not be called an S_N2 reaction since only one molecule **9.12** is involved. The reaction is then completed by an S_N2 reaction in which water reacts with α-lactone **9.13**, producing hydroxyacid **9.11** again with inversion of configuration at the stereocentre.

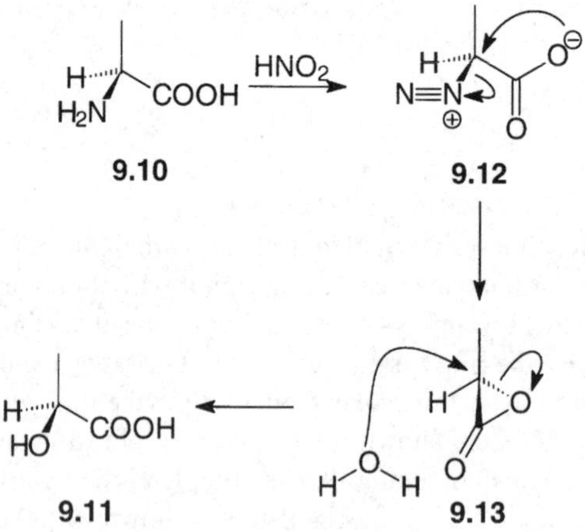

Scheme 9.6

The carboxylate ion in **9.12** is a neighbouring group which participates in the reaction and alters the stereochemical outcome. Neighbouring group participation in substitution reactions is quite common and should be expected whenever a nucleophile is suitably located within a molecule to form a three- to six-membered ring intermediate.

9.3 Elimination reactions

An elimination reaction occurs whenever two groups are expelled from a molecule. The most common case is when the two groups are located on adjacent carbon atoms (a 1,2-elimination) as shown in **Scheme 9.7**. Only 1,2-eliminations will be considered here. In the most general case of a 1,2-elimination reaction, the starting material contains two stereocentres both of which are destroyed in the

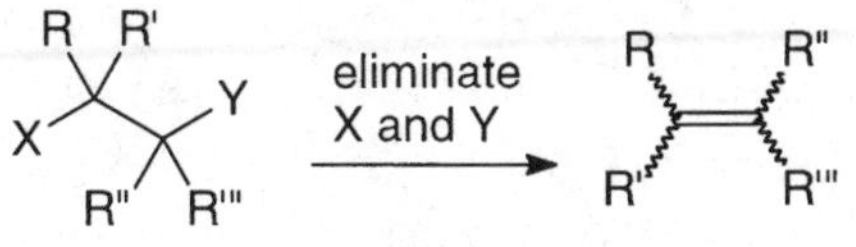

Scheme 9.7

course of the reaction. The product will be achiral, but contains an alkene which may exist as *cis*- and *trans*-isomers. Depending upon the mechanism of the elimination reaction and the relative configuration of the stereocentres in the starting material, the product may be formed as a single diastereomer or as a mixture of the two diastereomers. The most common eliminations occur by mechanisms called E2 and E1 respectively. The E stands for elimination, and the 1 or 2 refers to the molecularity of the rate determining step. These two mechanisms will be discussed in turn in the following sections, after which a third type of 1,2-elimination (syn-elimination) will be discussed.

Whatever the mechanism, elimination never occurs to give a bridgehead alkene in a bicyclic system unless each of the rings contains at least six atoms. This is called **Bredt's rule** and occurs because the bridgehead carbon atom cannot be planar as would be required if it was part of an sp^2 hybridized alkene system. Thus in the example shown in **Scheme 9.8**, bromide **9.14** undergoes an elimination reaction to give only alkene **9.15**. The bridgehead alkene **9.16** is not formed. The isomeric alkyl bromide **9.17** cannot undergo an elimination reaction at all, as this would have to produce the bridgehead alkene **9.16**. Thus bridge-head alkyl halides such as **9.17** do not undergo S_N1, S_N2 or elimination reactions. In fact, they do not react much like alkyl halides at all; their chemistry more closely resembles that of alkanes. This is an example of the stereo-chemistry of a system dramatically affecting its chemistry.

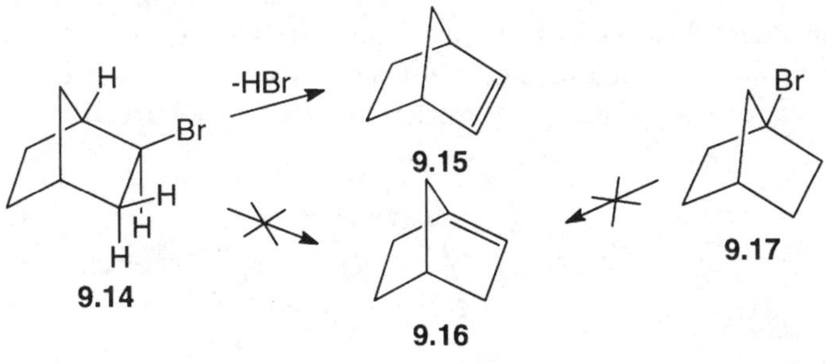

9.15

9.14

9.16

9.17

Scheme 9.8

9.3.1 E2 eliminations

In the E2 mechanism, the two groups which are eliminated (the two bromine atoms in the example shown in **Scheme 9.9**) must adopt an antiperiplanar conformation. The two groups are then eliminated in a concerted process,

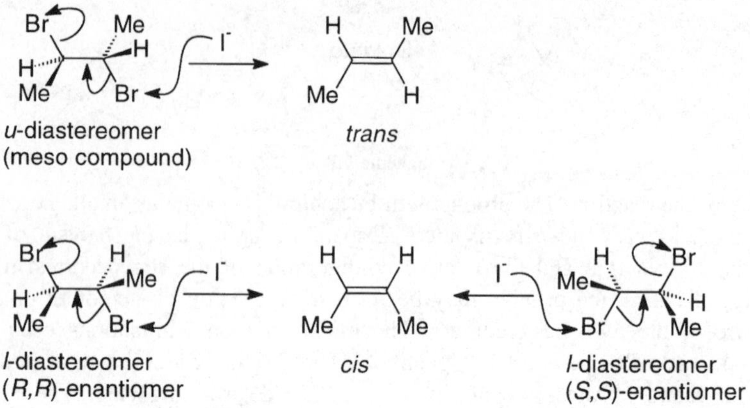

u-diastereomer
(meso compound)

trans

l-diastereomer
(*R*,*R*)-enantiomer

cis

l-diastereomer
(*S*,*S*)-enantiomer

Scheme 9.9

producing the alkene product in a single step. This mechanism is often called an anti-elimination. As a direct consequence of the antiperiplanar arrangement of the groups being eliminated and the concerted nature of the elimination, an E2 reaction is stereospecific. Thus each diastereomer of the starting material gives a single, different diastereomer of the product. In the example shown in **Scheme 9.9**, the *u*-diastereomer of the dibromide (which, in this case, is an achiral meso compound) gives only *trans*-2-butene, whilst the *l*-diastereomer of the dibromide produces only *cis*-2-butene. The *l*-diastereomer is chiral and can exist as a pair of enantiomers; however, both enantiomers will give the same diastereomer of the product.

The reason why an E2 reaction requires the two groups being eliminated to be antiperiplanar is stereoelectronic in nature. During the elimination of X and Y, the two electrons occupying the C–Y σ-bond are used to form the π-bond of the alkene and for this to be possible there must be an interaction between this σ-bond and the σ*-orbital of the C–X bond. This is only possible if the two groups being eliminated are antiperiplanar as shown in **Figure 9.1**.

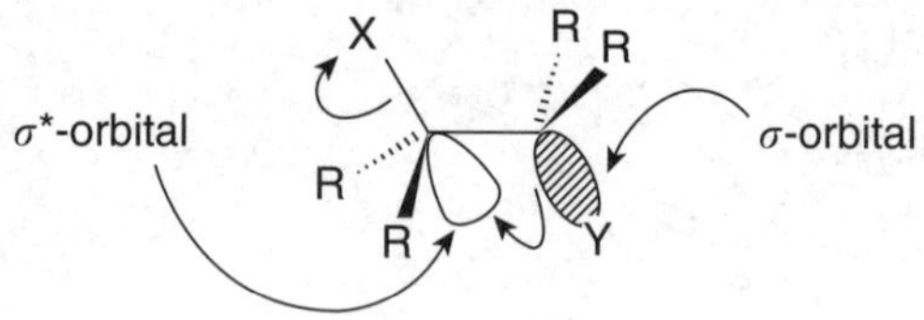

σ*-orbital

σ-orbital

Figure 9.1 The stereoelectronic requirement for an E2 elimination.

In cyclohexane systems, it is only possible for the two groups being eliminated to be antiperiplanar if the two groups are both in axial positions. In the isomer of hexachlorocyclohexane shown in structure **9.18**, there are no hydrogen and chlorine atoms *trans* to one another on adjacent carbon atoms. The minimum energy conformation of compound **9.18** has all of the chlorine atoms in equatorial positions, and the other chair conformation would have all of the hydrogen atoms in equatorial positions. Thus there is no conformation of compound **9.18** in which an adjacent hydrogen and chlorine are both in axial positions and hence compound **9.18** cannot eliminate HCl via an E2 mechanism.

Another example of the stereochemical consequences of an E2 elimination in cyclohexane derivatives is shown in **Scheme 9.10**. The minimum energy conformation of compound **9.19** is as shown in **Scheme 9.10**, since this allows two

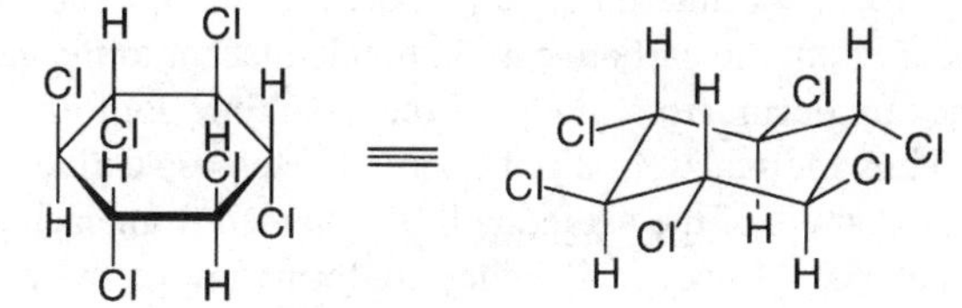

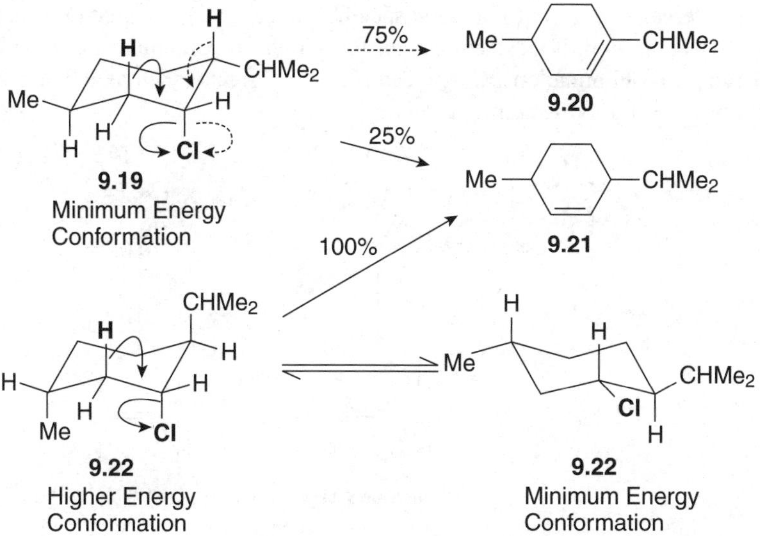

Scheme 9.10

of the three substituents to adopt equatorial positions. Compound **9.19** undergoes rapid elimination of HCl, as in the minimum energy conformation there are two hydrogen atoms axially located adjacent to the axial chlorine atom. Thus a mixture of the two regioisomeric alkenes **9.20** and **9.21** is produced. The minimum energy conformation of the diastereomeric cyclohexane derivative **9.22**, however, has all three substituents in equatorial positions, thus elimination of HCl cannot occur through this conformation. The other chair conformation of compound **9.22** is at a much higher energy as all three substituents are now in axial positions, so this conformer has only a very low population. However, in this high energy conformer, there is a single hydrogen axially orientated adjacent to the now axial chlorine atom. Thus compound **9.22** undergoes elimination of HCl much more slowly than compound **9.19**, but does so to give alkene **9.21** as the only product.

9.3.2 E1 eliminations

The mechanism of an E1 elimination is shown in **Scheme 9.11**. The reaction proceeds through the same intermediate (a carbenium ion) as an S_N1 reaction. Rotation can occur about the carbon–carbon bond adjacent to the carbenium ion and elimination can occur from both of the resulting conformations of the carbenium ion. Thus the reaction is not generally stereoselective; a mixture of the *cis*- and *trans*-isomers of the alkene will be obtained from each diastereomer of the starting material. However, if either conformation of the carbenium ion has a significantly lower energy than the other conformation, then elimination will occur preferentially through the lower energy conformation and the reaction will be stereoselective but not stereospecific. Since an E1 elimination proceeds via a carbenium ion, if a system cannot form a planar carbenium ion it will not undergo an E1 elimination. Bridgehead bicyclic systems (such as **9.9** and **9.17**) again provide a good example of this.

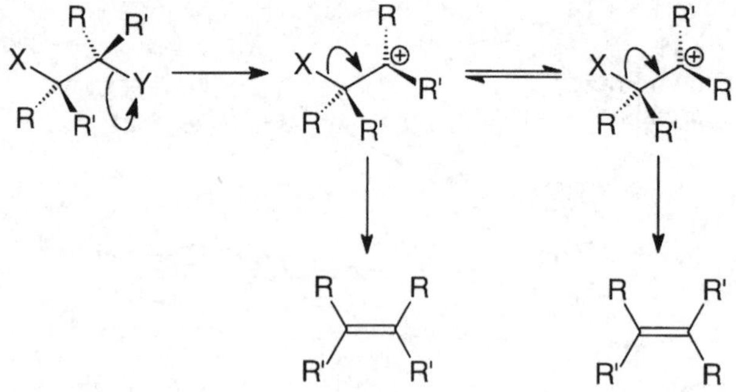

Scheme 9.11

9.3.3 Syn-eliminations

If an anti-elimination is impossible, then syn-elimination may occur provided the two groups being eliminated are synperiplanar. An example of a syn-elimination is the elimination of HCl from cyclopropane **9.23** as shown in **Scheme 9.12**. As the Newman projection of compound **9.23** shows, the chlorine and hydrogen atoms are exactly synperiplanar but the deuterium atom is not antiperiplanar to the chlorine. Hence, an anti-elimination of DCl cannot occur but syn-elimination of HCl can.

Another class of elimination reactions that occur by a syn-elimination are pyrolytic eliminations, which occur when appropriate substrates are heated in the absence of base. The best known examples are the eliminations of esters

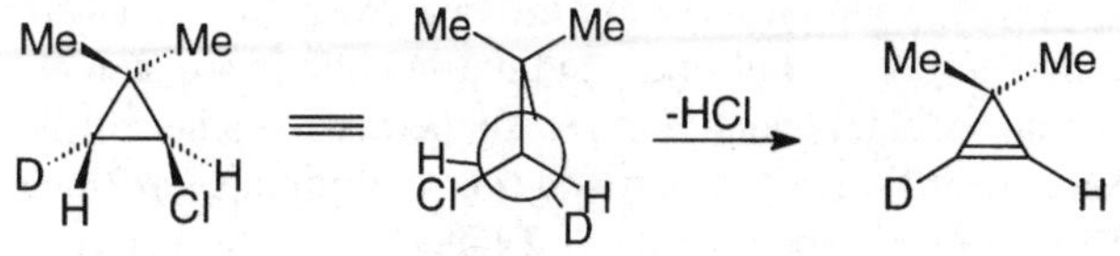

9.23

Scheme 9.12

(and their sulphur analogues), sulphoxides (or selenoxides) and amine oxides as shown in **Scheme 9.13**. The elimination of selenoxides is of particular synthetic utility, since the reaction occurs spontaneously at room temperature. In each case, the elimination occurs in a single step and in the transition state three pairs of electrons are moving around a five or six-membered ring. It is the requirement for the formation of this cyclic transition state that accounts for the observation of a syn-elimination, and the reactions are stereospecific, with each diastereomer of the starting material giving a single diastereomer of the alkene product.

If the elimination is occurring through a six atom cyclic transition state (as in the elimination of esters), then the transition state will adopt a chair conformation (**Scheme 9.13**), and the two groups being eliminated need not adopt an exactly synperiplanar conformation; a synclinal conformation is also compatible with the formation of the chair transition state. However, for elimination to occur through a five-membered ring transition state then the two groups being eliminated must be exactly synperiplanar.

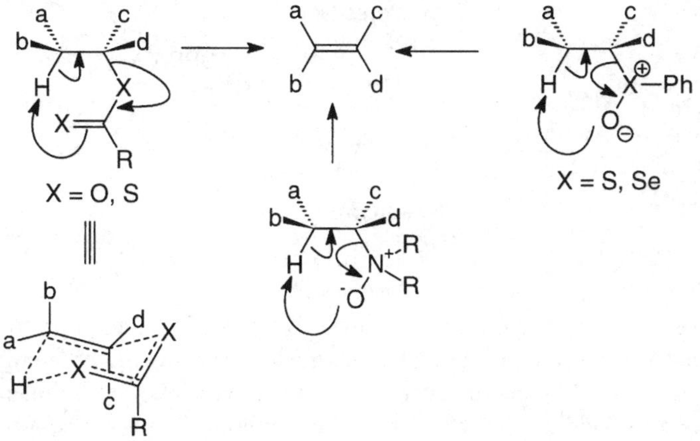

Scheme 9.13

9.4 Addition reactions to alkenes

Addition reactions to alkenes can be either syn-additions or anti-additions. In a syn-addition, both groups being added are attached to the same face of the double bond, whilst in an anti-addition the two groups are attached to opposite faces of the double bond. Both syn- and anti-additions are stereospecific with each diastereomer of the alkene giving a single diastereomer of the product as shown in **Scheme 9.14** for the anti-addition of bromine to 2-butene, and in **Scheme 9.15** for the syn-epoxidation of 2-butene.

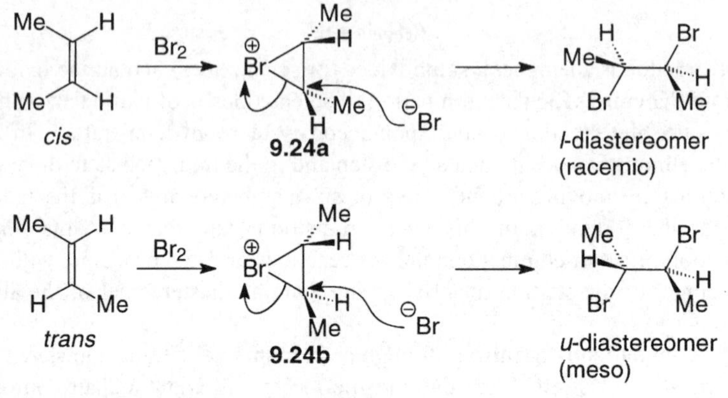

Scheme 9.14

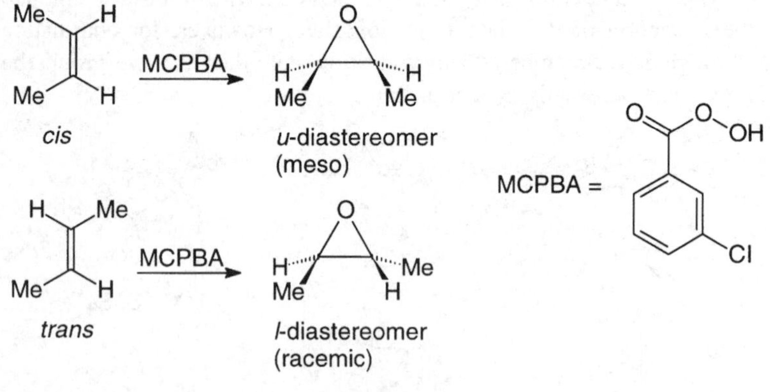

Scheme 9.15

A large number of reagents are known to undergo anti-addition to alkenes and, in most of these cases, a cyclic intermediate such as the bromonium ions **9.24a,b** are formed as intermediates. The formation of this cyclic intermediate occurs by a syn-addition, that is both carbon–bromine bonds of the bromonium ions **9.24a,b** are formed on the same face of the alkene. These cyclic inter-mediates then undergo an S_N2 reaction to give the final products and it is the inversion of configuration that is known to occur during an S_N2 reaction (cf. section 9.2.1) that accounts for the observed overall anti-addition. The product if chiral will be racemic, since the two faces of the achiral alkene are enantiotopic and so cannot be distinguished by an achiral reagent (cf. Chapter 7, section 7.2). It is thus equally likely that the initial addition will occur on either face of the alkene and it is this that determines which enantiomer of the product is formed.

The common reactions which occur by a syn-addition to the alkene are epoxidation with *meta*-chloroperbenzoic acid (MCPBA **Scheme 9.15**), cyclopropanation (addition of a carbene), *bis*-hydroxylation (using potassium permanganate or osmium tetroxide), hydrogenation (using H_2 and a catalyst) and hydroboration (with HBR_2). In each of these reactions, the addition is thought to be a concerted process in which the two new σ-bonds are formed simultaneously as illustrated for hydroboration in **Scheme 9.16**. It is this concertedness which ensures a syn-addition, since a concerted anti-addition is difficult or impossible to achieve for steric reasons. Stereoelectronic effects also require concerted additions to be syn-additions, as will be discussed in section 9.6. Again, the product if chiral will be racemic since the two faces of the alkene are enantiotopic, so an achiral reagent is equally likely to react on either face.

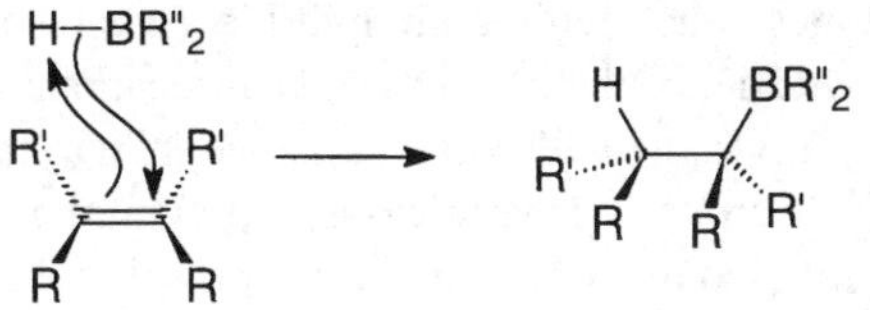

Scheme 9.16

For an anti-addition to a cyclohexene, the initial product is always the *trans*-diaxial conformer since this is the only way in which the two groups can add anti to one another. The diaxial adduct can usually undergo a chair–chair conformational exchange to give the more stable diequatorial conformer as the observed product as shown in **Scheme 9.17** for the addition of bromine to cyclohexene. However, if 3-*tert*-butyl-cyclohexene **9.25** is brominated, then the strong preference of the *tert*-butyl group to adopt an equatorial position (cf. Chapter 8, section

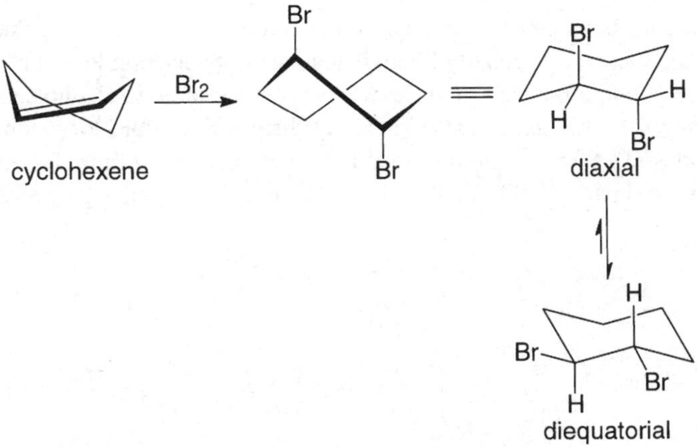

cyclohexene

diaxial

diequatorial

Scheme 9.17

8.7.1a) ensures that the observed product is the conformation in which the two bromine atoms are still in axial positions as shown in **Scheme 9.18**.

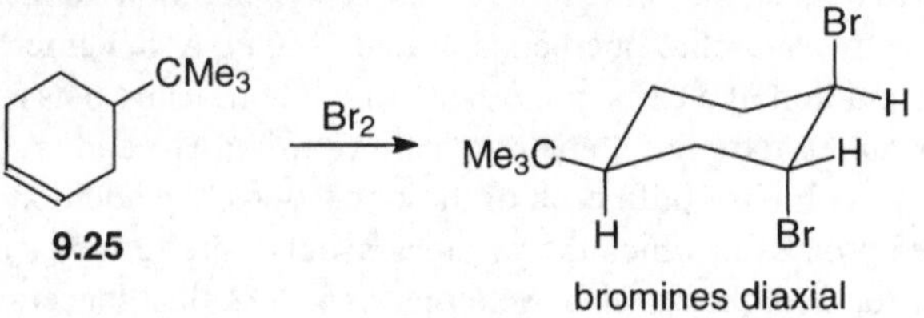

Scheme 9.18

9.5 Addition of nucleophiles to aldehydes and ketones

Aldehydes and ketones react with both hydride (from sodium borohydride, lithium aluminium hydride, or their derivatives) and carbanions to give alcohols in a process which involves the addition of a nucleophile to the carbonyl bond as shown in **Scheme 9.19**. In both general cases, a prochiral aldehyde or ketone is converted into a chiral alcohol. There are two types of carbanion which are not included in **Scheme 9.19**; these are enolates and phosphorus ylides, which are considered separately in sections 9.5.3 and 9.5.4.

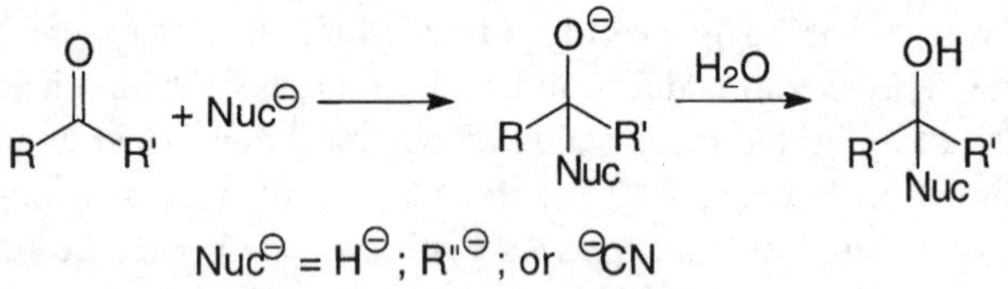

$$Nuc^{\ominus} = H^{\ominus};\ R''^{\ominus};\ or\ {}^{\ominus}CN$$

Scheme 9.19

Provided no stereocentre is present in the ketone, the two faces of the ketone are enantiotopic, so it is equally likely that an achiral nucleophile will react from either face giving a racemic mixture of the two enantiomeric alcohols as shown in **Scheme 9.20**. This scheme also illustrates the preferred trajectory along which nucleophiles react with a carbonyl bond; i.e. perpendicular to the C–C=O plane and at approximately 109° to the carbonyl bond. A reaction pathway/energy

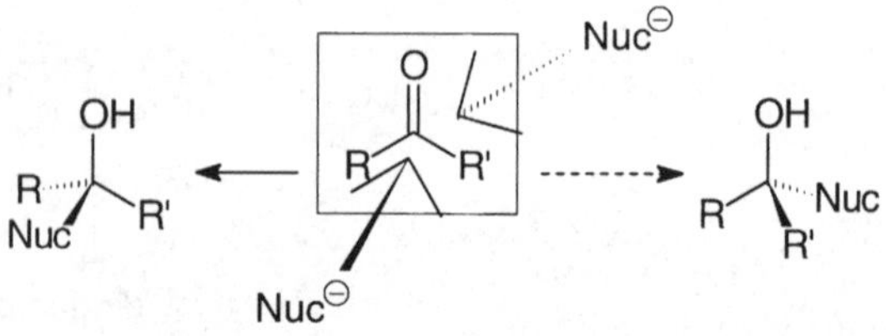

Scheme 9.20

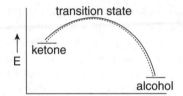

Figure 9.2 Reaction pathway/energy diagram for the addition of a nucleophile to a prochiral aldehyde or ketone.

diagram for this process is given in **Figure 9.2** which shows that the transition states leading to the two alcohols are of equal energy. Thus there is the same energy barrier to formation of either alcohol, which explains why a racemic mixture is produced.

However, if the ketone contains a stereocentre, then the two faces of the ketone are not enantiotopic but diastereotopic and the two alcohol products are not enantiomers, but diastereomers (**Scheme 9.21**) and hence will have different

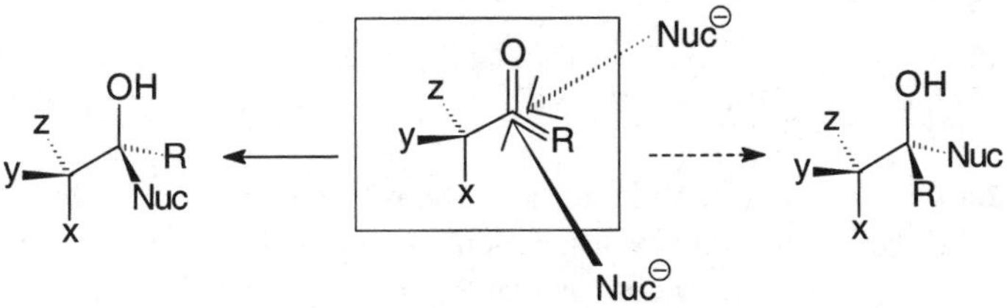

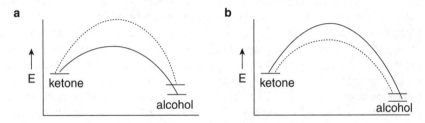

Scheme 9.21

energies. The transition states leading to the alcohols will also be diastereomeric and hence have different energies. Thus, the two alcohols may be formed in unequal amounts. The addition of a nucleophile to an aldehyde or ketone may be carried out under kinetic or thermodynamic control. In a kinetically controlled reaction, it is the relative energies of the transition states leading to products that determine which product is formed. For a thermodynamically controlled reaction, however, it is the relative energies of the products themselves which are important. Thus a kinetically controlled reaction may give the same product as a thermodynamically controlled reaction (**Figure 9.3a**), or a different product

Figure 9.3 Reaction pathway/energy diagrams for the addition of a nucleophile to a chiral aldehyde or ketone.

(Figure 9.3b). For addition of nucleophiles to chiral carbonyl compounds, both the transition states and the products are at different energies, so unequal amounts of the alcohols may be formed whether the reaction is occurring under kinetic or thermodynamic control. However, the ratio of the two alcohols may be different depending upon whether the reaction is carried out under kinetic or thermodynamic control. This analysis is true whether the ketone is racemic or enantiomerically pure. In sections 9.5.1 and 9.5.2, the stereochemistry of addition reactions which are carried out under kinetic control will be discussed, whilst sections 9.5.3 and 9.5.4 will consider two examples of thermodynamically controlled addition reactions to aldehydes and ketones.

It is possible to predict which of the two diastereomeric alcohols obtained by the addition of a nucleophile to a chiral ketone will be formed in excess; however, two different models are needed. If the ketone contains functional groups capable of forming a chelate with the metal counter ion to the nucleophile, then the reaction will follow **chelation control**. If no chelating groups are present then the stereochemistry will be predicted by the **Felkin–Anh model**.

9.5.1 Chelation control

The addition of a nucleophile to a carbonyl compound will be chelation controlled if the ketone contains a functional group in the α- or β-position to the carbonyl which has lone pairs available, and the metal counterion to the nucleophile is coordinatively unsaturated and is thus capable of forming a chelated complex as shown in structures **9.26** and **9.27**. In these cases, a rigid

9.26   **9.27**   X= OH, NH$_2$, NR$_2$, OR, C=O, SH etc

cyclic structure is formed in which the front and back faces of the ring are different. The nucleophile then attacks the carbonyl from the less hindered face of the chelate, resulting in the formation of one diastereomer of the product alcohol. An example of this is the reduction of benzoin **9.28** by sodium borohydride as shown in **Scheme 9.22**. The sodium ion first forms a chelate, and the hydride anion (as a borohydride complex not shown in **Scheme 9.22**) then attacks the chelate from the least hindered face giving a single diastereomer (the *u*-diastereomer) of diol **9.29**. This reaction is stereoselective since only one of the three possible stereoisomers of compound **9.29** is formed, but is not stereospecific since both enantiomers of benzoin **9.28** give the same diastereomer of diol **9.29**.

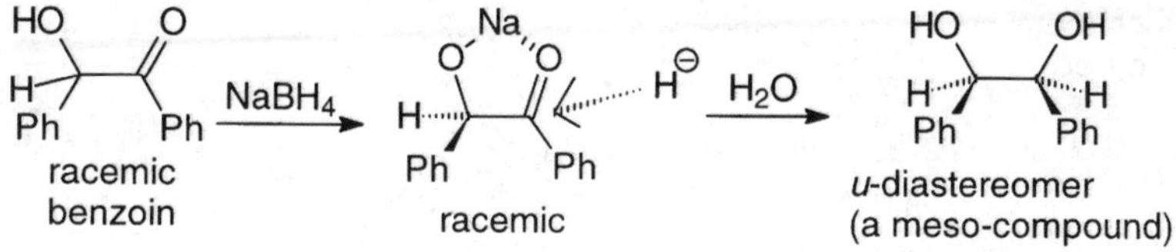

9.28   **9.29**

Scheme 9.22

9.5.2 The Felkin–Anh model

The Felkin–Anh model can be used to predict the stereochemistry of nucleophilic attack at ketones which do not contain a chelating group or when the nucleophile is associated with a counterion that is not capable of forming a chelate. Although the Felkin–Anh model successfully predicts the outcome of these additions in most cases, it may fail when an electron withdrawing substituent (such as a halogen) is one of the groups attached to the stereocentre. Variations of the Felkin–Anh model to cope with these situations have been developed and are included in the further reading.

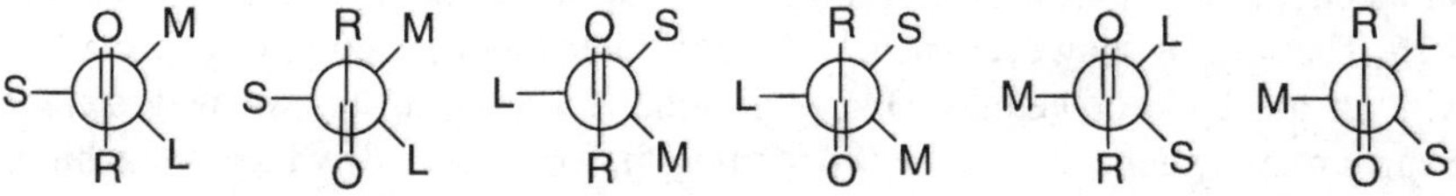

9.30a   **9.30b**   **9.31a**   **9.31b**   **9.32a**   **9.32b**

Consider the Newman projections **9.30–9.32a,b** of the six conformations of a chiral ketone in which one of the substituents is orthogonal to the carbonyl bond. In these projections, L, M and S represent the largest, medium and smallest groups respectively attached to the carbon adjacent to the carbonyl group. The preferred conformations will be **9.31a,b**, since in these conformations the largest substituent on the stereocentre (L) is orthogonal to the carbonyl oxygen and R group. Attack of a nucleophile will occur preferentially on conformer **9.31b**, rather than conformer **9.31a**, since in the former the nucleophile need only approach close to the small substituent whilst attack on conformer **9.31a** would require that the nucleophile approach close to either the medium or large substituents as shown in **Scheme 9.23**. This is the Felkin–Anh model and it correctly predicts the major diastereomer observed in many non-chelated reactions. An example is shown in **Scheme 9.24**.

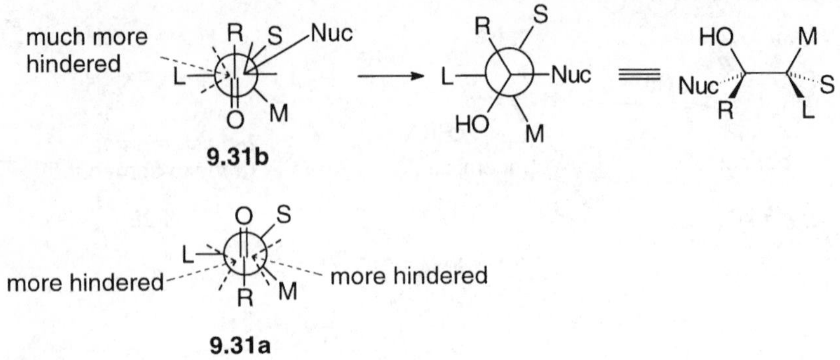

9.31b

9.31a

Scheme 9.23

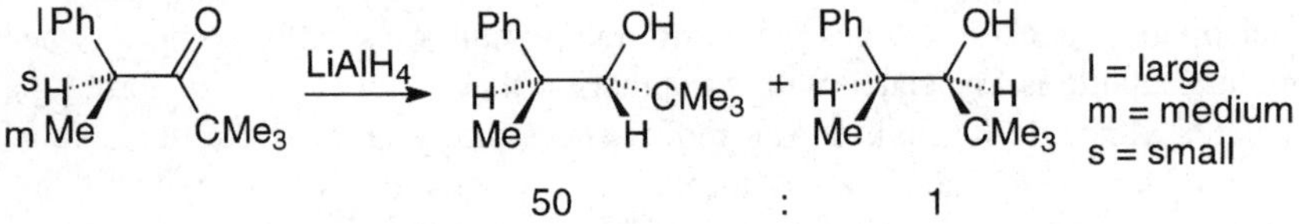

l = large
m = medium
s = small

50   :   1

Scheme 9.24

9.5.3   The aldol reaction

Strictly, the aldol reaction refers to the reaction between the enol or enolate of an aldehyde or ketone and a second aldehyde or ketone as shown in **Scheme 9.25**. The term is, however, widely used to refer to the reaction of any enolate with an aldehyde or ketone. During an aldol reaction, up to two new stereocentres may be generated from two starting materials both of which are achiral. Unlike the addition reactions discussed in sections 9.5.1 and 9.5.2, the aldol reaction is reversible, and the observed product is usually that corresponding to thermodynamic control.

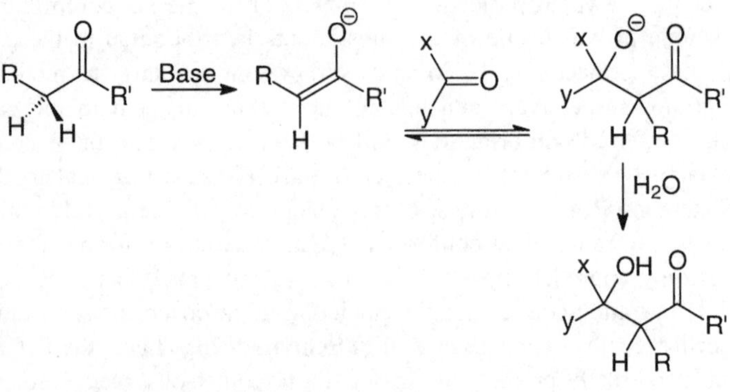

Scheme 9.25

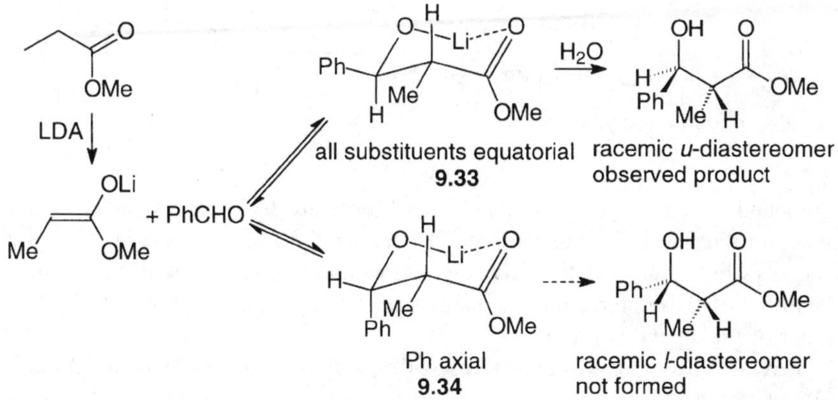

Scheme 9.26

In the majority of cases, the initial product of an aldol reaction is a six-membered ring cyclic chelate in which the metal counterion to the base is coordinated to both the carbonyl and hydroxyl oxygens of the product. This cyclic chelate adopts a chair conformation, and the most stable product (i.e. the thermodynamically controlled product) is the one in which as many substituents as possible are located in equatorial positions on the six-membered ring. An example of such a reaction is shown in **Scheme 9.26.** The aldol reaction between the lithium enolate of methyl propionate and benzaldehyde will initially give a mixture of the two possible chelates **9.33** and **9.34**. However, since the reaction is reversible, these two chelates will equilibrate and the most stable chelate **9.33** will predominate. Protonation of **9.33** then leads to the *u*-diastereomer of the product.

In this section it has only been possible to introduce briefly the stereochemical aspects of the aldol reaction. This is one of the central reactions in organic chemistry and a large amount of work has been done on the stereochemistry of this reaction. The reader may realize that there are other factors that can affect the stereochemical outcome of the reaction; these include the geometry of the enolate (*E* or *Z*), the reaction temperature (since at low temperatures the chelates may not equilibrate, so the kinetically controlled product may be obtained) and the nature of the metal. In some cases, the metal will not be able to form a chelate, so the mechanism described in **Scheme 9.26** cannot operate in these cases. More detail on the stereochemistry of the aldol reaction will be found in the further reading.

9.5.4   The Wittig reaction

The Wittig reaction is the reaction between an aldehyde or ketone and a phosphorus stabilized carbanion (an ylid) to generate eventually an alkene and a phosphine oxide. In the general reaction (**Scheme 9.27**), both the carbonyl

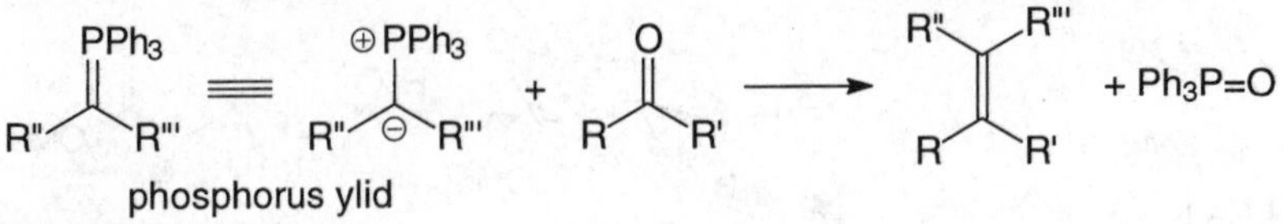

Scheme 9.27

compound and the ylid are prochiral, and there are two diastereomers of the alkene product (the *E*- and *Z*-isomers). Neither the starting materials nor the products are chiral, but the reaction intermediates are, and it is the stereochemistry of these chiral intermediates that directly determines which diastereomer of the alkene is produced.

There are two types of phosphorus ylides, stabilized ylides such as **9.35** and **9.36** where one of the groups adjacent to the carbanion is capable of stabilizing the negative charge by delocalization, and unstabilized ylides where no such groups are present. Wittig reactions with unstabilized ylides usually give predominantly the *Z*-isomer of the alkene, whilst Wittig reactions using stabilized ylides produce mainly the *E*-isomer of the alkene.

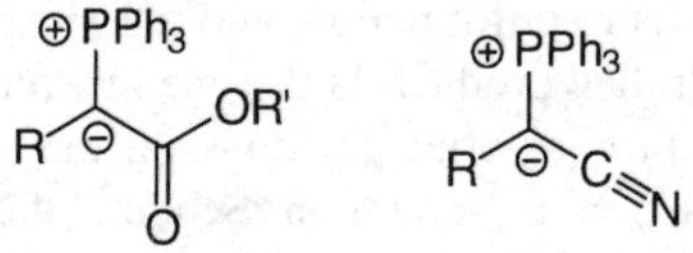

9.35   **9.36**

The mechanism of the Wittig reaction leading to a disubstituted alkene is shown in **Scheme 9.28**. The first step is the nucleophilic addition of the phosphorus stabilized carbanion to the carbonyl compound producing a betaine **9.37**. During this step, two new stereocentres are generated so there are two possible diastereomers (both of which will be a pair of enantiomers) of the betaine **9.37a** and **9.37b**. Isomer **9.37a** is thought to be both the kinetic and the thermodynamic product of the reaction since it minimizes steric repulsions between R and R′ in both the betaine and the transition state leading to the betaine. Unstabilized ylides are highly reactive and, for these species, this first step of the mechanism is irreversible so the subsequent reactions can only occur from intermediate **9.37a**. Stabilized ylides, however, are much more stable, as a result of which the formation of the betaines **9.37** is thought to be reversible and an equilibrium will be established between **9.37a** and **9.37b**.

The second step of the reaction mechanism involves the formation of phosphaoxetanes **9.38a** and **9.38b**, a process which involves a 180° rotation around the newly formed carbon–carbon bond followed by phosphorus–oxygen bond formation. The result of the 180° rotation, is that the R and R′ groups which were antiperiplanar in betaine **9.37a** become synperiplanar and eclipse one another in phosphaoxetane **9.38a**. Conversely, the less stable betaine **9.37b** leads to the more stable phosphaoxetane **9.38b** in which the R and R′ groups are on

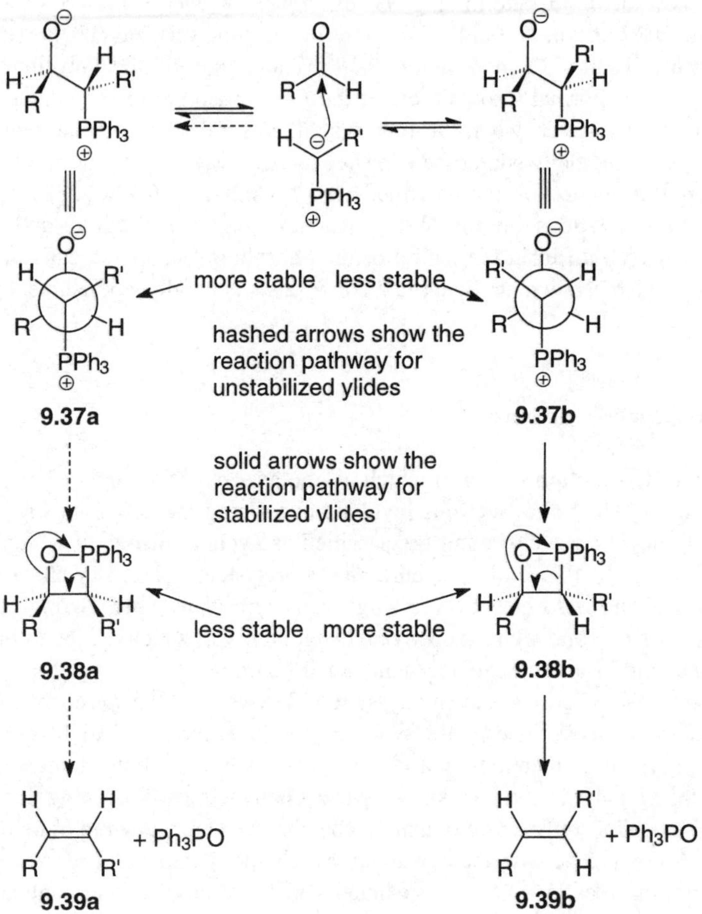

Scheme 9.28

opposite sides of the four-membered ring. For unstabilized ylides, since the first step of the mechanism is irreversible, the reaction has no option but to form phosphaoxetane **9.38a**. For stabilized ylides, however, an equilibrium is established between betaines **9.37a** and **9.37b**, and this allows the thermodynamically more stable phosphaoxetane **9.38b** to predominate.

The final step of the reaction mechanism is a syn-elimination (cf. section 9.3.3) of triphenylphosphine oxide from the phosphaoxetanes **9.38a** and **9.38b** generating the Z-alkene **9.39a** and the E-alkene **9.39b** respectively. Thus, in summary, unstabilized ylides irreversibly form the more stable betaine **9.37a** which leads through the less stable phosphaoxetane **9.38a** to the Z-alkene **9.39a**. Stabilized ylides reversibly form an equilibrium mixture of betaines **9.37a** and **9.37b** in which **9.37a** will predominate but which through **9.37b** can lead to the more stable phosphaoxetane **9.38b** and hence to the E-alkene **9.39b**.

The key factor in determining the difference in energy between the inter-mediates **9.37a,b** and **9.38a,b** is the size of the four variable substituents, two hydrogens, R and R′ in **Scheme 9.28**. Where two of the substituents are hydrogen, it is normal to obtain either the *E*- or *Z*-isomer of the alkene almost exclusively. However, when the four substituents are all of similar size then a much lower specificity should be expected as there will be little or no difference between the energy of the intermediates **9.37a,b** and **9.38a,b**. There are a number of variations on the Wittig reaction (e.g. the Wadsworth–Emmons reaction) which differ in the nature of the phosphorus stabilized carbanion, but the mechanism shown in **Scheme 9.28** is also generally applicable to these cases.

9.6   Pericyclic reactions

A **pericyclic reaction** is one in which all the new bonds are created simultane-ously and in which the electrons involved in forming the new bonds move in a circle. Pericyclic reactions can be classified as **cycloadditions** in which a new ring is formed by the coming together of two π-systems: **electrocyclic reactions** in which a ring is formed from a single π-system or in which a ring opens to give a π-system; and **sigmatropic rearrangements** in which a σ-bond migrates from one end of a conjugated system to the other end.

All of these reactions are usually stereoselective, and the stereochemistry of the reaction is determined by the symmetry of the HOMO and LUMOs (section 9.2.1). Pericyclic reactions may occur thermally or photochemically, and examples of some of the best known pericyclic reactions illustrating the stereo-selectivity and stereospecificity that can be observed are shown in **Scheme 9.29**. In the following sections, the way in which the stereochemical outcome of cycloaddition reactions can be predicted will be discussed. The application of the same concepts to predict the stereochemistry of other pericyclic reactions is discussed in the further reading.

9.6.1   Diels–Alder reactions: [4π + 2π] cycloadditions

Synthetically, the most important pericyclic reaction is the Diels–Alder reaction, which is a cycloaddition between a diene (or other species which contains two conjugated π-bonds and hence four electrons in π-orbitals) and an alkene or other species which contains a π-bond (containing two electrons) and is referred to as the dienophile. The basic Diels–Alder reaction is shown in **Scheme 9.30**. As was discussed in section 9.2.1, during a chemical reaction the electrons in the HOMO of the nucleophile interact with the LUMO of the electrophile. For a cycloaddition reaction, it does not matter which of the two species is considered as the electrophile and which as the nucleophile although, in practice, most Diels–Alder reactions are carried out on systems containing an electron deficient

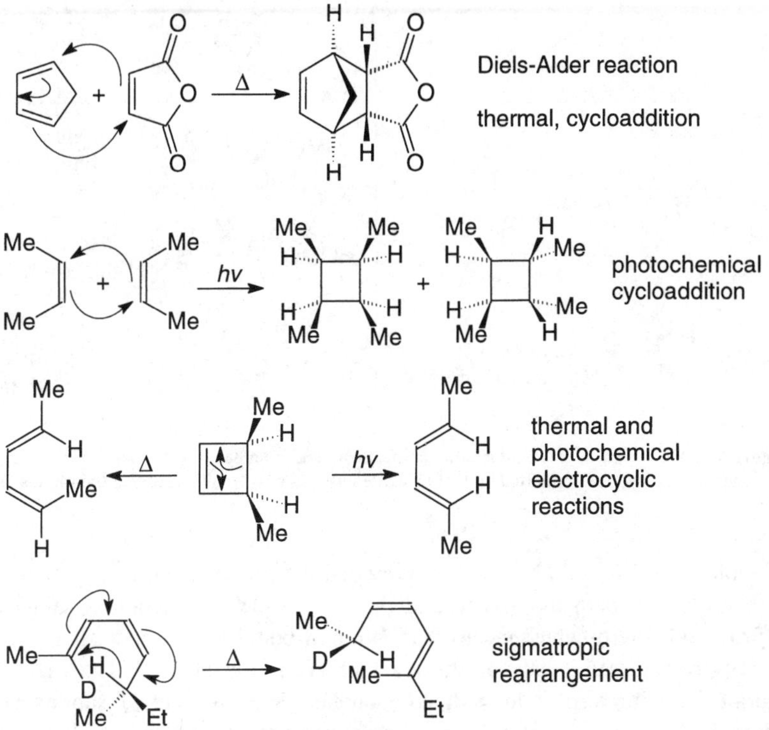

Diels-Alder reaction

thermal, cycloaddition

photochemical cycloaddition

thermal and photochemical electrocyclic reactions

sigmatropic rearrangement

Scheme 9.29

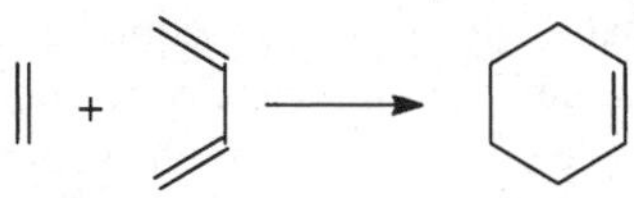

Scheme 9.30

alkene which is best considered as the electrophile, and an electron rich diene which is considered as the nucleophile. The key orbitals which then need to be considered are the LUMO of the alkene and the HOMO of the diene.

Both components of a Diels–Alder reaction contain π-bonds and, in such systems, the HOMO will always be a π-orbital and the LUMO a π^*-orbital. Molecular orbital diagrams of the π and π^* molecular orbitals of both ethene and butadiene are shown in **Figure 9.4**, where the HOMO and LUMOs are indicated. For a Diels–Alder reaction (or any other cycloaddition reaction) to take place, there must be a bonding interaction between the LUMO of the alkene and the HOMO of the diene at each atom where a new bond is being formed. Another way of saying this is that the lobes of the HOMO and LUMO which overlap to form the new σ-bonds must have the same phase, the phase being indicated by the shading of the orbitals. As can be seen in **Figure 9.5**, this is

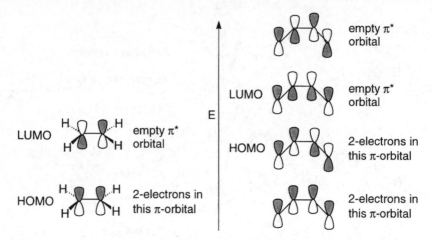

Figure 9.4 The π and π^* molecular orbitals of ethene and butadiene. Note that for butadiene the hydrogen atoms have been omitted and all orbital coefficients are shown as being the same size for clarity.

possible for a Diels–Alder reaction provided that the reaction is a syn addition with respect to both the alkene and the diene. Such a reaction is said to be **suprafacial**, a term which means that both new bonds are formed using lobes of orbitals on the same face of the reactant. Thus the Diels–Alder reaction is suprafacial with respect to both components. The opposite of suprafacial is **antarafacial**, which implies that the new bonds are formed using the lobes of an orbital on opposite faces of a molecule. Antarafacial cycloaddition reactions are uncommon since they are usually prevented by steric effects.

It is the requirement that a Diels–Alder reaction be suprafacial with respect to both components that directly determines the stereochemistry of the reaction. Thus groups which are located *cis* or *trans* to one another on either the alkene or diene retain their relative orientation in the product, and both new σ-bonds which are formed are always *cis* to one another. A number of examples of Diels–Alder reactions illustrating these stereochemical aspects are shown in **Scheme 9.31**. There are two other stereochemical aspects of a Diels–Alder reaction: the first of which is that the diene must be able to adopt the *s-cis* conformation since this is required to allow the two ends of the diene system to be correctly located to form bonds to the alkene.

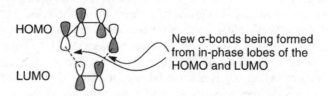

New σ-bonds being formed from in-phase lobes of the HOMO and LUMO

Figure 9.5 The orbital overlap during a Diels–Alder reaction.

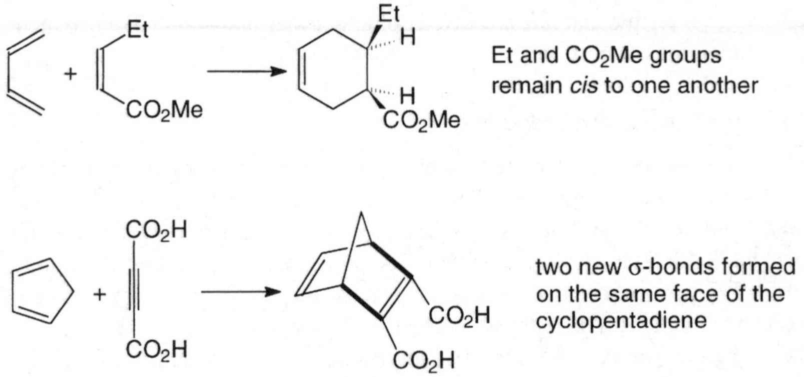

Et and CO$_2$Me groups remain *cis* to one another

two new σ-bonds formed on the same face of the cyclopentadiene

Scheme 9.31

The final stereochemical aspect of the Diels–Alder reaction is that, in some cases, two diastereomers of the product may be formed which are referred to as the **endo**- and **exo**-isomers. These two diastereomers arise from the two possible orientations in which the diene and alkene can react suprafacially as shown in **Scheme 9.32** for the reaction between cyclopentadiene and maleic anhydride.

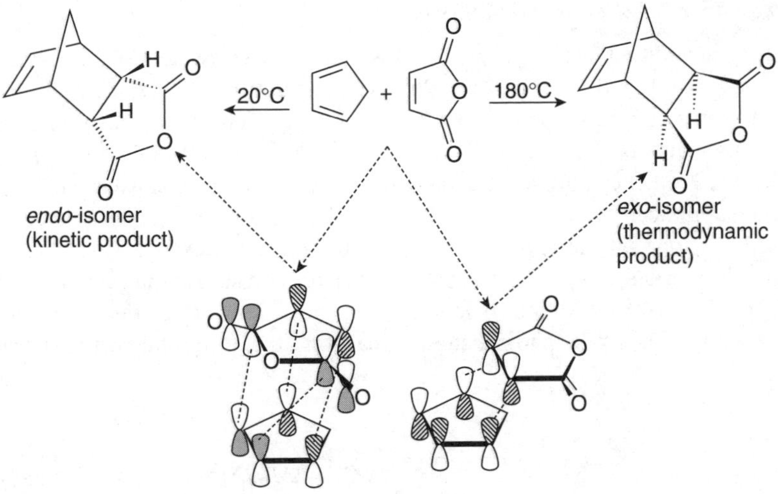

endo-isomer
(kinetic product)

exo-isomer
(thermodynamic product)

Scheme 9.32

Generally, the *endo*-isomer is the kinetic product of a Diels–Alder reaction despite the fact that it is the thermodynamically less stable product and is formed through the most sterically hindered transition state. The transition state leading to the *endo*-product is, however, often stabilized by favourable inter-actions between molecular orbitals not directly involved in the Diels–Alder reaction; these are shown by filled grey orbitals in **Scheme 9.32**. Such inter-actions are called **secondary orbital interactions**. The Diels–Alder reaction is, however, reversible and if it is carried out under thermodynamic control then the *exo*-isomer can be obtained.

9.6.2   1,3-dipolar cycloadditions

In the Diels–Alder reaction, the 4π electrons are provided by the four atoms of a diene. There are a number of species, however, which contain 4π electrons distributed over just three atoms: examples include ozone **9.39** and diazomethane **9.40**. This type of compound is called a 1,3-dipole and they undergo $[4\pi + 2\pi]$ cycloaddition reactions in exactly the same way (suprafacial with respect to both components) as a diene but give a product containing a five-rather than six-membered ring. The allyl anion **9.41** although not a 1,3-dipole also contains 4π electrons spread over three atoms and undergoes cycloaddition reactions with alkenes.

9.39   **9.40**   **9.41**

9.6.3   [2π + 2π] cycloaddition reactions

The cycloaddition reaction between two alkenes (or related species such as ketones) is referred to as a $[2\pi + 2\pi]$ cycloaddition reaction since each of the components contributes two π-electrons to the reaction. As is shown in **Figure 9.6**, this reaction, if carried out thermally, must be antarafacial with respect to one of the alkenes and suprafacial with respect to the other since the suprafacial/suprafacial interaction does not result in in-phase overlap between the HOMO of one alkene and the LUMO of the other. However, the antarafacial reaction is usually impossible for steric reasons as discussed in section 9.6.1.

    $[2\pi + 2\pi]$ cycloaddition reactions can, however, be carried out photochemically. During a photochemical reaction, the electromagnetic radiation

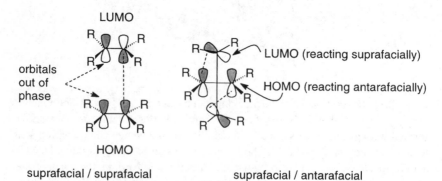

Figure 9.6 Orbital overlap during a $[2\pi + 2\pi]$ cycloaddition.

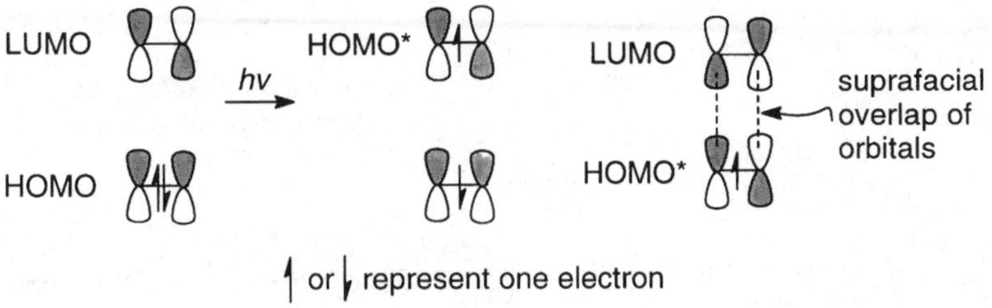

↑ or ↓ represent one electron

Figure 9.7 Orbital overlap during a photochemical, suprafacial [2π + 2π] cycloaddition.

promotes an electron from the HOMO to the LUMO of one of the reactants. This creates two singly occupied molecular orbitals (SOMOs) and the higher in energy of the SOMOs can be considered as the new HOMO (represented by HOMO*). As shown in **Figure 9.7**, this orbital does have the correct symmetry to overlap suprafacially with the LUMO of a second (unexcited) alkene. The stereochemical consequences of a photochemical [2π + 2π] cycloaddition are then the same as those of the Diels–Alder reaction: groups which are *cis* to one another in the starting materials remain *cis* to one another in the product.

An example of a photochemical [2π + 2π] cycloaddition illustrating the suprafacial nature of the reaction and the resulting stereochemistry is included in **Scheme 9.29**. The reader should beware, however, as many apparently [2π + 2π] cycloaddition reactions have actually been shown to occur by non-concerted radical mechanisms. Such reactions are not subject to the orbital overlap requirements of a cycloaddition and so may give products of different stereochemistry.

9.6.4   Cheletropic reactions

Cheletropic reactions are cycloaddition reactions in which both new σ-bonds are made to the same atom. The best known such reaction is the reaction between an alkene and a carbene to form a cyclopropane as shown in **Scheme 9.33**. This

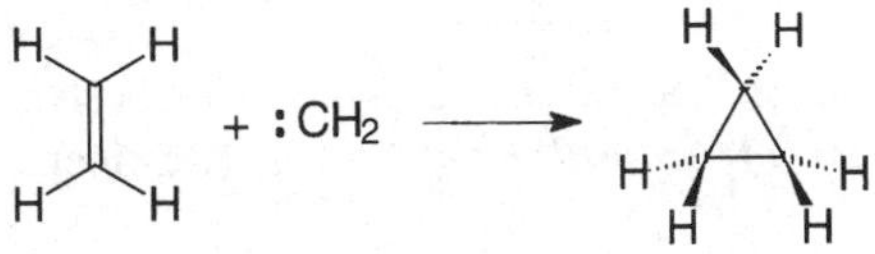

Scheme 9.33

reaction resembles a [2π + 2π] cycloaddition, yet it occurs thermally. The explanation for this reactivity is that since the LUMO of the carbene is localized on a single atom, it can react with the HOMO of the alkene in an orientation that would not be possible for the LUMO of an alkene as illustrated in **Figure 9.8**. The substituents on the carbene are initially parallel to the plane of the alkene, but as the new σ-bonds form they rotate to their final locations perpendicular to

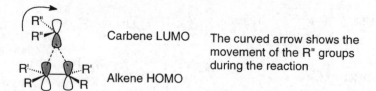

Figure 9.8 Orbital overlap during the cheletropic reaction between an alkene and a carbene.

the plane of the alkene. As **Figure 9.8** shows, the result is a syn-addition of the carbene to the alkene, arising from a suprafacial interaction between the HOMO and LUMO.

9.7   Further reading

General

Stereochemistry of Organic Compounds E.L. Eliel and S.H. Wilen. Wiley: London, 1994, chapter 12.
A Guidebook to Mechanism in Organic Chemistry P. Sykes. Longman: London, 1982.

Stereoelectronic effects in reactions

Stereoelectronic Effects in Organic Chemistry P. Deslongchamps. Pergamon: Oxford, 1983, chapters 5 and 6.
Stereoelectronic Effects Oxford Chemistry Primer Number 36, A.J. Kirby. Oxford University Press: Oxford, 1996.

Addition of nucleophiles to carbonyl compounds

Asymmetric Synthesis Vol. 2, E.L. Eliel (J.D. Morrison ed.). Academic Press: London, 1983, chapter 5.
Core Carbonyl Chemistry Oxford Chemistry Primer Number 47, J. Jones. Oxford University Press: Oxford, 1997.

The aldol reaction

Stereoselective Synthesis R.S. Atkinson. Wiley: Chichester, 1995, chapter 8.
Asymmetric Synthesis of Natural Products A. Koskinen. Wiley: Chichester, 1993, chapter 3.

Pericyclic reactions

Frontier Orbitals and Organic Chemical Reactions I. Fleming. Wiley: Chichester, 1976, chapters 4 and 6.
Pericyclic Reactions G.B. Gill and M.R. Willis. Chapman and Hall: London, 1974.
Introduction to Organic Photochemistry J.D. Coyle. Wiley: Chichester, 1986.
Cycloaddition Reactions in Organic Synthesis W. Carruthers. Pergamon: Oxford, 1990.

9.8   Problems

1. For each of the following substitution reactions, determine the absolute configuration of each stereocentre in both the starting material and the product. Do the reactions proceed with retention or inversion of configuration? In the third case, it is found experimentally that, even if the starting material is enantiomerically pure, the product is racemic. Explain why.

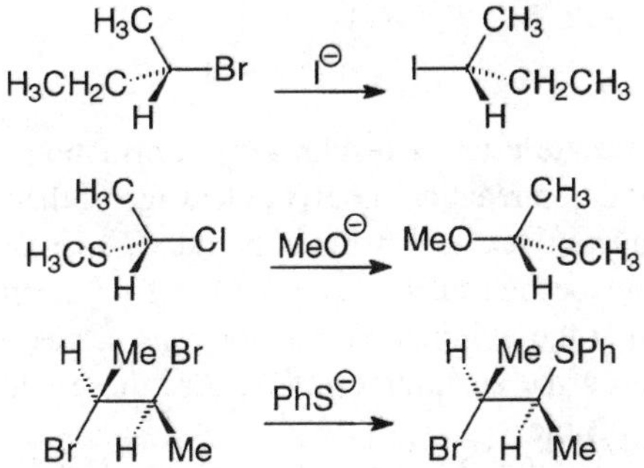

2. When enantiomerically pure alkyl iodide A is treated with radioactive iodide, racemic but radiolabelled A is obtained. It is found that the rate of racemization is twice the rate at which radioactive iodide is incorporated. Explain these results and explain the stereochemical significance of the kinetic data.

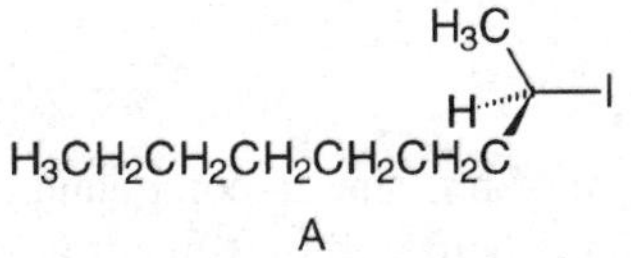

A

3. Treatment of compound A with a base generates carbanion B, which rearranges to sulphonate C. It is tempting to draw this reaction as an intramolecular substitution reaction as shown in the scheme below. However, if the reaction is carried out on compound A, 50% of which contains a ^{13}C atom at the carbon adjacent to the Ts group *and* in which the OCH_3 group has been deuterated, whilst the other 50% of A carries neither of these isotopic labels, then some of the product contains only one of the two labels. Suggest a mechanism that is consistent with these findings and explain why the intramolecular reaction shown below does not occur.

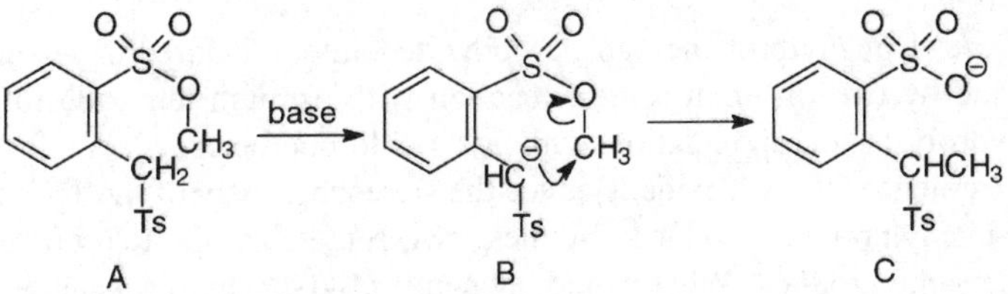

4. Explain why neither diastereomer of the bromo-ketone shown below elim-
   inates HBr when treated with base.

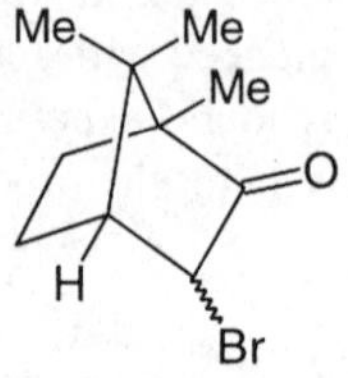

5. Pyrolysis of acetate ester A results in the formation of alkene B. Explain
   why this elimination reaction involves cleavage of the C–H bond rather than
   the C–D bond and explain why only the (E)-isomer of the product is
   formed. It may be helpful to build a model of the transition state. What
   would happen if the elimination reaction was carried out on the diastereomer of A and on the enantiomer of A? Are these reactions stereoselective
   and/or stereospecific?

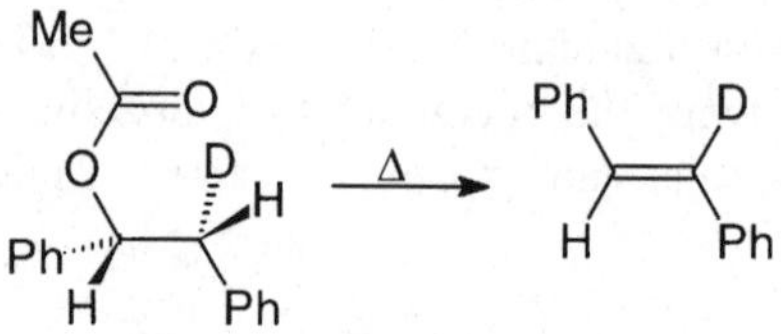

6. Explain why addition of bromine to compound A gives a product in which
   the two bromine atoms both occupy equatorial positions whilst addition of
   bromine to compound B gives a product in which the two bromine atoms
   are in axial positions.

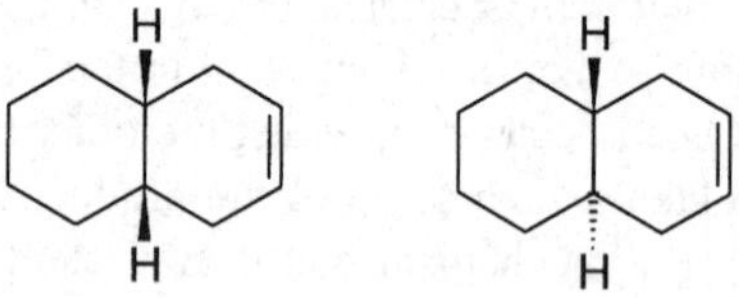

7. *Trans*-1-phenylpropene can be converted into 1,2-dihydroxy-1-phenylpro-
   pane by two different routes: reaction with osmium tetroxide followed by
   hydrolysis; or epoxidation with *meta*-chloroperbenzoic acid followed by
   treatment with hydroxide. Deduce the stereochemistry of the 1,2-dihydroxy-
   1-phenylpropane produced by these two routes. Do the two processes give
   the same product? What would happen if *cis*-1-phenyl-propene was used as

the starting material? Are these reactions stereoselective and/or stereospecific?

8. Predict the stereochemistry of the predominant diastereomer formed in each
   of the following reactions. Which of these reactions are stereoselective and/
   or stereospecific?

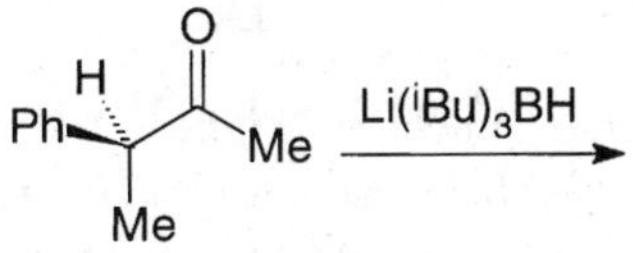

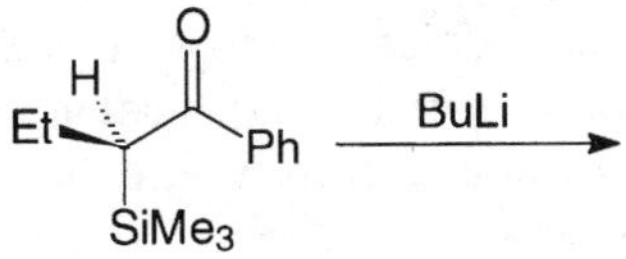

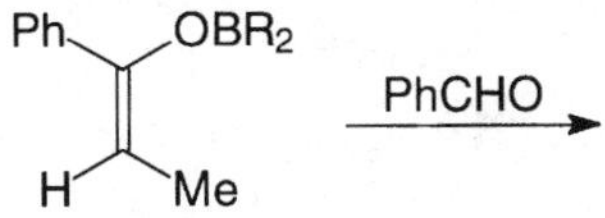

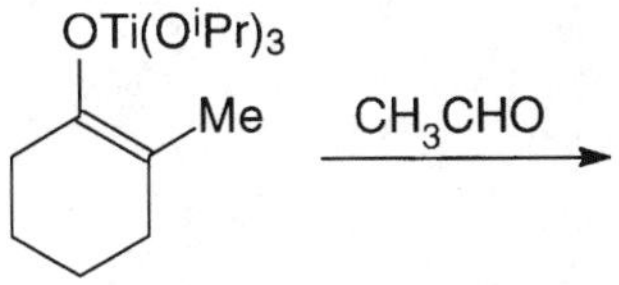

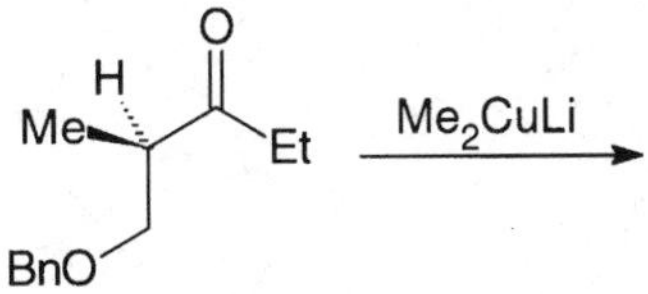

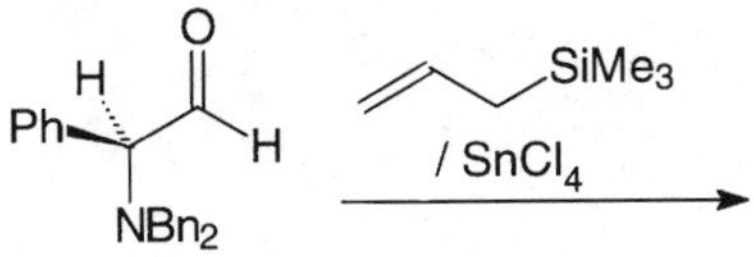

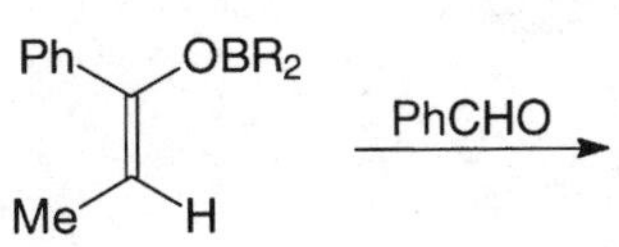

9. The Wittig reaction shown below produces two stereoisomeric products. What are the two products and what is the stereochemical relationship between them?

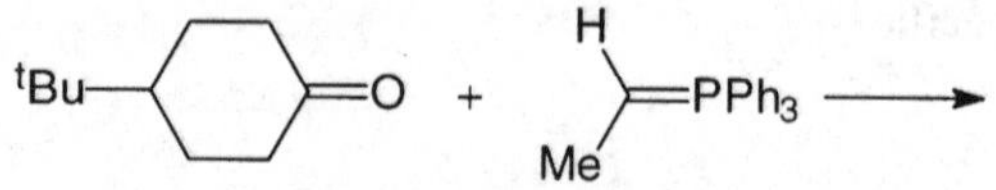

10. Show, using the orbitals drawn in **Figure 9.4**, that the stereochemical consequences of a Diels–Alder reaction remain unchanged if the alkene is treated as the nucleophile and the diene is considered as the electrophile.

11. Ketenes ($R_2C=C=O$) are unusual in that they will undergo thermally induced $[2\pi + 2\pi]$ cycloaddition reactions (involving the C=C double bond) with other alkenes. Suggest an explanation for this behaviour.

12. Place the dienes shown below in increasing order of reactivity in a Diels–Alder reaction.

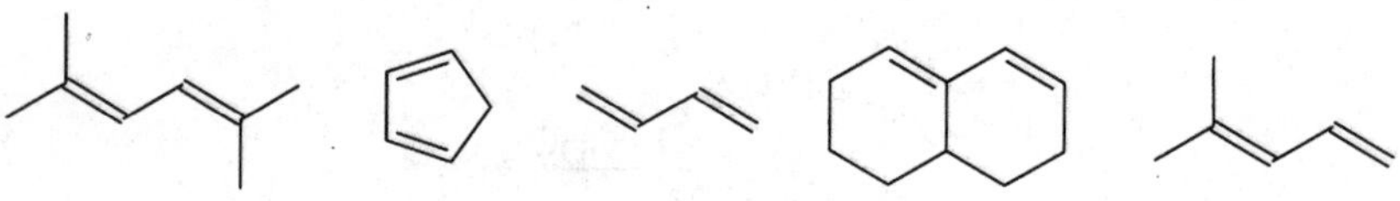

10 Asymmetric synthesis

10.1 Introduction

In Chapter 9, the stereochemical consequences of various reactions were discussed. Although many of these reactions create new stereocentres, if the starting material is achiral or racemic then the product of the reactions must also be achiral or racemic. It was also shown in Chapter 9, however, that if one stereocentre is already present in an enantiomerically pure starting material, then this can be used to control the configuration of other stereocentres which may be generated during subsequent reactions. This was most clearly illustrated for the addition of nucleophiles to chiral carbonyl compounds (cf. Chapter 9, section 9.5) but is equally applicable to the other reaction types. The remaining challenge in carrying out a synthesis of an enantiomerically pure product is then to introduce this first stereocentre into an achiral precursor in such a way that the product is not racemic. The methods for achieving this are the topic of this chapter. It is worth recalling at this stage that the two enantiomers of a compound (and more generally each stereoisomer of the compound) may have different biological properties (cf. Chapter 3, section 3.6), so if a product is destined for a biological application (pharmaceutical, agrochemical, foodstuff, etc.) it is important to be able to prepare the product as a single stereoisomer.

One way in which an enantiomerically pure compound can be obtained is by the resolution of the corresponding racemate. Various methodologies for achieving this task were discussed in Chapter 5 (section 5.3). Whilst in favourable circumstances a resolution can give a >50% yield of the desired enantiomer by *in situ* or subsequent racemization of the unwanted enantiomer, in most cases the chemical yield is <50% and the other enantiomer is waste which needs to be disposed of.

A second possibility is to make use of the enantiomerically pure compounds that nature provides. A wide range of naturally occurring, enantiomerically pure amino acids, carbohydrates, terpenes and other miscellaneous chemicals are commercially available and are often referred to as the '**chiral pool**'. A growing number of other enantiomerically pure compounds are also commercially available and can be used as chiral starting materials. These compounds are often the unwanted enantiomer from the resolution of an intermediate in another synthesis. The main problem with this synthetic approach is that it lacks generality. During the design and development of a new biological product, it is necessary to test many analogues of a lead compound to optimize the desired response and minimize side effects. If the synthesis of the lead compound started from (S)-alanine **10.1**, it might be desired to investigate the effect of replacing the methyl group with other alkyl groups. Unfortunately, whilst the ethyl to butyl analogues **10.2–10.4** are available, they are 20 times more expensive than (S)-alanine, and most other analogues are just not available.

Me

H······
H$_2$N   COOH

10.1

R

H······
H$_2$N   COOH

10.2 R = Et
10.3 R = Pr
10.4 R = Bu

In view of the limitations of preparing enantiomerically pure compounds either by resolution or using naturally occurring, enantiomerically pure starting materials, it would be very useful to have available methodology to convert achiral starting materials into chiral, non-racemic products. Such a process is called an **asymmetric synthesis** and is defined as the preparation of a non-racemic chiral product from achiral starting materials without the use of a resolution procedure. **Figure 10.1a** shows a reaction pathway energy diagram for the preparation of the two enantiomers of a chiral product from an achiral starting material in the absence of any other chiral species. It can be seen from **Figure 10.1a** that the reaction to form each enantiomer of the product has the same activation energy and the same Gibbs free energy of reaction. This is because the products are enantiomers of one another and the transition states are also enantiomeric. Thus the two enantiomers will form at the same rate and the product will be racemic (cf. Chapter 3, section 3.6.1).

In order to carry out an asymmetric synthesis, it is necessary to introduce some other non-racemic, chiral species into the reaction mixture. This chiral

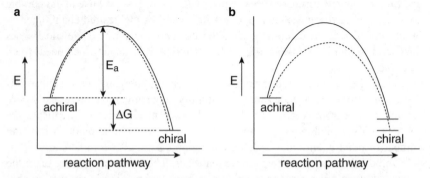

Figure 10.1 Energy diagram for the formation of the two enantiomers (solid and broken lines) of a chiral product from an achiral starting material: **a**, in an achiral environment; **b**, in a chiral environment.

species should be capable of interacting with the starting material so that the transition states for formation of the two products become diastereomeric (**Figure 10.1b**) and hence of different energy. In this way, the enantiomer resulting from the lower energy transition state will be formed preferentially and an asymmetric synthesis will be achieved. If the chiral species also interacts with the products, then these will also be diastereomeric and so have different energies as shown in **Figure 10.1b**. However, if there is no interaction between the chiral species and the products then the two possible products will be enantiomers (of the same energy), even though the transition states leading to them are diastereomeric.

In principle, the chiral species can be introduced in a large number of ways, for example, a chiral solvent could be used or the reaction could be carried out in the presence of circularly polarized light (cf. Chapter 3, sections 3.4.2 and 3.6.1), or in a rotating magnetic field. In practice, however, only two approaches have so far found synthetic utility: the chiral species may be attached to the achiral starting material (**chiral auxiliary approach**), or the chiral species may be attached to a reagent used to carry out the chemical reaction (**chiral reagent approach**). In the latter case, the reagent may be used stoichiometrically or catalytically. In this chapter, it is not possible to discuss all of the chiral auxiliaries and chiral reagents that have been developed. This is a very large and rapidly growing area and whole books have been written on this topic, some of which are listed in the further reading. Rather, we will examine a small number of the best known and most versatile chiral auxiliaries and reagents, and concentrate on understanding the stereochemical principles which control the way in which they work.

10.2 Use of chiral auxiliaries

Scheme 10.1 outlines diagrammatically the key features of the chiral auxiliary approach to asymmetric synthesis. An achiral starting material containing a prochiral centre **10.5** is reacted with an enantiomerically pure auxiliary **10.6** to form an adduct **10.7**. The effect of this reaction is that the two groups or faces of the prochiral centre which were enantiotopic in compound **10.5** become diastereotopic in **10.7** (cf. Chapter 7, sections 7.1–7.3). A chemical reaction is then carried out on adduct **10.7**, so that the prochiral centre is converted into a stereocentre forming adduct **10.8**. Since the two groups or faces of the prochiral centre are diastereotopic in adduct **10.7**, an achiral reagent can distinguish between them and react selectively with one of the groups or faces. Finally, the chemical bond between the auxiliary and the product is cleaved, releasing the enantiomerically pure product **10.9** and reforming the chiral auxiliary **10.6**, which in favourable cases can be recycled.

An example of the use of a chiral auxiliary in the asymmetric alkylation of the enolate of an aldehyde is shown in **Scheme 10.2**. Propanal **10.10** is the achiral

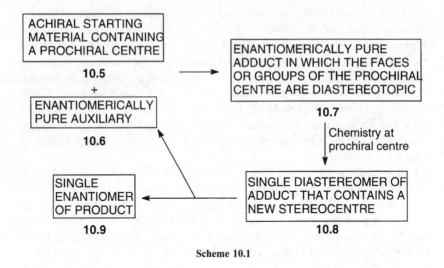

Scheme 10.1

starting material and contains two prochiral centres, one of which is indicated and has two enantiotopic hydrogen atoms attached to it. Reaction of propanal with (S)-1-amino-2-(methoxymethyl)pyrrolidine (abbreviated to SAMP) **10.11** results in formation of the chiral imine **10.12** in which the two indicated hydrogen atoms are now diastereotopic. Treatment of imine **10.12** with a strong

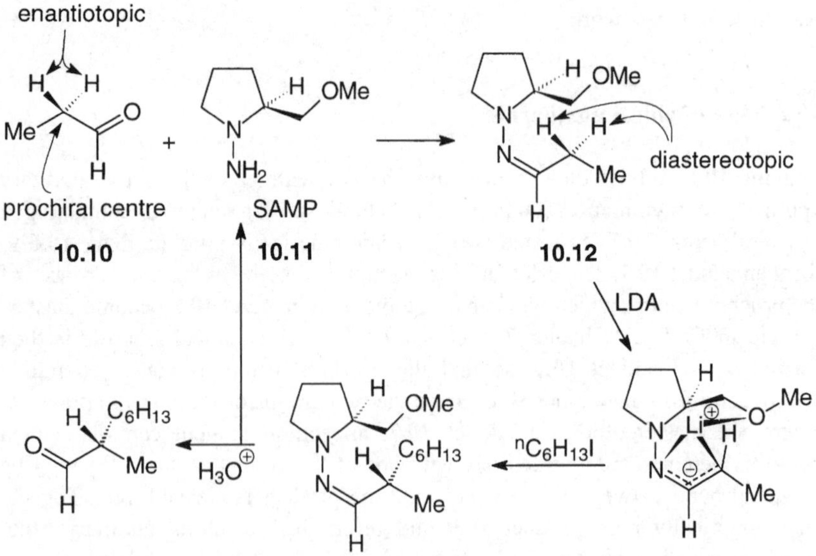

Scheme 10.2

base (LDA) results in the formation of carbanion **10.13** in which the lithium atom can complex to the ether oxygen and to both ends of the imine-anion, forming a six-membered ring chelate which effectively blocks the *si*-face (top face as drawn in **10.13**) of the carbanion. Alkylation of the carbanion by hexyl iodide then occurs preferentially on the unobstructed *re*-face, generating adduct **10.14**. Finally, hydrolysis of imine **10.14** generates the enantiomerically pure aldehyde **10.15** and regenerates the SAMP auxiliary.

Overall, the effect of the chemistry shown in **Scheme 10.2** is to replace the *pro-S* hydrogen of propanal **10.10** with a hexyl group, converting the prochiral centre into a stereocentre and giving enantiomerically pure product **10.15**. This is a very general reaction for the asymmetric alkylation of aldehydes. Propanal **10.10** could be replaced by any other achiral aldehyde or ketone with a prochiral methylene group adjacent to the carbonyl and the hexyl iodide could also be replaced by any other alkylating agent. In each case, the absolute configuration of the product could be predicted in advance since it is always the *pro-S* hydrogen of the aldehyde that is replaced.

A second example of the use of a chiral auxiliary is shown in **Scheme 10.3**. The chemistry is very similar to that shown in **Scheme 10.2** but the enolate of a carboxylic acid derivative rather than an aldehyde is formed and alkylated. Auxiliary **10.16** is called the Evans auxiliary and is very versatile. In addition to the enolate alkylation shown in **Scheme 10.3**, asymmetric aldol reactions can be carried out with this auxiliary, and the stereochemistry at both new stereocentres can be controlled by the choice of appropriate reaction conditions. In addition, the auxiliary can be attached to α,β-unsaturated acids, giving adducts **10.17** which undergo asymmetric Michael additions and asymmetric cycloaddition reactions.

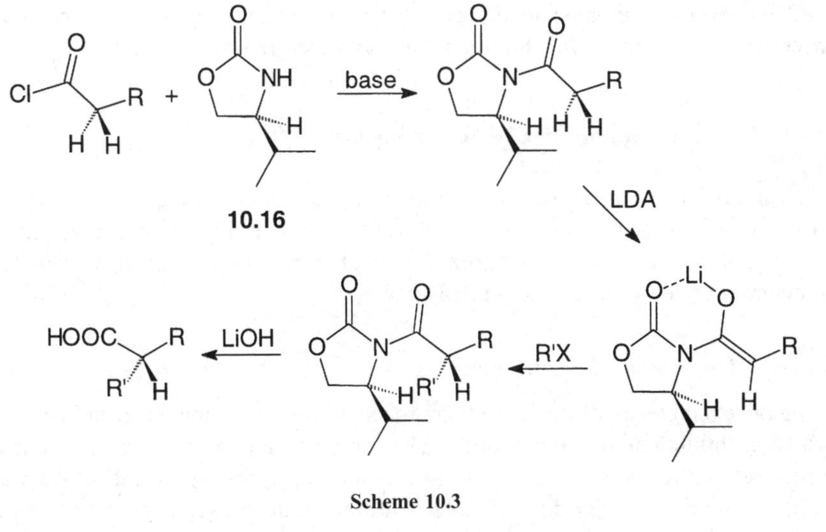

Scheme 10.3

10.17

It is worth noting that both chiral auxiliaries **10.11** and **10.16** contain a group which is capable of forming a chelate with the enolate intermediate. This is a very common occurrence with chiral auxiliaries since cyclic compounds have far fewer available conformations than acyclic compounds. The chiral auxiliaries are designed to form a cyclic intermediate in which one face of the prochiral centre is more hindered than the other. Reaction then occurs preferentially on the less hindered face.

The chiral auxiliaries used in **Scheme 10.2** and **Scheme 10.3** are both prepared from naturally occurring amino acids, (S)-proline in the case of SAMP **10.11** and (S)-valine in the case of auxiliary **10.16**. Thus both are readily available from the 'chiral pool'. There are a large number of other chiral auxiliaries but not all are so readily available and the auxiliary may need to be resolved prior to use. One of the advantages of the chiral auxiliary approach is that, if the reaction is not completely stereoselective, then the two stereoisomers which are initially formed (e.g. of **10.14** in **Scheme 10.2**) will be diastereomers of one another and so easily separated by crystallization, chromatography or any other purification methodology. Only after subsequent removal of the chiral auxiliary do the products become enantiomers of one another. The main disadvantage of the use of chiral auxiliaries is that at least two extra chemical steps will be needed compared to the corresponding racemic synthesis, since it is necessary to introduce the chiral auxiliary and then remove it again.

10.3   Use of enantiomerically pure reagents

The alternative to attaching the chiral group to the substrate of a chemical reaction is to attach it to one of the reagents. The reagent then becomes chiral and is able to differentiate between the enantiotopic faces or groups around a prochiral centre in the achiral substrate.

10.3.1   Use of stoichiometric reagents

A good example to illustrate this approach is the reduction of acetophenone **10.18** by lithium aluminium hydride. The carbonyl carbon of acetophenone is a prochiral centre, and the molecule has enantiotopic *re*- and *si*-faces. Achiral lithium aluminium hydride is unable to differentiate between these two faces and so adds hydride equally well to either face, giving a racemic mixture of alcohol **10.19** as shown in **Scheme 10.4**. If, however, the lithium aluminium hydride is first reacted with an enantiomerically pure ligand such as (*R*)-binaphthol **10.20** (which contains a stereogenic axis, cf. Chapter 3, section 3.8.1), then the chiral reducing agent **10.21** is obtained. When reagent **10.21** interacts with acetophenone, the two enantiotopic faces of the carbonyl become diastereotopic and hydride is added selectively to the *si*-face, producing the (*R*)-enantiomer of alcohol **10.19** with an enantiomeric excess of 95% as shown in **Scheme 10.5**.

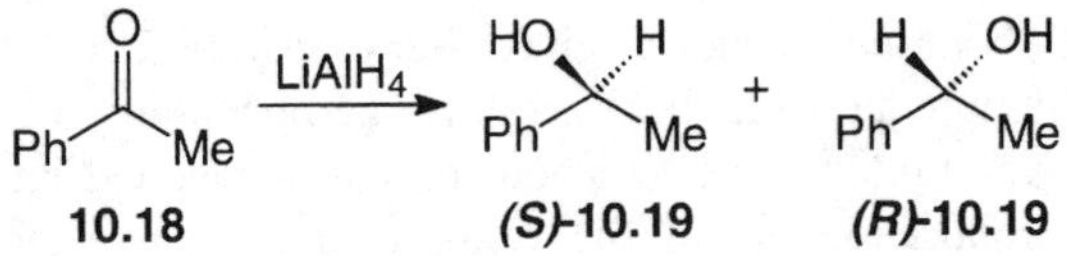

Scheme 10.4

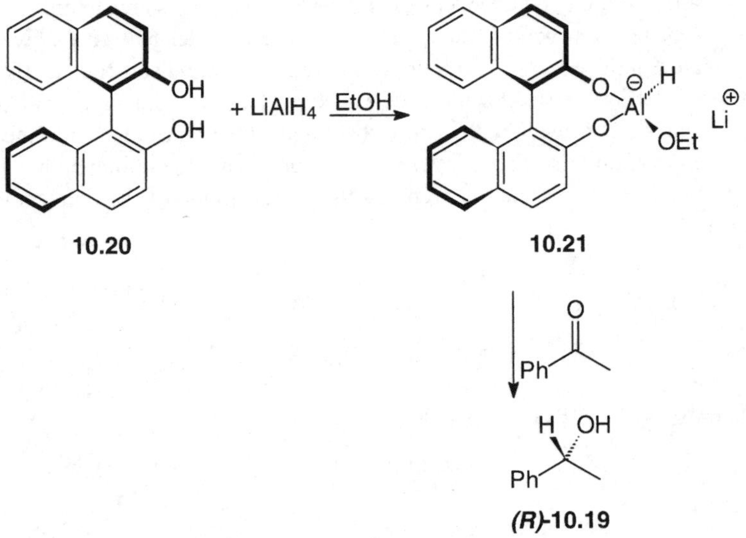

Scheme 10.5

This is a general procedure which can be used to reduce a wide variety of ketones, although particularly good results are obtained with α,β-unsaturated carbonyl compounds including aryl ketones. The mechanism of this reaction is not known but a model which explains the observed asymmetric induction is shown in **Figure 10.2**. The model relies on the formation of a six-membered ring complex involving both the lithium and aluminium atoms. The complex adopts a chair conformation, and the larger of the two groups attached to the carbonyl adopts an equatorial position.

Another example of a chiral reagent is shown in **Scheme 10.6**. In this case, the chiral reagent **10.22** is an enantiomerically pure derivative of borane and is

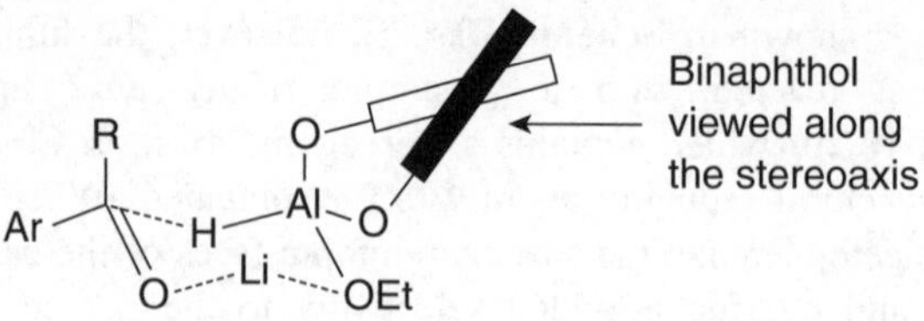

Figure 10.2 A model to account for the asymmetric induction during the reduction of ketones by complex **10.21**.

capable of hydroborating alkenes to give, after oxidative removal of the borane, enantiomerically pure alcohols. In general, the carbon atoms at both ends of the alkene will be prochiral centres and both may be converted into stereocentres during the hydroboration, as in the example shown in **Scheme 10.6**. The addition of any borane to an alkene is a *syn*-addition (cf. Chapter 9, section 9.4) and this determines the relative configuration of the product. The absolute configuration of the two stereocentres is determined by the stereocentres present in the α-pinene from which reagent **10.22** is prepared. Reagent **10.22** gives the best enantiomeric excesses with trisubstituted or *trans*-disubstituted alkenes. However, reaction of **10.22** with a second equivalent of α-pinene gives reagent **10.23**, which reacts well with *cis*-disubstituted alkenes. Both enantiomers of α-pinene are readily available from natural sources unlike the binaphthol **10.20** used in **Scheme 10.5**, which is prepared by the resolution of the corresponding racemate.

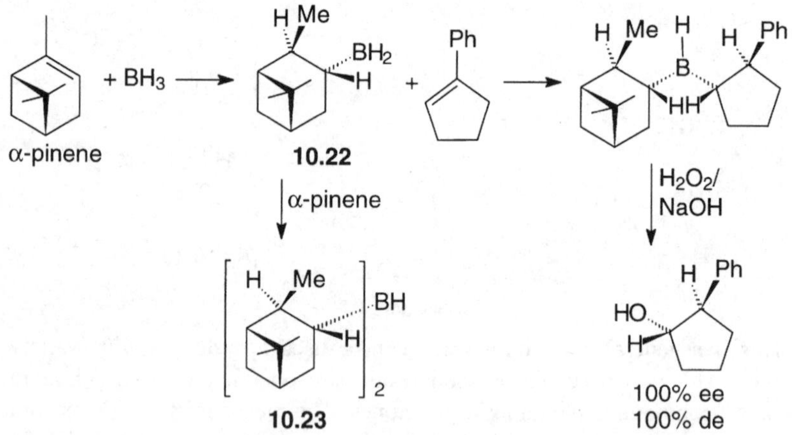

Scheme 10.6

Like chiral auxiliaries, the chemistry associated with chiral reagents is generally applicable. Thus compound **10.21** will reduce ketones other than acetophenone, and compounds **10.22** or **10.23** will hydroborate most alkenes. In both cases, the stereochemistry of the product is predictable in advance. Unlike the use of chiral auxiliaries, no extra steps are introduced into the synthesis, an achiral reagent is simply replaced by a chiral one. However, the product is not always enantiomerically pure, and since enantiomeric products are obtained directly they may not be easy to separate. Of the common purification techniques, only crystallization is sometimes able to increase the enantiomeric excess of a compound (cf. Chapter 5, section 5.3). Finally, the method can be expensive since a stoichiometric amount of the chiral reagent is needed. In view of this, much research is being devoted to the development of catalytic, chiral reagents.

10.3.2 Use of enantiomerically pure catalysts

In this approach, a chemical reaction is carried out at a prochiral centre within an achiral molecule using a stoichiometric amount of an achiral reagent and a catalytic amount of an enantiomerically pure catalyst. Compared to the use of a stoichiometric amount of an enantiomerically pure reagent (section 10.3.1), this method is more economical since only a catalytic amount of the often expensive chiral species is needed. The other advantages and disadvantages of the approach are the same as described for stoichiometric chiral reagents.

An example of the use of a chiral catalyst is shown in **Scheme 10.7**, in which the achiral ketone **10.24** is reduced to the corresponding alcohol **10.25** by sodium borohydride in the presence of a catalytic amount (10%) of the chiral catalyst **10.26**. The chiral catalyst **10.26** is readily prepared from the naturally occurring amino acid (S)-proline and induces the addition of hydride to the si-face of the ketone. It is informative to compare the chemistry in **Scheme 10.7** with that in **Scheme 10.5**, since in both cases an achiral ketone is reduced to an enantiomerically pure alcohol. In **Scheme 10.5** a stoichiometric amount of the chiral species is required and that has to be prepared by a resolution procedure. In contrast, only 10% of catalyst **10.26** is required to accomplish the same transformation in **Scheme 10.7** and the catalyst is readily available from the 'chiral pool'.

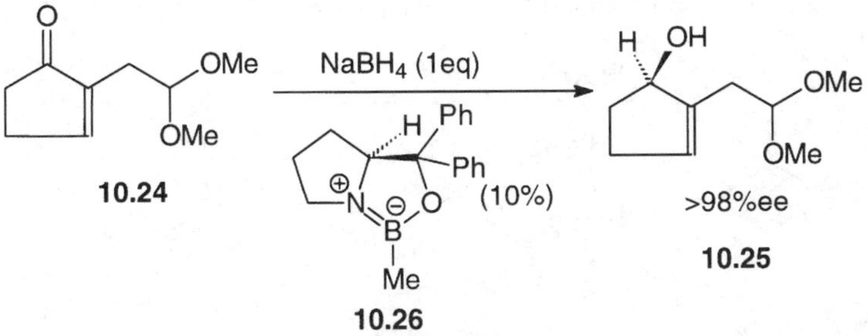

Scheme 10.7

Another reaction which can be accomplished with the aid of an enantiomerically pure catalyst is the asymmetric hydrogenation of prochiral alkenes. An enormous amount of work has been done in this area and many hundreds of catalysts have been investigated. Most success has been achieved in the hydrogenation of achiral α,β-didehydroamino acid derivatives as shown for the (chiraphos)$_2$Rh catalyst in **Scheme 10.8**. The enantiomeric excess obtained in this reaction is very dependent upon the nature of the groups on the alkene, amine and acid but, in favourable cases, enantiomerically pure amino acid derivatives can be obtained and the process is used industrially for the synthesis of amino acids.

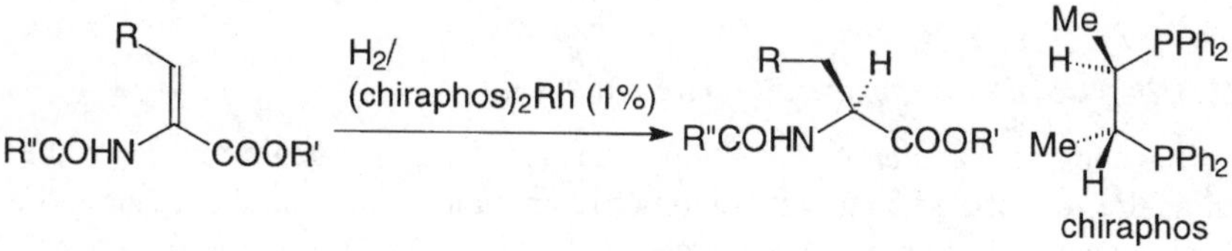

Scheme 10.8

In Chapter 5 (section 5.3.3), the Sharpless epoxidation of allylic alcohols was introduced as a method for the kinetic resolution of racemic allylic alcohols. The chemistry is equally applicable to achiral allylic alcohols, in which case it constitutes an asymmetric synthesis of epoxy alcohols as shown in **Scheme 10.9**. The process requires a stoichiometric amount of *tert*-butyl hydroperoxide but, provided 4 Å molecular sieves are added to the reaction mixture to remove any water present, is catalytic in both titanium(IV) isopropoxide and diisopropyl tartrate. The reaction gives best results with *cis*-disubstituted and trisubstituted allylic alcohols, and since both enantiomers of tartaric acid are readily available, either enantiomer of the epoxy alcohol product can be prepared. Since its discovery in the 1980s, the Sharpless epoxidation of allylic alcohols has become a cornerstone of asymmetric synthesis in both research laboratories and industrial production. A wide range of allylic alcohols are substrates for the reaction, which generally gives epoxy alcohols with very high enantiomeric excesses, and the epoxy alcohols are highly versatile synthetic intermediates that can be transformed into a wide range of other chemicals.

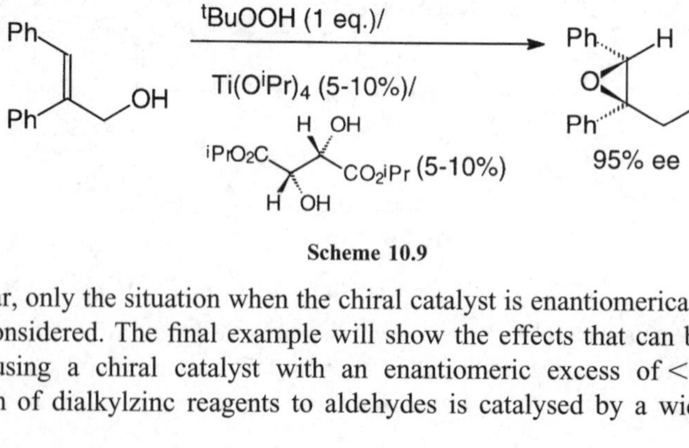

Scheme 10.9

So far, only the situation when the chiral catalyst is enantiomerically pure has been considered. The final example will show the effects that can be observed when using a chiral catalyst with an enantiomeric excess of <100%. The addition of dialkylzinc reagents to aldehydes is catalysed by a wide range of

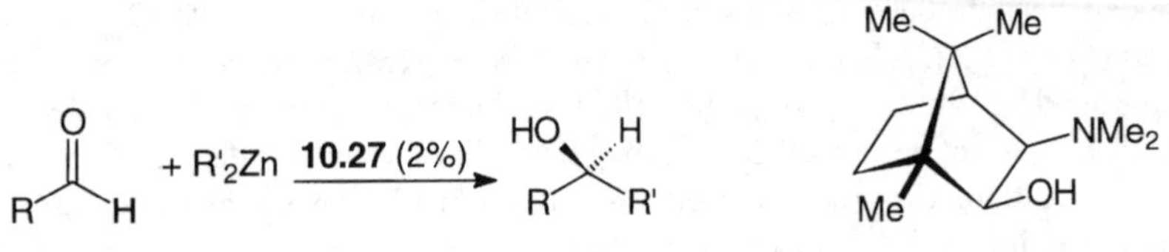

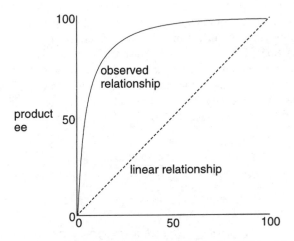

Scheme 10.10

bidentate chiral ligands, one of the best studied of which is DAIB **10.27** as shown in **Scheme 10.10**. If the reaction is carried out with benzaldehyde (R = Ph), diethylzinc (R' = Et) and enantiomerically pure DAIB, then the product 1-phenyl-propanol is obtained with an enantiomeric excess of 99%. It would seem reasonable that if the enantiomeric excess of the DAIB was reduced to 50%, then the enantiomeric excess of the product should be reduced to 49.5%, i.e. that there would be a linear relationship between the enantiomeric excess of the catalyst and the enantiomeric excess of the product. However, as **Figure 10.3** shows, this is not what is actually observed; rather even a catalyst with a very low enantiomeric excess (*ca.* 20%) produces a product with an enantiomeric excess > 90%.

The explanation for this unusual asymmetric amplification effect is that the catalyst/diethylzinc complex exists as a dimeric species, and if the catalyst is not enantiomerically pure then there are three possible dimers which may be formed **10.28–10.30** as shown in **Scheme 10.11**. Dimer **10.28** contains two molecules of (−)-DAIB, whilst dimer **10.29** contains two molecules of (+)-DAIB. These two dimers are enantiomeric and will have equal catalytic activity but will produce

Figure 10.3 The variation of product ee with catalyst ee in the DAIB catalysed addition of diethylzinc to benzaldehyde.

opposite enantiomers of the chiral alcohol. The third dimer **10.30**, however, contains one molecule each of $(-)$-DAIB and $(+)$-DAIB and is an achiral, meso compound. It turns out that dimer **10.30** is far more stable than either **10.28** or **10.29** and is not a catalyst for the reaction. Thus, when a catalyst of less than 100% enantiomeric excess is used, the minor enantiomer is virtually completely converted into achiral dimer **10.30**. Only the excess of the major enantiomer forms the catalytically active dimer **10.28** or **10.29** and catalyses the formation of the chiral alcohol.

(-)-DAIB
+ $\xrightarrow{\text{Et}_2\text{Zn}}$
(+)-DAIB

[(-)-DAIB-Et$_2$Zn]$_2$ + [(+)-DAIB-Et$_2$Zn]$_2$
10.28   **10.29**

+ [(-)-DAIB-(+)-DAIB-(Et$_2$Zn)$_2$]
10.30

Scheme 10.11

This process is more than just a curiosity of asymmetric catalysis, it is relevant to the origin of enantiomerically pure compounds in nature. It was shown in Chapter 3 (section 3.6.1) that there are a variety of processes which may have caused an achiral compound to be converted into a chiral compound with a very low enantiomeric excess. An asymmetric amplification process similar to the one discussed above could then have used the compound with a very low enantiomeric excess to form a compound with a much higher enantiomeric excess.

10.3.3 Use of enzymes

Enzymes are naturally occurring chiral catalysts; their structure and application to the kinetic resolution of racemates was discussed in Chapter 5 (section 5.3.3). There are thousands of different enzymes which catalyse a wide range of chemical reactions, but probably the two most useful for asymmetric synthesis are the oxidoreductases which catalyse the reduction of an aldehyde or ketone to the corresponding alcohol or the reverse reaction as shown in **Scheme 10.12**, and esterases which catalyse the esterification of a carboxylic acid or the hydrolysis of an ester (**Scheme 10.13**).

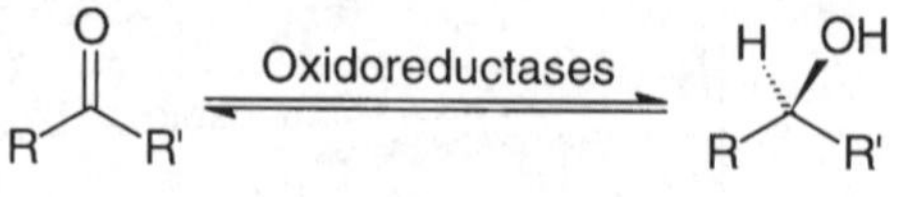

Scheme 10.12

RCOOR' $\xrightleftharpoons{\text{Esterases}}$ RCOOH + R'OH

Scheme 10.13

Enzyme catalysed asymmetric synthesis has all of the advantages and disadvantages of asymmetric synthesis using synthetic chiral catalysts. There are two other factors that need to be considered, however; the first being that, whilst the best known enzymes are commercially available and inexpensive, less well known enzymes are not commercially available and may be difficult to obtain. The easiest enzymes to use are those within commercially available yeasts such as bakers' yeast and brewers' yeast. The yeasts are readily available from supermarkets and can be used directly in chemical reactions as shown for the reduction of ethyl acetoacetate in **Scheme 10.14**. The second point is that if an enzyme is found which carries out the desired reaction but gives the wrong enantiomer of the product, then it will not be possible to obtain the enantiomeric enzyme since only one enantiomer is present in nature.

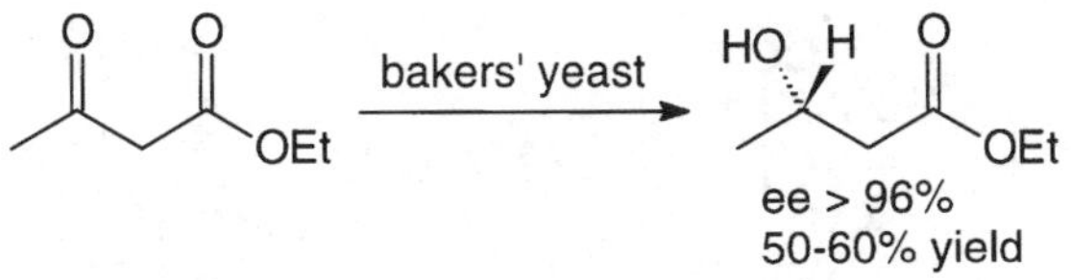

ee > 96%
50-60% yield

Scheme 10.14

It is possible to combine an enzyme catalysed asymmetric synthesis with *in situ* racemization of a starting material to provide a means for converting a racemic starting material into an enantiomerically and diastereomerically pure product. An example of this is seen with β-keto ester **10.31**. This compound contains a stereocentre between two electron withdrawing carbonyl groups and so undergoes rapid racemization (cf. Chapter 5, section 5.1.3) as shown in **Scheme 10.15**. If compound **10.31** is treated with bakers' yeast, then the yeast reduces only the (*R*)-enantiomer of the ketone and also produces exclusively the alcohol with the (*S*)-configuration at the new stereocentre **10.32**. Thus only one of four possible stereoisomers is obtained in greater than 50% yield from a racemic starting material. This type of kinetic resolution combined with asymmetric transformation can also be achieved with synthetic chiral catalysts and with chiral reagents.

Enzymes are also often used to carry out a slightly different type of asymmetric transformation, the desymmetrization of a meso compound as shown in **Scheme 10.16**. Compound **10.33** is an achiral meso compound which contains two stereocentres. The two ester groups are enantiotopic and an enzyme (or any other chiral species) can distinguish between them just as any two enantiotopic groups or faces can be distinguished by a chiral species. Thus treatment of compound **10.33** with an esterase enzyme results in selective hydrolysis of one of the ester groups giving enantiomerically pure acid **10.34**. The same type of desymmetrization of a meso compound has also been achieved using chiral reagents, and synthetic chiral catalysts.

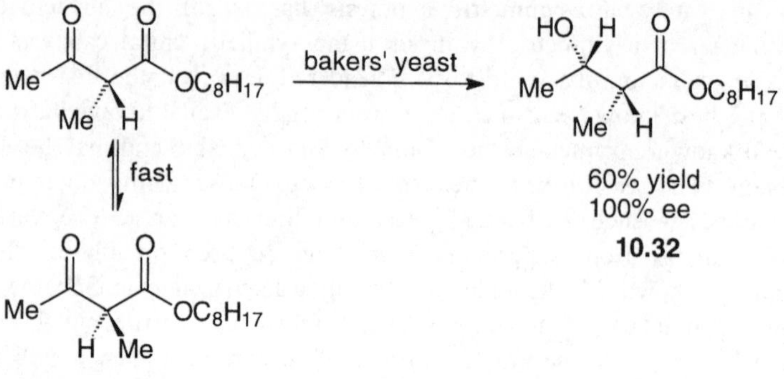
bakers' yeast
fast
60% yield
100% ee

10.31

10.32

Scheme 10.15

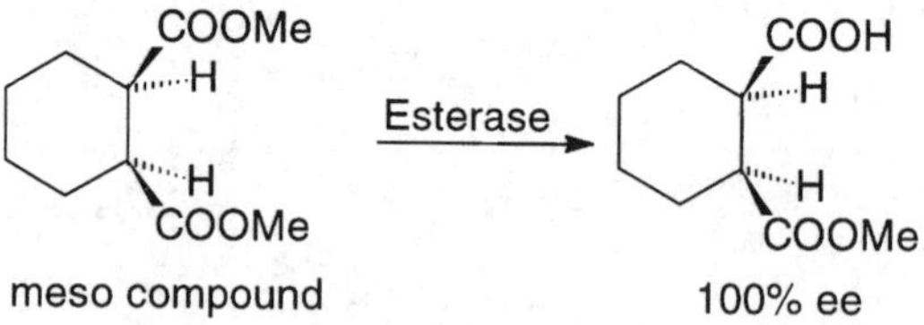
Esterase
meso compound
100% ee

10.33

10.34

Scheme 10.16

10.4 Further reading

All aspects of asymmetric synthesis

Asymmetric Synthesis G. Procter. Oxford University Press: Oxford, 1996.

Asymmetric Synthesis (R.A. Aitken and S.N. Kilényi eds). Chapman and Hall: London, 1992.

Enantioselective Reactions in Organic Chemistry O. Cervinka. Ellis Horwood: London, 1995.

Asymmetric Synthesis Vol. 2 (J.D. Morrison ed.). Academic Press: London, 1983.

Asymmetric Synthesis of Natural Products A. Koskinen. Wiley: Chichester, 1993.

Principles of Asymmetric Synthesis R.E. Gawley and J. Aubé. Pergamon: Oxford, 1996.

Asymmetric catalysis

Asymmetric Catalysis in Organic Synthesis R. Noyori. Wiley: Chichester, 1994.

Enzyme catalysed reactions

Enzymes in Synthetic Organic Chemistry C.-H. Wong and G.M. Whitesides. Pergamon: Oxford, 1994.

Biotransformations in Preparative Organic Chemistry H.G. Davies, R.H. Green, D.R. Kelly and S.M. Roberts. Academic Press: London, 1989.
H. Stecher and K. Faber. *Synthesis*, 1997, 1.

The chiral pool
Asymmetric Synthesis Vol. 4 (J.D. Morrison and J.W. Scott eds). Academic Press: London, 1983, chapter 1.

The Evans auxiliary
D.J. Ager, I. Prakash and D.R. Schaad. *Aldrichimica Acta*, 1997, **30**, 3.

Asymmetric amplification
M. Avalos, R. Babiano, P. Cintas, J.L. Jiménez and J.C. Palacios. *Tetrahedron Asymmetry*, 1997, **8**, 2997.

10.5 Problems

1. Draw reaction pathway/energy diagrams for the reactions shown in **Scheme 10.2** and **Scheme 10.3**. Your diagrams should show each reaction intermediate and transition state and clearly indicate any differences between the pathways leading to the two enantiomeric products.
2. Which of the following compounds will give an enantiomerically pure product when used instead of propanal in the reaction sequence shown in **Scheme 10.2**.

a Ph–CH₂–CHO **b** Me–CH₂–C(=O)–Ph **c** Me–CH(Me)–CHO

d Ph–CH₂–C(=O)–OCH₃ **e** F₃C–C(Me)(Me)–CHO **f** Me–CH₂–C(cyclopropyl)–CHO

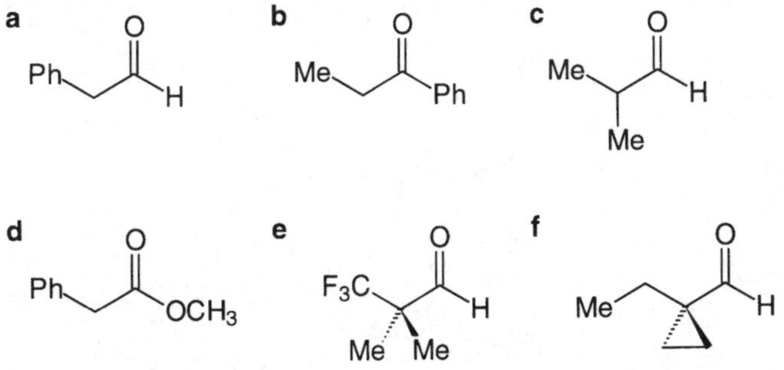

3. Despite not containing a prochiral methylene group adjacent to the carbonyl group, aldehyde A does give an enantiomerically pure product when subjected to the reaction sequence shown in **Scheme 10.2**. Explain why a non-racemic product is obtained, predict the stereochemistry of the product (assuming hexyl iodide is used as the alkylating agent) and draw a reaction pathway/energy diagram which illustrates the origin of the asymmetric

induction. How would the structure of the product be affected if the enantiomer of compound A was used in the same reaction sequence and if the
racemate of A was used?

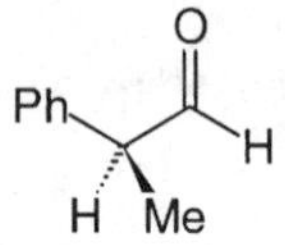

A

4. Account for the asymmetric induction observed in the reaction shown below.
The metal counterion has an important role to play in this and other reactions
of Evans oxazolidinones; suggest what this role might be.

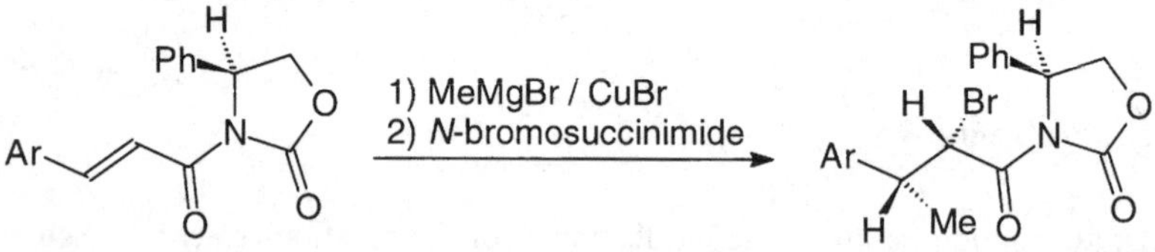

1) MeMgBr / CuBr
2) N-bromosuccinimide

5. What would be the stereochemical consequences for the reaction shown in
Scheme 10.10 if dimer **10.30** was much less stable than dimers **10.28** and
10.29.
6. **Scheme 10.9** shows how Sharpless epoxidation methodology can be used to
prepare one stereoisomer of an epoxy alcohol. How could the same methodology be used to prepare each of the other stereoisomers of the same epoxy
alcohol?
7. Suggest reagents to accomplish each of the following transformations.

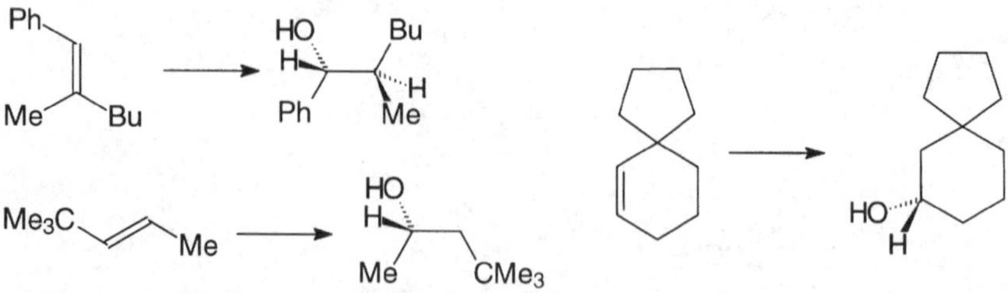

8. The enzyme oxynitrilase is readily available (from almonds) and catalyses
the decomposition of (R)-2-hydroxy-2-phenylacetonitrile to benzaldehyde
and hydrogen cyanide as shown in reaction A below. Which of reactions
B–D would it be reasonable to expect this enzyme to catalyse? The synthetic
catalyst E also catalyses the asymmetric addition of HCN to aldehydes
producing (R)-cyanohydrins. What advantages (if any) would the use of
synthetic catalysts such as E have in carrying out reactions B–D?

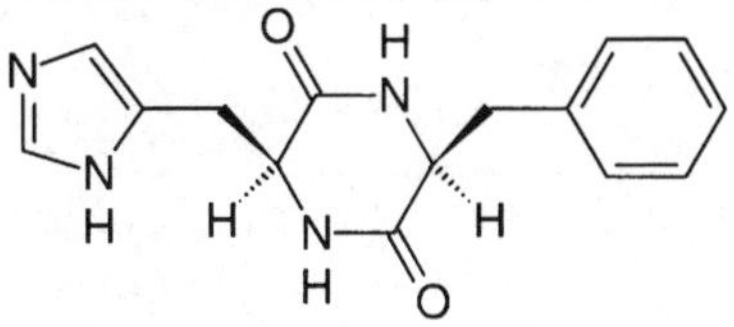

E

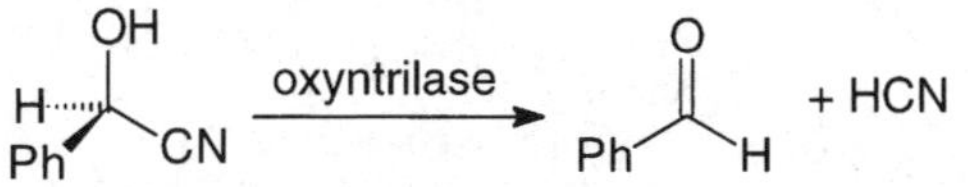

Reaction A

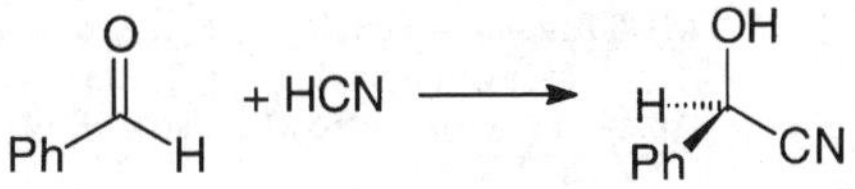

Reaction B

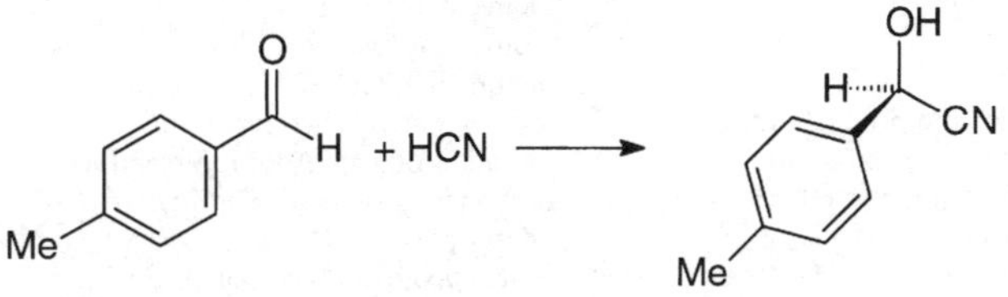

Reaction C

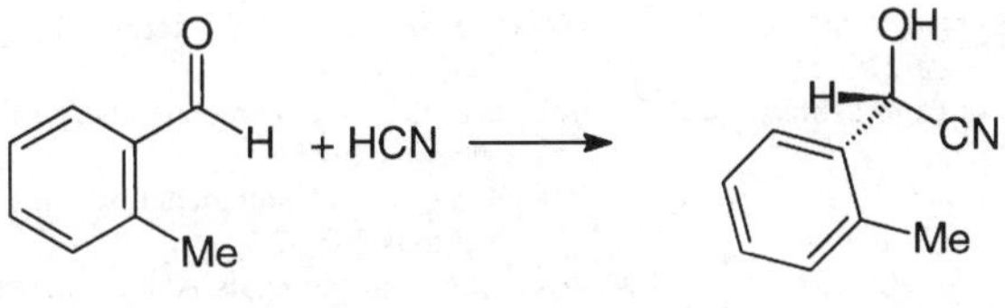

Reaction D

9. Reaction of anhydride A with (S)-proline methyl ester results in the formation of acid B. Draw a reaction pathway/energy diagram for this reaction which illustrates the origin of the stereoselectivity.

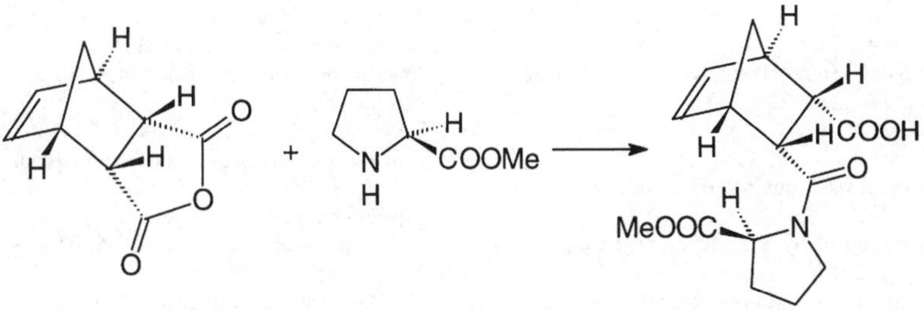

A

B

Index